THE VASCULAR FLORA OF OHIO

Volume One

THE *Monocotyledoneae*
CAT-TAILS TO ORCHIDS

This work is a project of the Ohio Flora Committee
of the Ohio Academy of Science, sponsored by the Academy
and by the National Science Foundation

BY E. LUCY BRAUN

THE
Monocotyledoneae

CAT-TAILS TO ORCHIDS

With Gramineae by Clara G. Weishaupt

Original drawings by Elizabeth Dalve' and Elizabeth King

THE OHIO STATE UNIVERSITY PRESS

PREFACE

Throughout the preparation of this volume of THE VASCULAR FLORA OF OHIO, I have kept in mind two principal objectives: (1) completeness and accuracy, and (2) usability, not alone by the student or botanist, but also by the amateur and the fieldworker in any of the natural sciences. I have defined (in parentheses) some of the less familiar technical terms where they are used; I have used gross characters wherever these suffice, resorting to more minute features only in those technical genera where only these can be relied on for species-determination. In many places in the text, mention is made of more or less familiar garden or greenhouse plants related to the Ohio species under consideration. Often, non-taxonomic information is added (usually in family- or genus-text) when this adds to the interest of the plants or stimulates field-observation. The book attempts to answer the questions: What is it? Where is it? How can I identify it? How can I recognize it in the field?

The text of the Gramineae, the Grass Family, was prepared by Clara G. Weishaupt, who has been studying Ohio grasses for many years.

My thanks are extended to my colleagues in various parts of Ohio for the loan of herbarium specimens, and especially to Clara G. Weishaupt, curator of the herbarium of the Ohio State University, who loaned me many hundreds of specimens. Specimens are the basis of all records, of all dots on the maps.

Illustrations have been drawn from fresh specimens whenever possible, a procedure which necessitated many trips afield. To those who by phone or letter told me when and where a wanted species was in flower, especially to C. A. Eulett and J. W. Culbertson, I extend thanks; also, I thank those who sent fresh specimens through the mails, most important of which were *Peltandra* sent from northeastern Ohio by Ervin M. Herrick, *Arisaema stewardsonii* from far eastern Ohio by Allison W. Cusick, and *Trillium cernuum* from northern Michigan by E. G. Voss, species which could not have been illustrated from dried specimens, and several species from E. S. Thomas.

For some of the drawings, color-transparencies were used in conjunction with herbarium specimens, the former giving the perspective, the latter the size; some of these were my own slides, others were loaned by J. A. Herrick, J. W. Culbertson, and Karl Maslowski.

Herbarium specimens were used for all species of which fresh material was not available, and of a large proportion of the Pondweeds, Rushes, Grasses, and Sedges.

A very few of the illustrations—other than those of magnified parts—were redrawn from Gleason's Illustrated Flora, by permission of the New York Botanical Garden; these are indicated by NY beside the drawing. Enlarged drawings of small parts—of capsules, achenes, perigynia, grass-spikelets, etc.— are original or are redrawn from sources indicated in family- or genus-text. These are gratefully acknowledged. The artwork, including the transcribing of names and of dots on the distribution maps, is by Mrs. Elizabeth Dalvé assisted by Mrs. Elizabeth King.

Acknowledgement is made of the aid of specialists in the identification of doubtful specimens: E. C. Ogden, for *Potamogeton;* J. R. Swallen, for grasses;

H. K. Svenson, for *Eleocharis;* F. J. Hermann, for *Carex, Juncus,* and *Luzula.* Without their help, these genera could not have been treated adequately.

Preliminary mimeographed lists of the "Grasses of Ohio," by Clara G. Weishaupt, and of the "Monocots of Ohio" (exclusive of the grasses), by E. Lucy Braun, circulated among Ohio botanists, stimulated collecting and resulted in many additional county records. For such collections thanks are extended, especially to Ervin M. Herrick, Arthur S. Brooks, Allison W. Cusick, and Floyd Bartley. The continuing help of my sister, Annette F. Braun, throughout the preparation of this book is invaluable. Suggestions and comments by those who read all or parts of the manuscript are here acknowledged.

Finally, I express my gratitude to the National Science Foundation for the support of this project through a grant—G-9232—which made the work in its present form possible, and to the Ohio Academy of Science, which organization administered the grant.

E. Lucy Braun

Cincinnati, Ohio
March, 1965

CONTENTS

INTRODUCTION

This volume of THE VASCULAR FLORA OF OHIO includes all native and naturalized species of the Monocotyledoneae—the monocots—that occur in Ohio (or did occur within the period represented by Ohio herbaria); many local adventive species and garden escapes are also included; mention is made (usually in family- or genus-text) of many ornamentals, and of food or other economic plants.

The Monocotyledoneae is one of the two subclasses of the Angiosperms or "flowering plants" and contains about one-fifth of its species; the other four-fifths belong to the Dicotyledoneae, commonly referred to as dicots. The designation monocot refers to one of the most distinctive features of the subclass; namely, embryo with one cotyledon or "seed-leaf," in contrast with the embryo of a dicot, which has two cotyledons. The usually parallel-veined leaves and three- or six-parted flowers (except in certain reduced kinds) generally facilitate recognition of monocotyledonous plants. Further characterization of the Monocotyledoneae is given on p. 17.

In size, monocots range from the smallest of all flowering plants (see Lemnaceae, p. 302 and Wolffia, p. 305) to large palm trees. Everywhere, the number of herbaceous species far exceeds the woody species; woody species are most abundant in tropical latitudes, and decrease rapidly in number in temperate latitudes. In Ohio, only bamboo (*Arundinaria*) and a few species of *Smilax* are woody plants, although *Yucca* has a short woody stem hidden by its crowded leaves (many warm-climate species are trees) and may be classed as woody.

Relatively few of the Ohio species of monocots have green foliage in winter; most of these (except sod- or mat-forming grasses and sedges) are tabulated on p. 13. Some retain their dry or shriveled fruits, which give clues to the identity of the species (for example, see pp. 390, 440). A large proportion have bulbs or rhizomes buried in the soil; very few are annuals.

The largest of all families of the Angiosperms is the Orchidaceae, with about 20,000 to 25,000 species. Other very large families are the Compositae and Leguminosae (dicots), the former sometimes claimed to be the largest of all families; but the frequent discovery of new species of orchids in the little-known parts of the Tropics results in the ever-increasing size of this highly developed family of monocots. Other very large families of monocots are the Gramineae (about 6,000 species) and the Cyperaceae (about 4,000 species).

In the Ohio flora, over 650 species of monocots are recognized; many varieties and a number of forms and hybrids swell the number to over 700 taxa (*sing.*, taxon, a taxonomic unit of any rank). Of these, almost two-thirds are grasses and sedges. Although a majority of the other one-third of Ohio's monocots have easily recognized, usually colored perianths, and therefore are often more showy, the grasses and sedges (and rushes) should not be neglected by the layman, for among these plants are some of the most graceful and beautiful forms of inflorescence (pp. 172, 186, 280). While determination of species in the Gramineae and Cyperaceae is often difficult, genera, for the most part, are easily recognized, and, except for the botanist, may furnish a satisfactory cognomen. Furthermore, the dominance of members of these families in meadows, marshes, and along swampy pond- or lake-borders means they are ever with us and should become recognized entities rather than unknowns. Certainly a lake-border is enhanced by the showy and graceful inflorescences of Wool-grass (p. 210).

The more than 650 species of Ohio's monocots are classified into 168 genera and 22 families, 74 of the genera in the Gramineae, 12 in the Cyperaceae, the two families thus containing slightly more than half our genera.

Nomenclature, for the most part, is that of *Gray's Manual of Botany*, ed. 8 (Fernald, 1950), although in some cases Ohio's plants are better classified in the works of Gleason (*Illustrated Flora of the Northeastern United States and Adjacent Canada*, 1952) and of Hitchcock for the grasses (*Manual of the Grasses of the United States*, 1951); other departures from Fernald's nomenclature may occur in genera or families where the aid of specialists was enlisted (acknowledged in the Preface and in the text of the genus or family).

It is not possible to designate or to list species included in this volume that were not previously known in Ohio because of changes in the interpretation of species (for example, in *Polygonatum*, in the herbaceous *Smilax*, etc.), recognition of additional varieties, etc., since the publication of Schaffner's "Revised Catalog of Ohio Vascular Plants" in 1932. However, a few outstanding examples should be noted:

> Potamogeton tennesseensis, p. 30
>
> Butomus umbellatus, p. 57
>
> Calamagrostis insperata, p. 113
>
> Eleocharis wolfii, p. 193
>
> Carex, p. 220, several species; see nos. 9,
>   12, 64, 96, 125, 142
>
> Arisaema stewardsonii, p. 296
>
> Erythronium rostratum, p. 357
>
> Streptopus roseus, p. 371
>
> Listera cordata, p. 437
>
> Epipactis helleborine, p. 429

Among these are species at the margin of range (as *Streptopus* and *Listera*) whose occurrence in Ohio does not change the distribution pattern; species known to be extending their ranges (as the introduced *Butomus* and *Epipactis*); species described from Ohio specimens (*Calamagrostis insperata*); and species whose Ohio occurrence is far from other stations (as the *Erythronium*). This last type of distribution is of particular interest because of the problems involved in explaining such distribution; such occurrences are of as much interest to persons outside of Ohio as to Ohio residents. Some of the species which have long been known in Ohio but which occur in only one or two counties (as *Nothoscordum*) are also disjuncts, and of particular interest in distribution studies.

Distribution maps of all species native to Ohio are included here; a dot in a county means that a specimen from that county has been examined. Sight records are not included. A dot, however, does not necessarily mean that the species is still growing in the county indicated—it may now be extinct there (for example, see *Nothoscordum bivalve*, p. 349, *Trillium recurvatum*, p. 377, *Cypripedium reginae*, p. 413, and *Habenaria leucophaea*, p. 419). Inclusion of records of occurrence in the past is desirable because of the more complete pattern of distribution thus afforded, and because of the aid they give in reconstructing a picture of original conditions and of past migrations.

Herbarium slips reporting specimens in various Ohio herbaria and those actually recording specimens examined (indicated by rubber stamp) are deposited as a card file with the herbarium at the Ohio State University. Timelimits in completing this work did not permit examining all specimens reported, but in so far as possible, all reported specimens that would have added a county record (a dot on the map) were borrowed and examined. Records are, with few exceptions, from herbaria of the following Ohio institutions and my own herbarium:

> Antioch College, Yellow Springs
> Baldwin-Wallace College, Berea
> Bowling Green State University, Bowling Green
> Kent State University, Kent
> Marietta College, Marietta
> Muskingum College, New Concord
> Oberlin College, Oberlin
> Ohio Agricultural Experiment Station, Wooster
> Ohio State University, Columbus
> Ohio University, Athens
> University of Cincinnati, Cincinnati

A study of herbarium specimens of a species of plant is, in a sense, a study of individuals that make up the species-population in the area under consideration. It discloses variations apparently related to unlike conditions in different parts of the geographic range (for example, in *Maianthemum canadense*, p. 367). It may show that varieties of a species which are fairly distinct in one part of its range are intergrading in another (see *Luzula*, p. 331, and *Carex flava* and *C. cryptolepis*, p. 270). It may raise questions about the nature of variation and thus suggest problems in cytotaxonomy (see *Arisaema atrorubens* and *triphyllum*, p. 296, *Stenanthium gramineum*, p. 341, *Trillium flexipes*, p. 381, and *Smilax ecirrhata*, p. 385). Such an approach has been useful in *Polygonatum*, p. 371, *Triglochin*, p. 47, and *Acorus*, p. 302.

Thus, we are dealing not alone with identification of plants, with the placing of specimens in their proper pigeon-holes, but also, in the newer systematics, with population-variation, hybridization and introgression (the result of infiltration of germ-plasm of one species into another), and genetic races.

The geographic ranges of species, even within a single state—Ohio—raise the question of why species are where they are. Ranges may, in many cases, be correlated with physical factors of the environment, with climate, topography or physiographic areas, underlying rock (because of its influence on soil), or

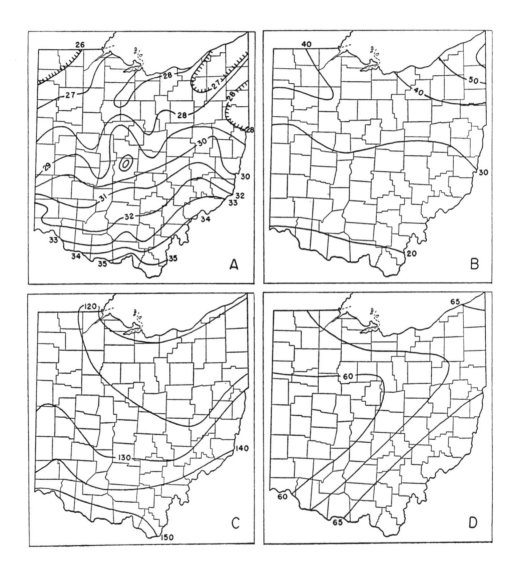

FIG. 1.—A. Average January temperature, ranging from below 26° F. in the northwest and below 27° F. in the northeast, to 33°–35° F. along the Ohio River. B. Average annual snowfall, in inches; note heavy snowfall of northeastern and northwestern Ohio. C. Length of available growing season, i.e. number of days without frost in 4 out of 5 years. Note shorter growing season of northern Ohio, except in the immediate vicinity of Lake Erie, and longer season along the Ohio River. D. Average relative humidity in July (8 P.M.) showing drier lobe over west-central Ohio. (A, from Ohio Department of Natural Resources, 1950; B, C, D, from Atlas of American Agriculture, 1918, 1922.)

supply of ground water. In other cases, an explanation based on the influence of existing environment seems inadequate; past conditions of climate, and migrations related to these, must be considered along with present-day environment. A comparison of distribution maps with the maps included here that

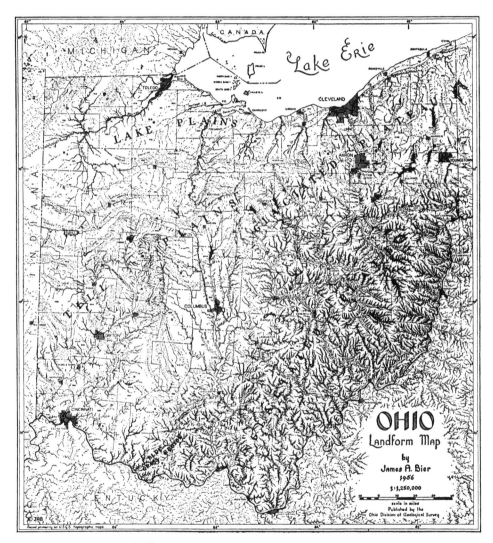

Fig. 2.—Relief map, showing principal physiographic areas of Ohio.

illustrate certain climatic conditions of Ohio (Fig. 1), physiographic areas (Fig. 2), geologic formations and lithological character of underlying rock (Fig. 3) will in many instances show positive correlation. The soils of the Allegheny Plateau, both of the unglaciated and of the glaciated sections, are prevailingly circumneutral (in mesophytic coves and ravines) or acidic (especially on slopes and ridges occupied by oak forest). Local areas of calcareous soil occur, particularly along the southeastern margin of Ohio. In areas of strong

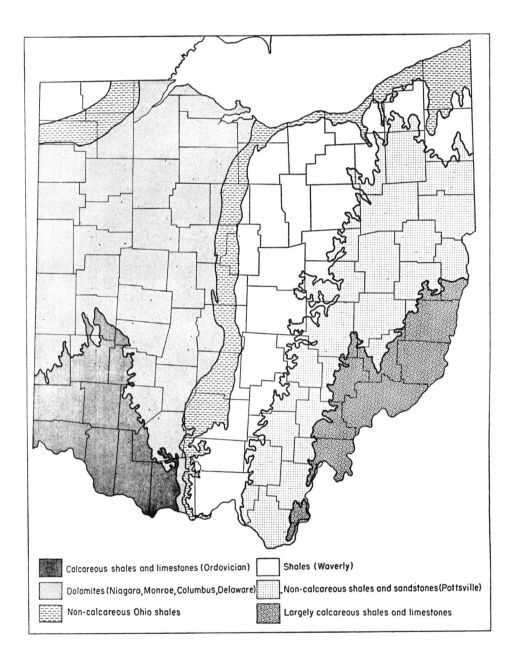

FIG. 3.—Map showing areas of calcareous and non-calcareous rock (after Stout, 1943).

relief (as in the Plateau), calcareous outcrops may be small in extent. The soils of the Till Plains area are largely derived from calcareous materials; however, in the Illinoian drift area, deep leaching of these old glacial deposits has produced acid soils on most of the flat areas. Wherever in the Till Plains streams have cut deeply, bedrock (Ordovician or Silurian) may be exposed. A small area west of the Plateau and east of the Illinoian Till Plain is underlain with Silurian dolomites, often not deeply covered with soil; here environmental conditions differ from those of the rest of Ohio; prairie communities are numerous, and xerophytes of calcareous soils occur.

Often, a combination of factors is involved in determining range or occurrence of species; for example, a southern species may have a soil-preference; *Hexalectris spicata* (p. 441) and *Agave virginica* (p. 395) are southern plants of limestone soil; *Tradescantia subaspera* (p. 313) and *Corallorhiza wisteriana* (p. 443) are less restricted southern species. Furthermore, all woodland herbaceous species are more or less dependent on type of forest, on the dominant tree-species, and hence on shade and on nature of humus; *Disporum lanuginosum* (p. 368) and *Goodyera pubescens* (p. 435) are essentially Alleghenian in range in Ohio, yet they occupy different habitats, the former in humus-rich soil of mixed forests, the latter in acidic soil of oak woods. Some species, for example *Poa cuspidata*, prevailingly Alleghenian in range, occur locally west of that physiographic area, in acidic soils of the Lake Plains or of the Illinoian Till Plain of southern Ohio. *Clintonia borealis* (p. 363) is a northern species of the Hemlock–White Pine–Northern Hardwoods Forest region, and in Ohio is found only where a lobe of that forest region extends into Ohio in the area of lower temperature and heavier snowfall.

When a definite distribution pattern seems evident, and still the correlations suggested above cannot be made, it is well to consider the possibility of glacial and postglacial migrations, and to compare the distribution map with the glacial map of Ohio (Fig. 4). The absence of *Trillium grandiflorum* from extreme southwestern Ohio, though it is abundant just north of the Wisconsin glacial boundary (map, p. 378); the westward extension of range of *Lilium canadense* (p. 353) in the area of Illinoian drift; the limits of range of *Polygonatum biflorum* (in the restricted sense), p. 373, and of *Dioscorea quaternata* (p. 392); and the ranges of the two variants of *Smilax ecirrhata* (p. 385), are a few examples of ranges apparently related to glaciation. Prairie species are frequent in Ohio, usually in west-central Ohio (the Prairie Peninsula area), in northwestern Ohio (Lake Plains), or in unglaciated southern Ohio, particularly in Adams County, sometimes in only one of these areas, sometimes in two or in all three. Such distribution patterns are usually explained as resulting from eastward migrations during the Xerothermic Interval some 5,000 to 3,000 years ago; or, south of the glacial boundary, perhaps from earlier migrations.

The different ranges of the several species of a genus may raise interesting questions as to environmental requirements; for example, *Trillium* (p. 375), *Uvularia* (p. 346), and *Lilium* (p. 355). Two species of *Trillium* are northern and very rare in Ohio; one is western and very local in western Ohio (or perhaps extinct); one (*T. nivale*) is a limestone–soil plant; one is Alleghenian in range;

two are rather general; and a third (*T. grandiflorum*), although apparently general, is strangely absent from extreme southwestern Ohio. In *Uvularia*, *U. grandiflora* is general (a plant of rich forest soil), and *U. perfoliata* is essentially

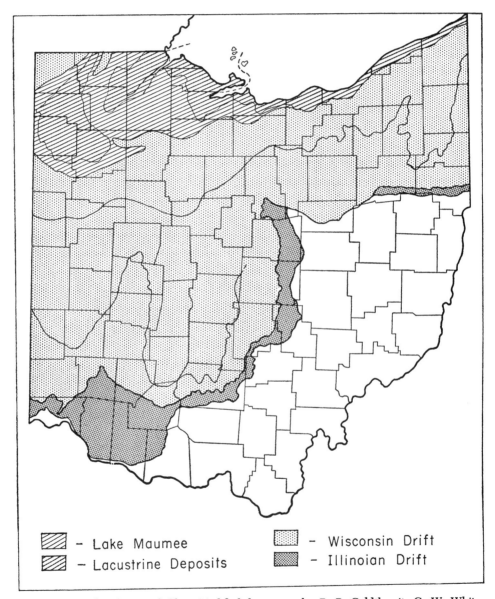

Fig. 4.—Glacial map of Ohio. Modified from map by R. P. Goldthwait, G. W. White, and J. H. Forsyth, published as separate map by U.S. Geological Survey, 1961. Lines within the area of Wisconsin drift indicate limits of readvance of ice and slight differences in clay-content of drift.

Alleghenian. In *Lilium*, the ranges of *L. canadense* and *L. michiganense* scarcely overlap, the former chiefly Alleghenian, the latter more western.

For swamp, bog, or streamside plants, the important factor is adequate water; distribution patterns may bear little relation to the maps of physical features. Large numbers of hydrophytes (aquatic and swamp plants) are wide-ranging because aquatic habitats occur more or less irrespectively of the moisture-climate, but they may display definite latitudinal range. Isolated occurrences of northern species of high-moisture requirements may be related to relic bogs or swamps—habitats dating back to postglacial times in which these isolated plants were stranded.

The recent construction of artificial lakes and innumerable farm ponds (principally in the last few decades) has resulted in the formation of habitats suitable to a large number of pond and swamp species. A surprising number have populated these areas. This has doubtless resulted in range-extension in Ohio, if not in the geographic range as a whole. Cat-tails (*Typha*) intermediate between our two species may occur along a pond-margin even though only one of the species, in typical form, is present. *Eleocharis quadrangulata*, reported by Schaffner (1932) only from northeastern Ohio, has appeared on the borders of artificial lakes in Adams and Brown counties in southern Ohio. Submerged aquatics, also, have appeared. The whole suggests the desirability of studying the recent migrations of hydrophytes, and the means of this migration. Doubtless, waterfowl have played an important part by carrying seeds embedded in mud on their feet southward in their fall migration.

For further discussion of the correlation of vegetation with environment, see *The Woody Plants of Ohio*, pp. 19–30 (Braun, 1961). In the same reference may be found suggestions concerning the use of keys, and definitions of descriptive terms.

KEYS

KEY TO FAMILIES,
OR IN SOME CASES, TO GENERA OF MONOCOTYLEDONS

a. Submersed or floating aquatics (plants with usually emersed leaves may have floating leaves in deep water; see aa).
    b. Plants without true leaves, very small, free-floating flat or flattish, lenticular, or globular fronds ....LEMNACEAE, p. 302
    bb. Plants with leaves; stems simple or branched.
        c. Leaves long (to 2 m), ribbon-like, arising from horizontal rhizomes, and floating vertically in the water ....*Vallisneria*, p. 60
        cc. Leaves shorter.
            d. Leaves radical, bladeless or phyllodial, at least some of them with expanded blade; inflorescence an emergent scape bearing whorls of flowers or of inflorescence-branches ....*Sagittaria, Alisma*, p. 49
           dd. Leaves cauline, sometimes low on stem, then with leaf-like bracts.
                e. Plants with unlike submersed and floating leaves, or with all leaves floating; leaves with stipules or expanded sheathing petiole-bases.
                    f. Floating leaves longer than wide, tapering to base or rarely truncate or subcordate at base; flowers in emergent crowded spikes or heads ....*Potamogeton*, p. 25
                    ff. Floating leaves wider than long, reniform; flowers 1-few on axillary peduncles ....*Heteranthera reniformis*, p. 317
                ee. Plants with submersed leaves only.
                    f. Leaves opposite or whorled.
                        g. Leaves with enlarged more or less toothed sheathing bases; flowers axillary, sessile ....NAJADACEAE, p. 42
                        gg. Leaves without enlarged bases, stipules may be sheathing.
                              h. Leaves opposite, linear-filiform, with sheathing stipules; flowers axillary, sessile, without perianth ....
                                        Zannichellia, p. 42
                            hh. Leaves, at least upper, in whorls of 3 or 4, linear, narrowed to base ....*Elodea*, p. 59
                    ff. Leaves alternate.
                        g. Leaves various, linear to lanceolate or narrow oblong; stipules free or adnate, sometimes inconspicuous; flowers in sessile or peduncled heads or spikes ...*Potamogeton*, p. 25
                        gg. Leaves linear, grass-like, translucent; flowers solitary from terminal spathes which appear lateral, ephemeral, pale yellow ....*Heteranthera dubia*, p. 316
aa. Emersed, amphibious, or terrestrial plants.
    b. Flowers inconspicuous, without perianth, or perianth-parts greenish or scale-like, abortive or represented by bristles.
        c. Flowers in axils of, or enclosed by dry scales or scale-like bracts and often hidden by them, or stigmas or stamens or both exserted.
            d. Plants with cauline leaves, *or* leaves reduced to bladeless sheaths or scales and flowers in dense spikes or spikelets.
                e. Stems (culms) with hard usually solid nodes and hollow cylindric or flattened internodes (rarely solid or with loose removable pith); leaves 2-ranked, sheaths usually split lengthwise opposite leaf-blade
                            GRAMINEAE, p. 61
                ee. Stems (culms) without hard solid nodes, solid but pithy, usually triangular in cross-section, rarely terete, flat or 4-angled; leaves or bladeless sheaths 3-ranked, sheaths closed, tubular ....
                CYPERACEAE, p. 175

  dd. Plants with all leaves basal.
    e. Leaves pellucid; flowers in tight involucrate head at tip of naked scape ....ERIOCAULACEAE, p. 309
    ee. Leaves opaque, grass-like; flowers in axils of hard scale-like bracts on naked scape; petals exserted, yellow, but soon falling ....
                       XYRIDACEAE, p. 309
 cc. Flowers not in axils of or enclosed by dry scales, or at least not hidden by them.
  d. Leaves linear or sword-like, many times longer than wide, flat or cylindric, or reduced to bladeless sheaths.
    e. Flowers in a dense apparently lateral spike on margin of 2-edged scape similar to leaves, and with leaf-like spathe extending beyond it
                         Acorus, p. 302
    ee. Flowers not in lateral spike.
      f. Inflorescence of large globose heads, the upper staminate, the lower pistillate .................................SPARGANIACEAE, p. 20
      ff. Inflorescence not of large globose heads, or if of heads, these all alike and composed of perfect flowers.
        g. Flowers in a thick dense spike divided into 2 contiguous or separated sections, the lower pistillate, the upper staminate
                       TYPHACEAE, p. 17
        gg. Flowers not in a thick dense spike.
          h. Flowers in slender loose spike-like raceme or in open raceme; carpels 3, separating at maturity from persistent axis, or into 3 inflated 1–2-seeded follicles; leaves terete or quill-like ..... JUNCAGINACEAE, p. 47
          hh. Flowers variously arranged; perianth of 6 similar parts; carpels united, forming a dehiscent capsule; leaves flat or cylindric or reduced to sheaths ....
                     JUNCACEAE, p. 317
  dd. Leaves broader, simple or compound, netted-veined; flowers on fleshy spadix subtended by or enclosed in spathe ....ARACEAE, p. 293
 bb. Flowers with herbaceous or colored perianth.
  c. Sepals herbaceous; petals white or colored, *or* both series green, greenish, or brownish and similar.
    d. Petals unlike sepals, white or colored.
      e. Stem with a whorl of 3 leaves at summit subtending a sessile or peduncled flower; fruit a large berry subtended by calyx or calyx-remnants ....*Trillium*, p. 375
      ee. Stem leafy or leaves all basal or nearly so.
        f. Leaves basal or appearing basal.
          g. Inflorescence an umbel; petals pink; plants of mud or shallow water ....BUTOMACEAE, p. 57
          gg. Inflorescence a raceme or panicle with flowers or branches whorled; petals white; fruit a ring or head of achenes; plants of swamps or shallow water ALISMATACEAE, p. 49
          ggg. Inflorescence a tight head on a naked scape; petals yellow, soon deciduous; plants of wet usually acid soils ....
                     XYRIDACEAE, p. 309
      ff. Leaves cauline; stem jointed; flowers regular or irregular, petals usually blue (rarely roseate) ...........COMMELINACEAE, p. 310
    dd. Petals and sepals green, greenish, or brownish, similar.
      e. Plants with grass-like, quill-like, or terete leaves, or leaves reduced to bladeless sheaths; carpels 3.
        f. Flowers in slender interrupted spike-like raceme or open raceme; carpels separating at maturity from persistent axis, or into 3 inflated 1–2-seeded follicles ..............JUNCAGINACEAE, p. 47
        ff. Flowers variously arranged, in spikes and umbellate clusters; perianth-parts, except sometimes at flowering time, dry and chaffy; carpels united forming a dehiscent capsule ....
                     JUNCACEAE, p. 317

ee. Plants with broad leaves; stems leaning or climbing, or with large ovate or rounded leaves crowded toward apex; dioecious.

   f. Stems herbaceous or woody, climbing by stipular tendrils (or without tendrils and leaves crowded); flowers in axillary umbels; perianth of 6 distinct similar, greenish or yellowish parts; ovary superior .... *Smilax*, p. 383

   ff. Stems herbaceous, without tendrils, twining or leaning; leaves strongly ribbed, petiole jointed near base; flowers in axillary panicles or interrupted spikes; perianth calyx-like, 6-cleft; ovary inferior ....DIOSCOREACEAE, p. 391

cc. Sepals not herbaceous, sepals and petals colored, similar or unlike, if greenish then ovary obviously inferior.

  d. Ovary superior.

     e. Flowers irregular, inflorescence a spike; perianth tubular, 2-lipped, blue; plants with basal leaves and 1-leaved flowering stem, growing on shores or in water .... *Pontederia cordata*, p. 316

     ee. Flowers regular or nearly so; perianth-segments distinct or united .... LILIACEAE, p. 335

  dd. Ovary inferior, visible below the perianth.

     e. Twining plants with cordate alternate leaves or the lower usually whorled, dioecious; fruit a 3-winged capsule .... DIOSCOREACEAE, p. 391

     ee. Not twining.

       f. Stamens 3 or 6; flowers regular or nearly so.

         g. Stamens 3; leaves 2-ranked, equitant, linear to lanceolate .... IRIDACEAE, p. 396

         gg. Stamens 6.

            h. Ovary only partially inferior (perianth adhering to base of ovary); perianth-segments united into a cylinder roughened without .... *Aletris*, p. 382

            hh. Ovary obviously inferior; flowers regular or nearly so AMARYLLIDACEAE, p. 392

       ff. Stamens 1 or 2; flowers very irregular, zygomorphic .... ORCHIDACEAE, p. 407

COMMON OR CONSPICUOUS SPECIES—NOT IN FLOWER

A number of the more conspicuous or more common species and species having green foliage throughout the winter may be identified in the field without use of the above key, and at seasons when floral characters are lacking.

Plants with solitary over-wintering leaves:

Tipularia, Crane-fly Orchid—Leaf bronzy green above, rich red-purple beneath (p. 440).
Aplectrum, Putty-root Orchid—leaf deep green, plicate (p. 440).

Plants with over-wintering rosettes:

Goodyera, Rattlesnake-plantain—orchids with dark green leaves with silvery white reticulations (p. 436).
Chamaelirium, Fairy-wand—a member of the Lily Family with bright green oblanceolate or spatulate leaves (p. 338).
Carex platyphylla—leaves glaucous, the 3-ranked arrangement evident in the rosette which develops in early summer and persists until early spring when flowering culms arise (p. 258). Other related species of *Carex* are similar.
Panicum, Panic grass—a number of species develop winter-rosettes of greatly shortened culms (p. 160).
Luzula carolinae, Woodrush—irregular rosette with wide grass-like leaves loosely hairy on margin near base (p. 332).

Plants with whorled leaves:

Lilium, Lily—plants with tall stems and narrow leaves in whorls, or some leaves alternate (pp. 352, 354).

Trillium—stem with whorl of 3 leaves at summit, these with principal veins longitudinal but otherwise netted-veined. Unless bearing fruit (a large berry subtended by calyx-remnants), leaves usually wither before mid-summer (pp. 376, 379, 380).

Medeola, Indian Cucumber-root—stems with terminal whorl of leaves or with two whorls (flowering and fruiting stems), slender and wiry, usually with some wooly hairs persisting (p. 376).

Isotria, Whorled Pogonia—an orchid with terminal whorl of slightly succulent leaves; stem fleshy, glabrous. The species does not flower freely and many stems with the whorl of leaves may be seen without finding a single one with the characteristic orchid flower or capsule (p. 426).

Woodland plants with more or less erect stems bearing alternate leaves:

Smilacina, Solomon's Plume—bright green or glaucous leaves and terminal panicle of berries (p. 366).

Polygonatum, Solomon's-seal—dull green or glaucous leaves and axillary peduncled berries (p. 372).

Uvularia, Bellwort—forked stems (not forked on young plants) with perfoliate leaves (the 2 commoner species) or sessile glaucous leaves (p. 344).

Disporum, Fairy-bells—forked stems with sessile ciliate leaves (p. 369).

Tradescantia, Spiderwort—narrow, 2-ranked leaves on jointed stem; clusters of flower-remnants or fruits in upper axils (p. 312).

Orchids—many species, usually recognizable as orchids by developing or dried capsules with remnants of perianth at apex; see figures.

Low plants, and plants with all leaves basal:

Iris cristata, Crested Dwarf Iris—arching "fans" of leaves 1–2 cm wide; often forming large patches (p. 400).

Allium tricoccum, Wild Leek—bulbous plants with 2–3 large bluish green leaves in early spring (p. 348).

Clintonia—plants with 2–4 large ascending, obscurely ciliate, basal leaves; fruiting plants with umbel of blue or black berries on naked scape (p. 364).

Orchids—a number of species have 2 large basal leaves; see *Cypripedium acaule* (p. 410), *Orchis* (p. 410), *Liparis* (p. 440), and species of *Habenaria* (p. 416).

Symplocarpus, Skunk-cabbage—coarse plants with large petioled basal leaves; growing in seepage swamps and swampy woods (p. 301).

Climbing or leaning plants, dependent for support:

Smilax—wide alternate leaves, stipular tendrils at least above (these often shriveling by mid-summer); both herbaceous and woody species (pp. 384, 386, 388).

Dioscorea, Wild Yam—vines with wide long-petioled leaves, at least the lowest in whorl, or mostly in whorls; petiole jointed near base, the basal part persistent on stem even after leaf-fall; pistillate plants may have axillary racemes of 3-winged capsules (p. 390).

Plants of pond or lake margins, with tall erect leaves,
or stems functioning as leaves:

Typha, Cat-tail—leaves bluish-green, lens-shaped in cross-section, loosely cellular—characters
which will identify cat-tails in the absence of fruiting stems (p. 17).

Acorus, Sweet-flag—flattened, bright green, strongly ribbed leaves (p. 301).

Iris shrevei, I. versicolor, I. pseudacorus—green or bluish green equitant leaves (pp. 402, 406).

Eleocharis quadrangulata, Four-angled Spike-rush—green 4-angled culms, at least some of
which bear a short (2–3 cm) brown spike (p. 190). Many smaller species with terete stems.

Scirpus, especially *S. americanus,* Three-square, and *S. validus,* Great Bulrush (p. 202).

Juncus, Rush, especially *J. effusus* and *J. balticus* (p. 320).

Sparganium, Bur-reed—leaves triangular in cross-section, deep green, loosely cellular and
spongy (p. 20).

FIG. 5.—Reference map for location of counties.

SYSTEMATIC TEXT

The characters given in keys and text are intended not only to aid in identification, but also to give an idea of the appearance of the plant and make it possible to distinguish it from other Ohio species. They would not, in many cases, suffice to distinguish these from species which may be found elsewhere. Including all characters would only complicate the text and make recognition of Ohio species more difficult.

In the following treatment, species native to Ohio are indicated in the text by **boldface** type. Ohio distribution of these native species is shown by maps with a dot in each county in which the species is known to occur; these records are based on specimens actually examined and do not include sight records (see p. 3 of Introduction). The names of introduced species, whether naturalized or adventive, are in SMALL CAPITALS; these species (with few exceptions) are not mapped.

Specific names may differ in different Manuals; correlations are facilitated by the synonomy included—such names are in *italics*; these include names in Fernald (1950), Gleason (1952), Hitchcock (1951), and Schaffner (1932).

MONOCOTYLEDONEAE

Embryo (and seedling) with one cotyledon; stems, as seen in cross-section, with woody fibers (vascular bundles) not arranged in a ring but appearing as scattered dots; parts of flower in twos, threes, fours, or sixes, never in fives; leaves mostly parallel-veined, sometimes (in wide leaves) with principal veins longitudinal, others reticulate, or in many of the *Araceae*, netted-veined.

The arrangement of families follows that found in current manuals. This is based, in large part, on floral features, especially of perianth and carpels, on presence or absence of endosperm, and other technical characters. In general, the sequence (with ramifications) is from taxa with rudimentary or primitive perianth and simple pistil of one carpel (or carpels distinct at maturity), to taxa with more developed perianth and three carpels, and then from superior to inferior ovary, and from regular to zygomorphic flowers.

TYPHACEAE.   Cat-tail Family

A family including only one genus.

1.   TYPHA L.   CAT-TAIL

Swamp or subaquatic plants of temperate and tropical latitudes of both Eastern and Western hemispheres. All have long linear upright leaves sheathing

Typha angustifolia

Typha latifolia

the base or lower part of the jointless vertical stem arising from a creeping rhizome, unisexual flowers, without true floral envelope, crowded in dense spikes, the staminate portion of the inflorescence above the pistillate, contiguous to it or separated by an interval of stem. Two well-marked species, *Typha latifolia* and *T. angustifolia*, and some intermediate forms combining characters of these two species occur in our area. These intermediate forms are variously assigned to *T. glauca* of Hotchkiss and Dozier (1949), and perhaps of Godron, or to *T. angustifolia* var. *elongata* (Dudley) Weigand, which Fernald (1950) considers to be synonymous with *glauca*.

Studies based on mass collections totaling 821 specimens, and on a large number of herbarium specimens, show that there has been introgression wherever *T. latifolia* and *T. angustifolia* occur together: some individuals (or clones), although closely similar to one of these species, have one or two characters of the other species; however, "there are forces, both external and internal, tending to maintain them as specific entities" (Fassett and Calhoun, 1952). In Ohio, there are many more or less recently constructed artificial lakes (state fishing lakes as well as private ponds) around which cat-tails have appeared. Waterfowl, which come in great numbers in late fall or early winter during southward migration, are doubtless responsible for introduction (perhaps by seed in the mud on their feet) of both species as well as of intermediate forms originally developed in more northern lakes. Characters most readily observed, and which will usually serve to distinguish our two species, are presence or absence of an interval of stem between staminate and pistillate portions of the spike, color of pistillate portion, shape of stigma, shape of aborted pistils (carpels), and presence or absence of thickening of tips of hairs of the pistillate flowers, shown in illustrations of separate flowers, and (below) of aborted pistils (after Fassett and Calhoun, 1952).

Leaves of our cat-tails, when split lengthwise, are seen to be made up mostly of air spaces, rectangular in section, separated by very thin white or pale pithy partitions, the green tissue confined to a layer near the surface. This structure gives extreme buoyancy to the long leaves, which may blow back and forth violently but seldom break.

Although cat-tails have usually been thought of as having little economic value, the leaves have been used to make rush-bottom chairs. Recent research at Syracuse University reveals a variety of possible uses.

At least in some areas, there seems to be a correlation between soil acidity or alkalinity and distribution of the two species. *T. latifolia* displays a wide range of tolerance, *T. angustifolia* is more often found in basic waters or even in saline marshes.

Pistillate and staminate portions of spike contiguous, mature pistillate portion dark brown, 1.5–2.5 cm in diam.; stigma flattened and broad at apex; hairs (from stipe) filiform, not thickened at ends; leaves 7–20 mm wide ....1. *T. latifolia*
Pistillate and staminate portions of spike separated by an interval of stem, mature pistillate portion reddish brown, 1.0–1.7 cm in diam.; stigma filiform; hairs (from stipe) minutely enlarged at tip; leaves 3–12 mm wide ....2. *T. angustifolia*

1.   **Typha latifolia** L.   COMMON CAT-TAIL

Almost throughout North America, and widely distributed in Europe and Asia; doubtless present in every county in Ohio, along the borders of lakes and ponds, in swamps, along seepage bands on banks, in roadside ditches. The figures illustrating inflorescence and pistillate flowers give key characters.

2.   **Typha angustifolia** L.   NARROW-LEAVED CAT-TAIL

Less wide-ranging in North America than the preceding, except in the East and South; also in California, Guatemala, Europe, and Africa. The occurrence of the more typical or representative Ohio specimens of the "pure species" is shown by vertical appendages on dots on the distribution map. Taller individuals, with wider leaves and thicker spikes have been referred to *T. angustifolia* var. *elongata* (Dudl.) Weig., or to *T. glauca* Godr.; for disposition, see discussion under *Typha*.

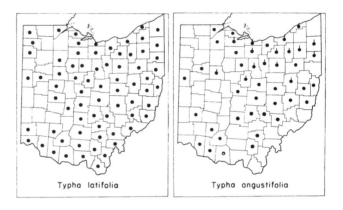

Typha latifolia   Typha angustifolia

SPARGANIACEAE.   Bur-reed Family

Like the preceding family, with only one genus.

1. SPARGANIUM L.   BUR-REED

Swamp or aquatic plants with linear flat, keeled, or (in transverse section) triangular leaves sheathing at base, alternate on a more or less elongate stem. Flowers unisexual, in compact globose sessile or short peduncled heads at or just above the nodes, the staminate above the pistillate; pistillate flowers with inconspicuous perianth of 3–6 linear or spatulate parts; fruit an achene. A genus of some twenty or more species widely distributed in temperate and colder latitudes of the Eastern and Western hemispheres, ten in eastern North America. The common name refers to the globose fruiting heads from which the achene-beaks or style bases project; achenes fall separately, the head breaking apart on the plant.

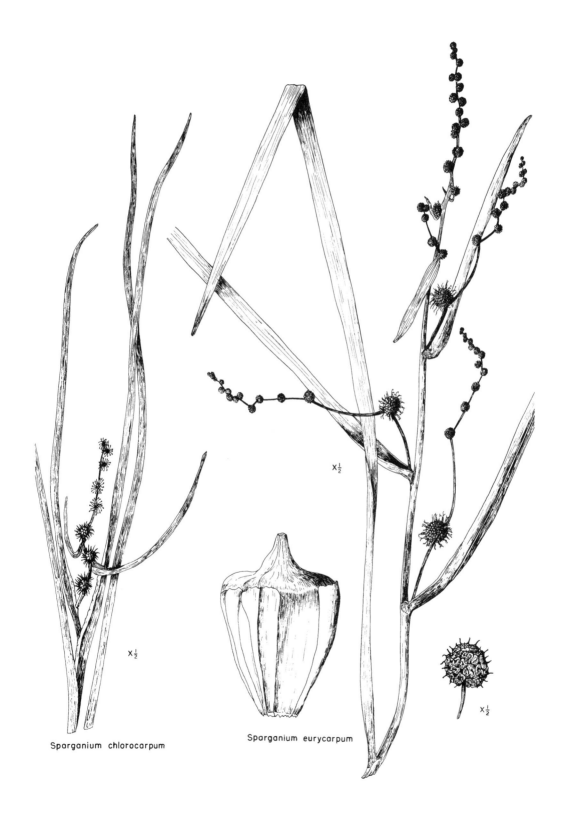

X½

X½

Sparganium chlorocarpum

X½

Sparganium eurycarpum

X½

X$\frac{1}{2}$

X$\frac{1}{2}$

Sparganium androcladum

Sparganium americanum

a. Achenes large (6–10 mm long and nearly as thick), angled, broadly obpyramidal, sessile; stigmas 2; fruiting heads 2–3.5 cm in diam. ....1. *S. eurycarpum*
aa. Achenes widest below summit, fusiform, on short stipe; stigma 1.
    b. Pistillate heads all appearing axillary, sessile or peduncled at nodes; leaves 6–12 mm wide; bracts slightly scarious margined toward base.
        c. Mature pistillate heads 2.5–3.5 cm in diam.; achenes straw-color to pale brown, lustrous; stigmas 2–4 mm long; leaves keeled ....2. *S. androcladum*
        cc. Mature pistillate heads 1.5–2.5 cm in diam.; achenes dull brown (sometimes pale), very finely pitted; stigmas 1–2 mm long; leaves flat or weakly keeled
                                          3. *S. americanum*
    bb. At least one of the pistillate heads supra-axillary, its peduncle fused with stem for a distance above axil; leaves flat or slightly keeled, 3–9 mm wide, bracts broadly scarious margined near base ....4. *S. chlorocarpum*

1. **Sparganium eurycarpum** Engelm.

Our most widely distributed species; distinguished from our other bur-reeds by the large fruiting heads (2–3.5 cm in diam.) made up of obpyramidal achenes 6–10 mm long, the angles a result of pressure while growing, or, earlier in the season, by the presence of 2 stigmas; inflorescence branching, the lower branches with 1–3 pistillate heads, the central axis often all staminate. Leaves usually flat, but often more or less keeled. The illustrations show the inflorescence at flowering time (early or mid-summer), a mature fruiting head (late summer), and one enlarged achene with appressed sepals.

2. **Sparganium androcladum** (Engelm.) Morong

S. lucidum Fern. & Eames, *S. simplex* var. *androcladum* of early ed., not *S. americanum* var. *androcladum* of ed. 7.

This and the next species are frequently confused, and some Ohio specimens of the two species appear to be more or less intermediate, with some characters of each species. The lustrous pale brown or straw-color fusiform achene is the most distinctive character. Inflorescence simple or branched, the branches rarely bearing pistillate heads; leaves often more or less strongly keeled.

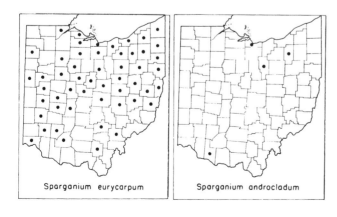

Sparganium eurycarpum   Sparganium androcladum

3. **Sparganium americanum** Nutt.

S. americanum, typ. var., and var. *androcladum* of ed. 7, *S. simplex* var. *nuttallii* Engelm.

More frequent in Ohio than the preceding species, and often confused with it. Recent studies (Beal, 1960) show that this is a complex polymorphic species, at least in some parts of its wide range. Ohio material includes many variations, some characters suggesting *androcladum*, and possibly a result of introgression from that species. Achenes dull brown, very finely pitted, fusiform, but with sides flattened or slightly constricted. Although the leaves of *S. americanum* are said to be flat, a number of specimens with fruit characters of this species have the leaves more or less strongly keeled. The bur-reeds of the *americanum-androcladum* complex need intensive field study and careful analysis of overlapping characters. The illustrations show inflorescence (bracts scarious margined toward base) in mid-July, fruiting head (October), an enlarged achene with surrounding sepals and long stipe, and one with sepals removed to show constriction.

4. **Sparganium chlorocarpum** Rydb.

S. diversifolium of Am. auth., not Graebn., *S. simplex* of ed. 6 of Gray's Manual, in large part.

Best recognized by the narrow leaves (2–7 mm wide) much overtopping the stem and unbranched inflorescence with sessile pistillate heads, at least the lowest of which is supra-axillary (because of fusion of peduncle with stem). Achenes slightly lustrous, greenish brown to brown, fusiform. More northern in range than our other species.

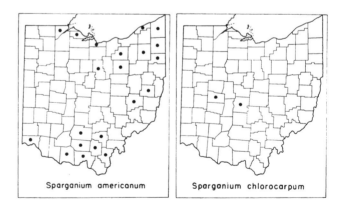

Sparganium americanum   Sparganium chlorocarpum

ZOSTERACEAE.   Pondweed Family

Aquatic, mostly submersed plants, with narrow leaves, or, in a few species, with some of the leaves floating and broader. Flowers inconspicuous, in axillary sessile or peduncled clusters or spikes. This and the next family often united under the name Najadaceae.

Leaves mostly alternate, rarely subopposite; flowers and fruits in peduncled spikes or more or less capitate clusters ....1. *Potamogeton*
Leaves all opposite; flowers and fruits sessile or nearly sessile, clustered in leaf-axils
2. *Zannichellia*

1. POTAMOGETON L. PONDWEED

A genus of some 80 species, cosmopolitan in distribution, found in ponds and slow streams, and the shallow borders of larger lakes. The pondweeds furnish one of the most important foods for wild waterfowl. Plants normally root in the bottom-mud, the flexuous stems ascending in the water and bearing slender, sometimes thread-like alternate leaves, or, in some species, larger but very thin leaves, and, in those species reaching the surface, bearing floating leaves very different in shape and texture from the submersed leaves; both submersed and floating leaves are stipulate, the stipules often relatively large and sheathing. Flowering is sporadic, although some species flower and fruit freely. Shallowing of water during dry periods often results in more abundant fruit production, for under such conditions more of the flowering spikes remain emergent, a position necessary for pollination, as all but a few species (all of ours except *pectinatus*) are wind-pollinated.

Flowers consist only of 4 stamens with 2-locular sessile anthers, and 4 pistils; a true perianth is lacking, its place taken by 4 sepal-like structures, outgrowths of the connectives of the sessile anthers. Each of the 4 pistils may develop into an achene-like, more or less compressed fruit, beaked by the persistent style, and often with a dorsal crest. The embryo is in the form of a hook, ring, or spiral (snail-like), the shape clearly visible in some species. The flowers and fruits are in short capitate, cylindric, or interrupted axillary spikes, at first sheathed by leaf-stipules, later emergent. Winter-buds or hibernaculae are produced in many species; these are greatly shortened and hardened branches made up of axes with crowded stipules and often a few shortened leaves (see *P. strictifolius*, p. 38). In some species, winter-buds are produced only late in the season, in others, more or less throughout the season. These winter-buds, when detached, sink to the bottom because of lack of air spaces in their tissues. They are an important means of vegetative propagation.

For purposes of identification, the submersed leaves and stipules furnish the best characters. Floating leaves, although the most conspicuous feature of those species producing submersed and floating leaves, vary greatly and cannot always be relied on. Stipules—their shape, position (adnate, free, sheathing, divergent), and texture—must be carefully examined. Fruit characters are very good, but because a large proportion of all pondweed specimens collected do not bear mature fruit, and because some species rarely fruit, our key is based on vegetative characters. Color and texture of the outer layer (exocarp) of the mature ovary, presence or absence of dorsal keel or keels, and shape and position of beak are diagnostic characters. In longitudinal section, the curvature of the seed and the inward projection of the endocarp (inner layer of ovary) may be seen. This inward projection, known as the endocarp loop, is solid in all but two of the broad-leaved Ohio species, as shown in the illustration of *P. epihydrus* (p. 29). In *P. praelongus* and *P. richardsonii,* there is a cavity in the loop (see illustration of *praelongus,* p. 38). Fruits of some of the species are illustrated by much enlarged drawings (some original, some after Fernald, 1932, or Ogden, 1943, all to the same scale).

Further details about the species can be found in our Manuals. The linear-leaved species are treated in great detail and illustrated by natural-size photographs in a monograph by Fernald (1932); here also will be found an interesting discussion of geographic distribution, and of the diagnostic value of various structures. The broad-leaved species are treated in a monograph by Ogden (1943). Both monographs contain distribution maps of all species. A later key by Ogden (1953) includes all North American species; it is unique in character, depending on elimination rather than selection, and permits use of any and all characters the specimen or fragment shows.

The distribution maps include Ohio specimens cited by Fernald and by Ogden in the above-mentioned monographs. A large proportion of our specimens have been annotated by Fernald or by Ogden (1961).

In addition to the 24 species of the Ohio flora (some represented by more than one variety), a number of hybrids are recorded.

a.   Plants with floating and submersed leaves.
   b.   Floating leaves large, coriaceous, mostly more than 4 cm long, or if smaller, then acute.
      c.   Petioles for about 1 cm adjacent to leaf blade somewhat contracted and curved, whitish or brownish; submersed leaves linear, 2 mm wide or less, not differentiated into blade and petiole ....1.   *P. natans*
      cc. Petioles of floating leaves uniform throughout, or but gradually tapering, without joint-like portion.
         d.   Submersed leaves linear or ribbon-like, less than 10 mm wide, lacunate in a median band; floating leaves tapering into petiole.
            e.   Submersed leaves ribbon-like, 2–10 mm wide, often plainly 2-ranked; floating leaves mostly opposite, elliptic to narrowly oblanceolate or spatulate, 3–8 cm long; stems flattened ....2.   *P. epihydrus*
            ee. Submersed leaves narrowly linear, less than 2 mm wide; floating leaves lanceolate to lance-oblong, tapering to apex and base, 2–4 cm long; stems terete ....3.   *P. tennesseensis*
         dd. Submersed leaves not linear or ribbon-like; floating leaves mostly alternate.
            e.   Submersed leaves large, 2–8 cm wide and to 20 cm long, arcuate or falcate-folded, variable in shape, tapering into petiole; transitional leaves usually present, floating often absent ....4.   *P. amplifolius*
            ee. Submersed leaves mostly smaller, not arcuate.
               f.   Submersed leaves sessile, from linear to narrowly elliptic, mostly less than 10 cm long; stem much branched ......5.   *P. gramineus*
               ff. Submersed leaves sessile or petioled, mostly longer; stem simple or little branched.
                  g.   Stems and petioles black-spotted; submersed leaves mostly 10–15 cm long; floating leaves rounded to cordate at base
                                              6.   *P. pulcher*
                  gg. Stems and petioles not black-spotted; floating leaves rounded to acute or tapering at base.
                     h.   Floating leaves often absent, transitional leaves present; submersed leaves mostly 2–5 cm wide ....
                                              7.   *P. illinoensis*
                     hh. Floating leaves well developed; submersed leaves decaying, absent on most specimens ....8.   *P. nodosus*

bb. Floating leaves smaller, mostly less than 3 cm long; submersed leaves linear, less than 2 mm wide.

    c. Floating leaves acute ....3. *P. tennesseensis*

    cc. Floating leaves obtuse, rounded at tip, or sometimes minutely emarginate.

        d. Stipules of submersed leaves free from leaf-base, not forming a connate sheath; winter buds numerous, very slender and long-pointed ..9. *P. vaseyi*

        dd. Stipules of submersed leaves adnate to base of leaf, forming a connate sheath with free stipular tips; winter buds rarely developed.

            e. Connate sheath longer than free tip; submersed leaves obtuse; upper stipules hyaline, scarcely fibrous; sides of fruit with deep central depression ....10. *P. spirillus*

            ee. Connate sheath about half as long as free tip, submersed leaves acute to subobtuse; upper stipules becoming fibrous; sides of fruit flat or slightly depressed ....11. *P. diversifolius*

aa. Plants with submersed leaves only.

    b. Leaves linear.

        c. Leaves often distinctly 2-ranked and crowded on sterile branches, 3–10 mm wide.

            d. Leaves auricled at base; stipules whitish, fibrous; inflorescence branching
                12. *P. robbinsii*

            dd. Leaves narrowed to base, ribbon-like; stipules hyaline; spikes absent on submersed plants, from axils of floating leaves on plants with both submersed and floating leaves ....2. *P. epihydrus*

        cc. Leaves not 2-ranked (or if so, internodes long), mostly less than 5 mm wide.

            d. Stems strongly flattened, winged, almost as wide as leaves, and constricted at nodes; leaves thin and grass-like, many-nerved ....
                13. *P. zosteriformis*

            dd. Stems terete, or if flattened, slender; leaves with 1–9 nerves.

                e. Stipules adnate to sheathing leaf-base; stems much branched; leaves very narrow or bristle-like; peduncles long, slender, whorls remote
                  14. *P. pectinatus*

               ee. Stipules free from leaf-base.

                  f. Stipules strongly fibrous, whitish.

                      g. Stem compressed; leaves 1.5–3.5 mm wide, 5–7 nerved; peduncles flattened, 1.5–5 cm long; spike interrupted cylindric ....15. *P. friesii*

                      gg. Stem compressed filiform; leaves 0.5–2.5 mm wide, 3-nerved, obtuse or mucronate; peduncles filiform-clavate, 1-9 cm long; spike interrupted cylindric ....16. *P. strictifolius*

                      ggg. Stem filiform; leaves 1–2.2 mm wide, 3-nerved, acute and cuspidate; peduncles clavate, short (0.5-1.5 cm long); spike capitate ....17. *P. hillii*

                  ff. Stipules membranous or delicately scarious.

                      g. Leaves usually without basal glands; peduncles short, 3–10 mm long; spike subcapitate to thick-cylindric, the 2–3 whorls approximate ....·.............18. *P. foliosus*

                      gg. Leaves usually with 2 small basal glands; peduncles filiform.

                          h. Peduncles 1.5–8 cm long; spike elongate, 6–12 mm long, interrupted, with 3–5 distant whorls; stipules connate, tubular ....19. *P. pusillus*

                        hh. Peduncles 3.5–4 cm long; spike subglobose, 2–8 mm long, scarcely interrupted, with 1–3 whorls; stipules flat or with inrolled free margins ...... 20. *P. berchtoldi*

bb. Leaves lanceolate, oblong, elliptic, but not linear.

    c. Leaves finely and sharply serrulate, at least in apical half ....21. *P. crispus*

    cc. Leaves entire.

        d. Leaves tapering to sessile or petioled base; floating leaves often present.

            e. Stipules mostly less than 3 cm long; leaves sessile, 3–8 cm long ....
                5. *P. gramineus*

            ee. Stipules mostly more than 3 cm long; leaves petioled.

    f.   Leaves falcate-folded, with more than 25 nerves ....
                                                 4. *P. amplifolius*
    ff. Leaves not folded, with less than 20 nerves ....7. *P. illinoensis*
dd. Leaves cordate or rounded at base, or sometimes narrowed to petiolar very slightly clasping base.
    e.   Leaves ovate- or lance-oblong, often with cucullate tip splitting under pressure, 10–20 cm long; stipules white, usually persistent and conspicuous, 3–10 cm long ....22. *P. praelongus*
    ee. Leaves narrowly lanceolate to ovate lanceolate, 3-10 cm long; stipules whitish, soon disintegrating into whitish fibers ... 23. *P. richardsonii*
    eee. Leaves ovate, or ovate lanceolate to orbicular, 1–6 cm long; stipules membranous, delicate, appressed to stem, soon disintegrating ....
                                             24. *P. perfoliatus*

1. Potamogeton natans L.   FLOATING PONDWEED

A northern species ranging from New England and the Maritime provinces of Canada through the Great Lakes region and westward to the Pacific states and to Alaska; also in Eurasia. Most easily recognized by the joint-like portion of the long petiole adjacent to the leaf-blade of floating leaves. Floating leaves 5–10 cm long, oblong-ovate to elliptic-ovate, rounded at base, rounded or slightly mucronate at apex. Submersed leaves very long and slender (10–20 or more cm long, 2 mm or less wide), early disintegrating. Stipules conspicuous, 4–10 cm long. The obovoid plump fruits rounded on back or slightly keeled, short-beaked.

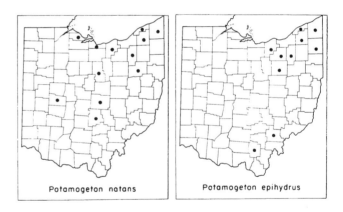

Potamogeton natans   Potamogeton epihydrus

2. Potamogeton epihydrus Raf.

A species made up of two intergrading varieties with overlapping ranges. The var. *ramosus* (Peck) House (var. *nuttallii* (C. & S.) Fern.) has the submersed leaves more prominently 2-ranked and crowded on new shoots than var. *epihydrus*, and with fewer veins (3–7 nerves instead of 7–13); it is very common in the East, but ranges westward to the Pacific Coast, as does also the typical variety, which is most frequent in the Great Lakes region. The opposite floating leaves and long (up to 2 dm) ribbon-like 2-ranked submersed leaves distinguish this species. Fruits flattened, 2.5–4 mm long, with 3 dorsal keels, the middle one most prominent.

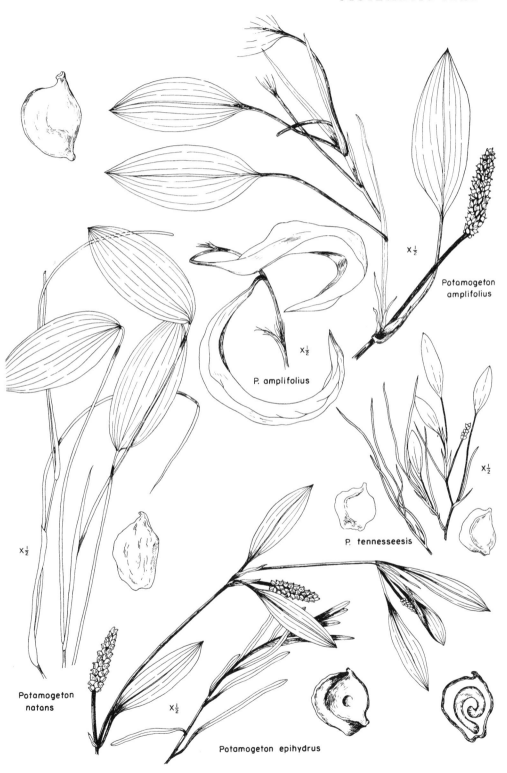

x½

Potamogeton
amplifolius

x½

P. amplifolius

x½

P. tennesseesis

x½

Potamogeton
natans

x½

Potamogeton epihydrus

3. **Potamogeton tennesseensis** Fern.

A southern species originally described from Tennessee, later found in West Virginia (Ogden, 1947), and now known to occur in Ohio (specimens determined by Ogden, 1961). Submersed leaves linear, 0.2–2.0 mm wide (or sometimes wider), with 1–3 veins (rarely 5), lacunate along mid-vein or from margin to mid-vein; stipules free from leaf-blade or adnate at base, a single branch sometimes with both free and adnate stipules. Floating leaves 2–5.5 cm long, 0.5–3.5 cm wide. Peduncle clavate, 3–8 cm long; spike cylindric. Fruit quadrate-orbicular, 2.5–3 mm long with low convex sides and acute, usually entire, dorsal keel.

4. **Potamogeton amplifolius** Tuckerm.   LARGE-LEAF PONDWEED

As the specific name suggests, the leaves (both submersed and floating) are large; transitional leaves usually present; floating leaves sometimes absent. Submersed leaves of two types—the upper arcuate or falcately folded (because margins are longer than midrib), broadly lanceolate to ovate, the lower lanceolate and early decaying, both types 1–2 dm long, tapering to petioles 1–6 cm long, widely acute at apex. Floating leaves with rounded or mucronate apex and rounded or cuneate base. Stipules large, thin and fibrous, those of floating leaves with 2 distinct keel-nerves. Peduncles up to 3 dm long, thickening upward, the spikes 4–8 cm long, 1–1.5 cm thick. Plants of wide geographic range, usually growing in deep water.

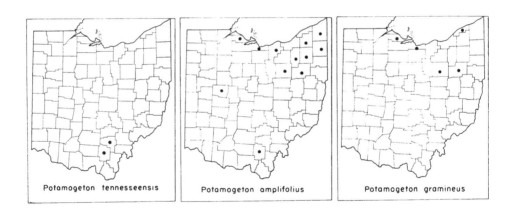

Potamogeton tennesseensis   Potamogeton amplifolius   Potamogeton gramineus

5. **Potamogeton gramineus** L.

P. heterophyllus of most auth., not Schreb.

An extremely variable species "characterized by a stem with many lateral compound branches bearing numerous small leaves." Three varieties are distinguished by Ogden (1943) as follows:

a. Principal submersed leaves narrowly elliptic to oblanceolate, (1–)1.5–9(–13) cm long, .2–1(–1.5) cm wide, 5–10 times as long as broad, or if more than 10 times, then not less than 6 cm long, sides not parallel; nerves (3–)5–9.

    b. Principal submersed leaves (1–)1.5–4.5(–6.5) cm long, .2–.6(–.8) cm wide; nerves
        5–7 ....var. *typicus*
    bb. Principal submersed leaves (3–)6–9(–13) cm long, .6–1(–1.5) cm wide; nerves
        7–9(–11) ....var. *maximus*
aa. Principal submersed leaves linear (1–)1.5–3.5(–5.3) cm long, .1–.25(–.3) cm wide,
    10–20(–30) times as long as broad, sides essentially parallel for most of their length,
    tapering at apex to an acute tip; nerves 3 ....var. *myriophyllus*

Floating leaves (if present) elliptic to ovate, 2–5 cm long, long-petioled. A wide-ranging species, more abundant north of our area; also in Eurasia; poorly represented in Ohio.

6. Potamogeton pulcher Tuckerm. Spotted Pondweed

A southern species, chiefly of the Coastal Plain and Mississippi Valley, with scattered interior occurrences. Submersed leaves narrower than those of *amplifolius*, with acute apex and tapering to short petiole or sessile. Floating leaves ovate or almost rotund, with subcordate base, and rounded but mucronate tip. The large cordate floating leaves, lanceolate submersed leaves tapering to short petiole, and black-spotted stems aid in recognition of this species.

7. Potamogeton illinoensis Morong.

P. lucens of Amer. auth., not L.; *P. angustifolius* of auth., not Berchtold & Presl.

A highly variable American species, long confused with the European *P. lucens,* ranging from southern Canada southward to peninsular Florida, Texas, and southern California, and also in the West Indies, Mexico, and Central America. Floating leaves often absent or transitional from submersed leaves, elliptic to ovate- or oblong-elliptic, with 2 or 3 rows of air cells (lacunae) along midrib, bluntly mucronate, petioles shorter than blades. Submersed leaves elliptic or oblong-elliptic to lanceolate, often somewhat arcuate. Fruits obovate to orbicular or ovate, with well-developed dorsal keel and less prominent lateral keels.

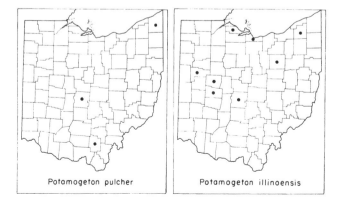

Potamogeton pulcher   Potamogeton illinoensis

Potamogeton gramineus

X½

X½

Potamogeton
pulcher

X½

Potamogeton illinoensis

X½

X½

Potamogeton nodosus

X½

P. nodosus

8. **Potamogeton nodosus** Poir.   FLOATING PONDWEED
   P. americanus C. & S.; *P. lonchites* Tuckerm.

A common and widespread species ranging over much of the United States except the extreme Southeast; also in Mexico, the West Indies, South America, Africa, and Eurasia. The most frequent of the large-leaved floating pondweeds in Ohio, occurring in all quarters of the state. Often confused with *P. natans* (which is northern), perhaps because of the superficial resemblance of floating leaves; readily distinguished from that species by the absence of the joint-like section of the petiole adjacent to the leaf-blade, and by the linear lanceolate to broadly lance-elliptic submersed leaves (these frequently disintegrating before abundant fruiting of the floating branches). Peduncles usually thicker than the stems. Fruits obovate, with prominent keels.

9. **Potamogeton vaseyi** Robbins

One of our four pondweeds with small floating leaves (6–15 mm long, 3–8 mm wide). Stems filiform, bearing numerous short lateral branches ending in small slender winter buds; submersed leaves linear-filiform, less than 0.5 mm wide, 3–6 (–8) cm long; floating leaves on very slender petioles about as long or longer than leaves. Plants with floating leaves bear flowers and fruit; spike interrupted cylindric, on peduncle 10–15 mm long. Plants without floating leaves do not bear flowers, and could be mistaken for sterile specimens of *P. pusillus*, from which they can be distinguished by the small winter buds on short branches, and the free stipules; those with floating leaves can be distinguished from *P. spirillus* of similar range (*P. diversifolius* is more southern) by stipules and flowering habit. *P. vaseyi* is a northern species, with range extending from Maine and eastern Pennsylvania through the mid and southern Great Lakes region to eastern Minnesota.

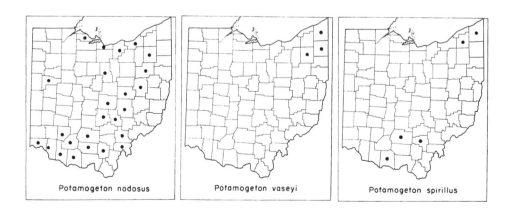

Potamogeton nodosus   Potamogeton vaseyi   Potamogeton spirillus

10. **Potamogeton spirillus** Tuckerm.
   P. dimorphus in part, of Gray, ed. 7, not Raf.

This species (which is northern) and the next (which is southern) have often been confused, partly because of similar appearance, partly because of

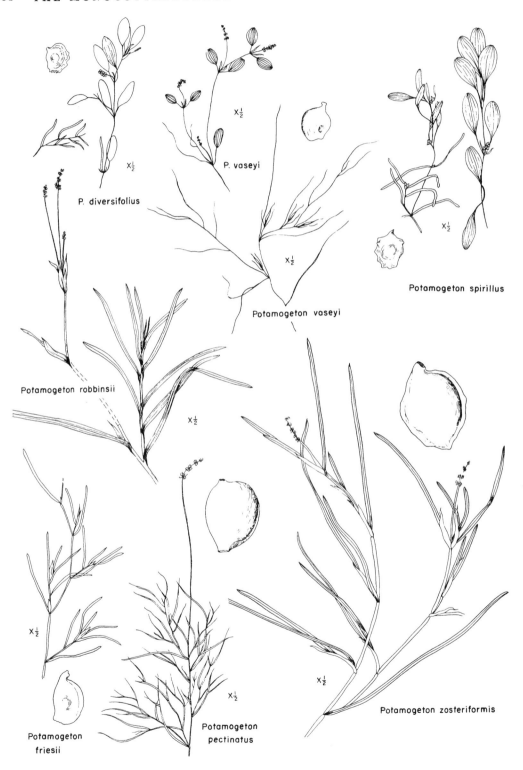

P. diversifolius

X½

X½

P. vaseyi

X½

Potamogeton vaseyi

X½

Potamogeton spirillus

Potamogeton robbinsii

X½

Potamogeton zosteriformis

X½

X½

Potamogeton
friesii

Potamogeton
pectinatus

X½

X½

confusion of nomenclature; both have been included in "*P. dimorphus.*" "It is characterized by its obtuse (usually round-tipped) and definitely linear pale-green, often curving, submersed leaves mostly definitely emarginate; the very few flowered and nearly sessile submersed spikes; the fruits large for the group, with dorsal keel toothless or with at most 7 remote low teeth, the sides rounded and without sharp keels" (Fernald, 1932).

11. Potamogeton diversifolius Raf.

P. dimorphus in part, of Gray, ed. 7, not Raf.

A species of southern range (in contrast to the two preceding), common in the Mississippi River drainage to southern Indiana and southwestern Ohio, southwestern Wisconsin and southeast Minnesota, westward to and across the Rocky Mountains, and on the Piedmont and Coastal Plain northward to Pennsylvania and New Jersey. "From *P. Spirillus* . . . it is at once distinguished by the small fruits with sharp lateral keels. . . . submersed leaves usually narrower and less rounded at tip; the free tip of the lower stipules much longer; the upper stipules longer and more fibrous; its submersed spikes usually more exserted; its emersed spikes on shorter peduncles; its fruits at most 1.5 mm in diameter, with sharp lateral keels (in *P. Spirillus* 1.3–2.2 mm in diameter and with rounded keelless sides) and with flatter sides" (Fernald, 1932).

12. Potamogeton robbinsii Oakes

Distinguished by the 2-ranked linear leaves auricled at base and crowded in feather-like arrangement on sterile branches, their margins pale and cartilaginous, usually minutely spined toward acute apex; whitish acute stipules closely wrapping the internode above and becoming finely fibrous; branched inflorescence on elongate stems bearing smaller and more remote leaves. A northern species ranging from New Brunswick and New England westward across Canada and northern United States to British Columbia and Oregon.

13. Potamogeton zosteriformis Fern.   FLAT-STEM PONDWEED. EEL-GRASS POND-WEED

P. zosterifolius of Amer. auth., not Schum.; *P. compressus* of Amer. auth., not L.

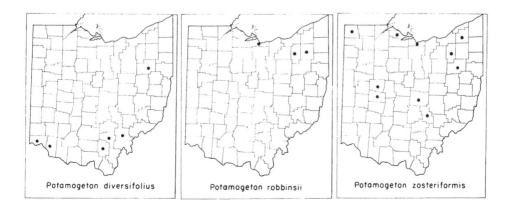

Potamogeton diversifolius   Potamogeton robbinsii   Potamogeton zosteriformis

Long confused by American students with the European species having similar vegetative characters but quite unlike fruit. Distinguished from all other of our species by the flattened winged stems constricted at the nodes. Leaves grass-like, linear, with 3 veins more prominent than the numerous fine ones; stipules pale, fibrous, 1–3 cm long; spikes cylindric, 1.5–2.5 cm long; peduncles 2–5 cm long; fruits large (4–5 mm long, 3–3.5 mm wide), truncate at base; winter-buds large, the green leaf-tips projecting beyond the pale fibrous stipular wrapping. A rather wide-ranging and abundant species of New England, New York, and the Great Lakes northward to James Bay and northern Alberta, west to British Columbia and northern California; entirely north of the Ohio River.

14. Potamogeton pectinatus L.   FENNEL-LEAVED PONDWEED

The long, narrowly linear leaves (3–10 cm) tapering to long bristle tips, and manner of branching distinguish this submersed pondweed. Stems filiform, up to 1 m long, little-branched below, intricately and dichotomously branched above, the branchlets sheathed by the adnate stipules; peduncles long and slender; spike with 2–6 whorls, the lower remote, the upper approximate; fruits plump, 3–4 mm long, short-beaked. A very wide-ranging species of shallow brackish or calcareous waters of coastal and interior areas; also in South America, Africa, and Eurasia. The tubers, borne on the slender rootstocks, are said to be an important food for ducks.

15. Potamogeton friesii Rupr.

A linear-leaved northern pondweed of which two Ohio specimens were cited by Fernald (1932). Stipules strongly fibrous, connate below, lacerate in age. Peduncles flattened and broadened upward. Winter-buds terminating short lateral branches and covered with pale fibrous stipules.

16. Potamogeton strictifolius Ar. Benn.

Rare and local in northern Ohio. A linear-leaved species with strongly fibrous stipules and filiform but compressed stems. Leaves firmer than those of the preceding species, with fewer nerves, peduncles more slender. Winter-buds on elongate and on short branches.

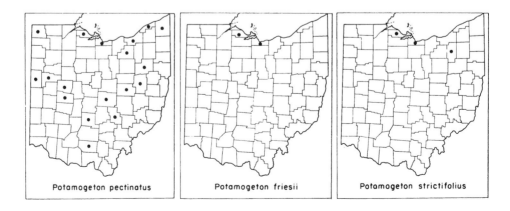

Potamogeton pectinatus   Potamogeton friesii   Potamogeton strictifolius

17. Potamogeton hillii Morong

A very local northern species of which a Portage County specimen was cited by Fernald (1932). A linear-leaved species with acute and cuspidate leaves, whitish stipules with strong persistent fibers; peduncles bearing few-flowered capitate spikes; winter buds unknown.

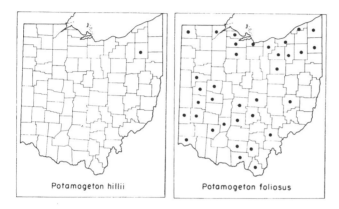

Potamogeton hillii   Potamogeton foliosus

18. Potamogeton foliosus Raf.

A common and widespread linear-leaved submersed pondweed segregated by Fernald into two varieties which "in their extremes . . . might easily be taken for two species, but so many transitions in size of foliage and in shape of fruit occur that it has been found impossible to make the sharp separation which should be required for fully differentiated species" (Fernald, 1932). The typical variety, *foliosus,* or var. *genuinus* of his monograph (*P. foliosus* var. *niagarensis* (Tuckerm.) Morong) is larger than var. *macellus* Fern. (*P. foliosus* of most authors), and more southern in range. The more northern var. *macellus* is "smaller throughout" and "in the East is rare south of the Potomac"; that is, in the East, the varieties are geographically separated, while in the interior, the ranges not only overlap but to a considerable extent coincide. In Ohio, some specimens can be referred to var. *foliosus,* some to var. *macellus,* but many are intermediate; varieties are not separated on our map. We have specimens of the typical variety, annotated by Fernald, from Auglaize, Champaign, Erie, Hamilton, Medina, Miami, and Montgomery counties, and of var. *macellus* from Champaign (same lake and date as typical variety), Cuyahoga, Erie, Fairfield, Licking, Logan, Summit, and Williams counties. Characters of var. *foliosus* and (in parentheses) of var. *macellus* are: primary leaves 4–10 (1–7) cm long, 1.4–2.7 (0.3–1.5) mm broad, 3–5 (1–3) nerved; midrib compound below middle (simple or subsimple) with 1–3 rows of lacunae on each side at base (without marginal lacunae or with a single row on each side below the middle); stipules 7–18 (3–11) mm long; fruits obliquely suborbicular, 2–2.5 mm long (obliquely obovoid, 1.8–2.3 mm long); winter buds sessile in the axils or on short branches (terminating mostly elongate branches), 1–1.6 cm long (0.8–1.2 cm long) (from Fernald, 1932). The short (3–10 mm) stout peduncles, slightly thickened upward, and subcapitate or thick-cylindric spikes are distinctive characters of the species.

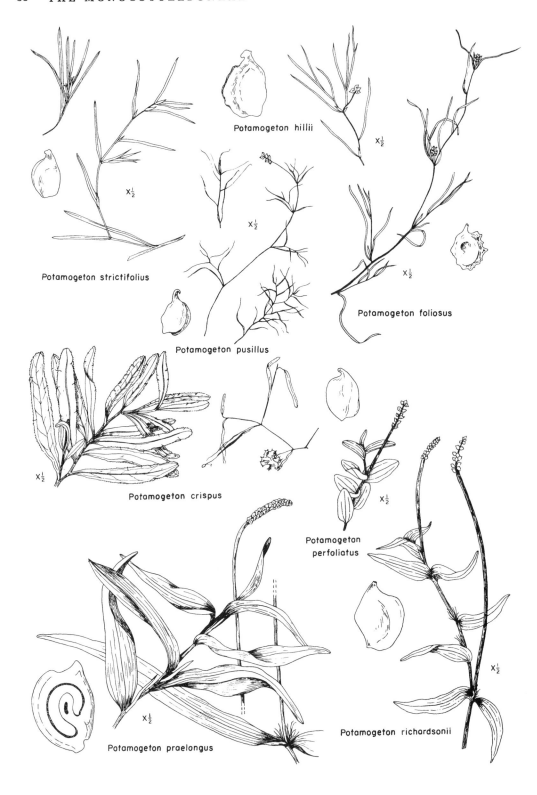

Potamogeton hillii

Potamogeton strictifolius

Potamogeton pusillus

Potamogeton foliosus

Potamogeton crispus

Potamogeton perfoliatus

Potamogeton praelongus

Potamogeton richardsonii

19. **Potamogeton pusillus** L.   SMALL PONDWEED

P. pusillus var. *capitatus* of Gray, ed. 7; *P. panormitanus* Biv., incl. var. *minor* Biv.

The interpretation and nomenclature of "pusillus" is confused. In his monograph, Fernald (1932) distinguished *panormitanus* and *pusillus*, and listed two varieties of the former, six of the latter. His maps and citations give five Ohio stations for *panormitanus* (Ashtabula, Auglaize, Erie, Ottawa, Summit counties) and two for *pusillus* var. *tenuissimus* (Franklin and Ottawa counties). In his manual (1950), we find that *panormitanus* is now *pusillus*, and the *pusillus* of the mongraph is now *P. berchtoldi* Fieber, with the var. *tenuissimus* (Mert. & Koch) Fern. to which our two records (Franklin and Ottawa counties) should be transferred. Furthermore, *pusillus* of Gray, ed. 7, is now *berchtoldi* var. *acuminatus* Fieber (of which we have an Ashtabula County specimen determined by Ogden), but *pusillus* var. *capitatus* of Gray, ed. 7, is now along with *panormitanus* included in *pusillus*.

Potamogeton pusillus

Our small pondweed, *pusillus*, is a very slender stemmed, freely branching plant with narrowly linear leaves and filiform peduncles.

The two very similar species may be separated by the following characters:

Leaves narrowly linear, .8–7 cm long, 1–3 mm wide; stipules slenderly tubular, 6–17 mm
   long; peduncle filiform, 1.5–8 cm long, spikes strongly interrupted, 6–12 mm long
                                                                                          P. pusillus
Leaves .8–8.5 cm long, .3–2.4 mm wide; stipules flat or with inrolled but free margins,
   3–14 mm long; peduncles filiform, .3–3(–4.5) cm long; spikes subglobose or slightly
   interrupted, 2–8 mm long ....*P. berchtoldi*

The one specimen that we have which was annotated by Fernald as *pusillus* var. *tenuissimus* (from Franklin County) has the spikes interrupted, and quite similar to those of specimens annotated as *panormitanus*. Our specimen of *berchtoldi* var. *acuminatus* has very short-cylindric and subcapitate spikes. On the distribution map, *pusillus* and *berchtoldi* are not distinguished.

20. Potamogeton berchtoldi Fieber

Confused with *P. pusillus* and not separated from that species by Gleason (1952). Differing in its somewhat shorter and more capitate spikes and stipules, with free instead of connate margins. Concerning the stipules characters, Gleason writes, "While free stipules or united ones may be characteristic of some species, there is little reason to infer that both types or gradations between them may not occur within a single species." It has been shown that such gradation does occur in *P. tennesseensis*. For differences between *pusillus* and *berchtoldi* (with six varieties, two in our range), and for nomenclature, see discussion under the preceding species, *P. pusillus*.

21. POTAMOGETON CRISPUS L.   CURLY-LEAF PONDWEED

A native of Europe, now widely naturalized. A submersed species, easily recognized by its oblong to linear-oblong, sharply serrulate sessile leaves, which at maturity are often wavy-margined or crisped. Midrib prominent, a pair of lateral nerves near leaf margin connected with midrib by oblique cross-veins. Fruits obliquely ovoid, tapering to a long beak, at base a somewhat recurved appendage. Winter-buds bur-like, somewhat isodiametric.

22. Potamogeton praelongus Wulfen.   WHITE-STEM PONDWEED

A widespread northern species found in Ohio in a few northern counties. "Plants characterized by large ovate-oblong leaves, cucullate at the tip, whitish stem, large conspicuous stipules, and with long peduncles bearing large fruits" (Ogden, 1943). Because of the cucullate or hooded leaf-tip, the tip sometimes splits under pressure resulting in two small apical teeth. The cavity in the endocarp loop (see fig.) is a character found (in our species) only in *praelongus* and *richardsonii* (Ogden, 1943).

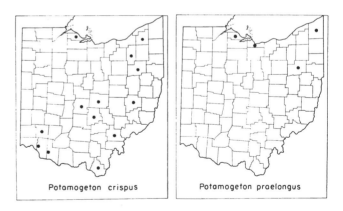

Potamogeton crispus   Potamogeton praelongus

23. Potamogeton richardsonii (Ar. Benn.) Rydb.   RED-HEAD PONDWEED

Comparable in range to the preceding species. Leaves perfoliate, mostly ovate-lanceolate, varying from ovate on lower part of stem to narrow-lanceolate above, mostly large (3–10 cm long) but on some specimens small (2–3 cm long) with crisped but entire margin, rather coarsely veined, with 3–7 nerves more prominent than the others; stipules thin but coarsely veined, soon disintegrating into whitish fibers; fruit with cavity in endocarp loop. Closely related to the next species, and often mistaken for it.

24. Potamogeton perfoliatus L.

The typical variety of Eurasia, northern Africa, and Australia occurs sparingly in the maritime provinces of Canada; the var. *bupleuroides* (Fern.) Farw. has a more extensive range, but is common only in the northeastern part of its range—New England and New York—becoming rare westward. "*P. perfoliatus* and *P. richardsonii* can scarcely be distinguished by leaf shape. They can easily be separated when mature fruits are present [*richardsonii* has a cavity in the endocarp loop, *perfoliatus* does not] but when sterile one must rely on the stipules which are delicate in *perfoliatus* and coarsely fibrous in *richardsonii*" (Ogden, 1961, *in litt.*). Probably the two species hybridize resulting in plants with stipules of intermediate character. Ohio specimens of *P. perfoliatus* var. *bupleuroides* have been cited by Ogden (1943) from Put-in-Bay and from Roscoe [Coshocton Co.], the latter one of Riddell's early collection. Another sheet of Riddell's is in the herbarium of the University of Cincinnati; this Ogden states (*in litt.*) "is intermediate between *Potamogeton richardsonii* and *P. perfoliatus*, but . . . closer to *P. richardsonii*," and that probably "all of the Ohio plants should be referred to *P. richardsonii*."

Potamogeton richardsonii

POTAMOGETON HYBRIDS RECORDED FOR OHIO

(Determined by E. C. Ogden)

P. praelongus × richardsonii
    Summit.

P. gramineus × richardsonii
    Stark, Lake.

P. nodosus × richardsonii ?
    Erie.

P. richardsonii × sp.
    Cuyahoga, Ashtabula.

2. ZANNICHELLIA L.   Horned Pondweed

Submersed, very slender and widely branching herbs, cosmopolitan in distribution. Flowers sessile, unisexual, without perianth, usually both staminate and pistillate in the same leaf-axil; achenes oblique, often falcate, narrowly keeled, and tipped with the short style, hence the name "horned pondweed."

1. **Zannichellia palustris** L.   Horned Pondweed
Ohio plants belong to the typical variety, which ranges almost throughout North America, and is found also in South America, Africa, and Eurasia. Distinguished from species of *Najas* by the longer (3–10 cm) and more slender leaves, and clustered achenes, 2–3 mm long, exclusive of beak.

Zannichellia palustris

NAJADACEAE.   Naiad Family

Submersed aquatics with narrow opposite leaves enlarged to a sheathing base; minute flowers (without perianth) and one-seeded fruits solitary in leaf- or branchlet axils.

1. NAJAS L.   Naiad

A cosmopolitan genus of some 35 species growing in quiet non-saline water. Five species in our area, including the introduced *N. minor*, which is becoming locally abundant. Leaves opposite, leaf-blades toothed or with minute marginal spinules (see illustration of *N. gracillima*), small (1–4 cm long), very slender (mostly less than 1 mm wide), and widening to a sheathing base. The sheathing base (not sheathing stipules) distinguishes *Najas* from other small linear-leaved aquatics. Flowers unisexual, without perianth; the single ovary of pistillate flower with 2 or 3 stigmas; fruit ellipsoidal or fusiform, 1.5–3.5 mm long (–7 mm in *N. marina*), the outer wall very thin and with age breaking and sloughing off (or can gently be teased off), then exposing the seed similar in shape and size to the fruit, but with its walls characteristically sculptured, the sculpturing produced by alternation of tiny air spaces (areolae) and their separating walls (shown much enlarged with the illustrations of four species).

The best characters to use in identifying species of *Najas* are shape (and teeth) of sheathing leaf-base, nature of leaf-margin (definitely toothed or with minute spinules, i.e., projecting transparent marginal cells), and especially the markings on seed-coat. These characters are shown in accompanying illustrations. Aspect and leaf-length are of little value, because of extreme variability.

a. Leaves (including teeth) 2–4 (–5) mm wide, teeth coarse, angular, giving sinuate appearance to leaf-margin, sheathing base rounding to blade, entire or with 1–2 teeth; fruit ellipsoidal, 4–7 mm long; seed finely pitted or reticulate; dioecious......1. *N. marina*
aa. Leaves (including teeth) less than 2 mm wide, teeth fine, sometimes not discernible without lens; monoecious.
    b. Sheathing leaf-base abruptly narrowing to linear blade, more or less truncate and toothed across shoulders or auricles; fruit slender.
        c. Teeth of leaf-margin minute, a rounded sinus between leaf-blade and conspicuously toothed sheath-auricles; areolae of seed-surface quadrangular, higher than wide ....2. *N. gracillima*
        cc. Teeth of leaf-margin evident, short-toothed sheath-auricles usually divergent from blade without noticeable sinus; areolae of seed-surface much wider than high, the longitudinal ribs between rows of areolae discernible on fruit as well as on seed...3. *N. minor*
    bb. Sheathing leaf-base rounding or tapering to linear blade, spinulose on margin, teeth of leaf-margin minute; fruit slender or broadly ellipsoid.
        c. Leaves tapering to slender pointed tip; style and stigmas long (1–2 mm), seed-coat finely and inconspicuously reticulate, the areolae more or less hexagonal ....4. *N. flexilis*
        cc. Leaves acute to obtuse; style and stigmas short (less than 1 mm); areolae of seed-coat squarish ....5. *N. guadalupensis*

1. **Najas marina** L.

Readily distinguished from our other species by its wider and conspicuously toothed leaves. Represented in Ohio herbaria by a single collection from Erie County.

2. **Najas gracillima** (A. Br.) Magnus

A very slender and delicate aquatic of ponds and shallow water of lake margins from Maine to Minnesota and southward to Virginia, Kentucky and Missouri. Leaves 1.5–3.5 cm long, less than 0.5 mm wide, the teeth mere cellular projections (see fig.). The shape and toothing of auricles of leaf-sheath and the quadrangular vertically elongated areolae of seed-surface distinguish this species.

3. NAJAS MINOR All.

Native of Eurasia and Africa, and according to our manuals only locally established in American waters; the first collections were made in 1934 in the Hudson River at the mouth of the Mohawk (Clausen, 1936). In recent years, specimens have been collected from several of Ohio's artificial lakes, in which it may be the most abundant species. Leaves crowded toward end of branches, 1–2.5 cm long and very narrow, teeth of leaf-margin more conspicuous than those of other species (except the rare *N. marina*); subtruncate sheath-summit

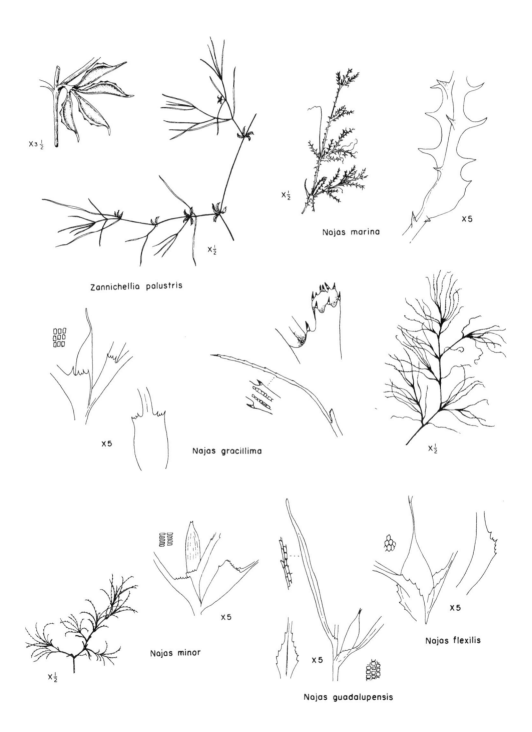

X3½

X½

Zannichellia palustris

X½

Najas marina

X5

X5

Najas gracillima

X½

Najas minor

X5

X½

Najas flexilis

X5

Najas guadalupensis

X5

X5

toothed and not separated from blade by a rounded sinus. Long-leaved plants resemble *N. gracillima,* but are distinguished by the transversely elongate areolae of seed-coat.

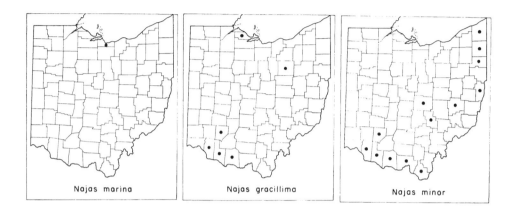

4. **Najas flexilis** (Willd.) Rostk. & Schmidt

A wide-ranging species of northern latitudes of North America and Europe. Leaves 1–3 (–4) cm long, gradually tapering to long slender often recurving tip, and to rounded sheathing base. Fruit sometimes broader in relation to length than in our other species; seed hard and lustrous, the hexagonal areolae visible only with high magnification. A distinctive bushy-branched habit is seen on many specimens, several to many short branches arising from the tuft of fine roots.

5. **Najas guadalupensis** (Spreng.) Magnus

Distinguished by its shorter (0.5–2 cm), often crisped and sometimes finely toothed leaves, and squarish areolae of seed surface. Ranging through much of tropical America, northward through the United States.

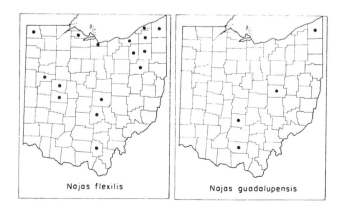

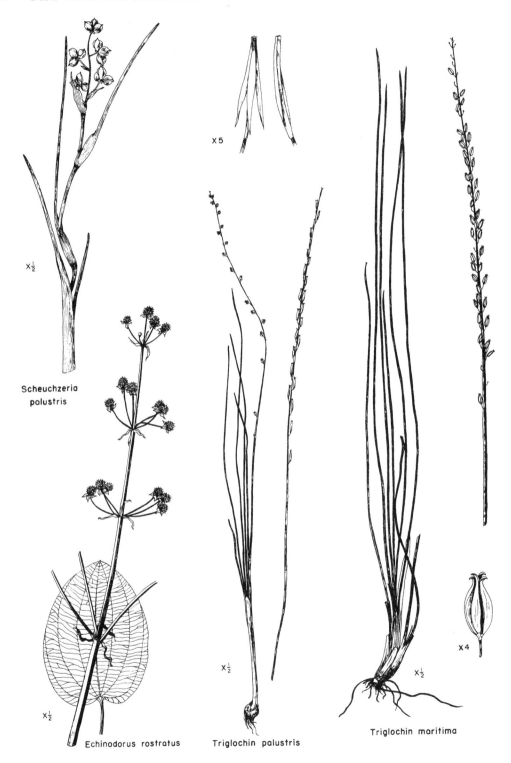

X 5

$\times \frac{1}{2}$

Scheuchzeria
palustris

$\times \frac{1}{2}$

$\times \frac{1}{2}$

$\times \frac{1}{2}$

$\times \frac{1}{2}$

$\times 4$

Echinodorus rostratus

Triglochin palustris

Triglochin maritima

JUNCAGINACEAE.   Arrow-grass Family

Swamp or bog herbs with elongate narrowly linear leaves arranged in two vertical ranks, their bases enlarged and sheathing. Flowering stems unbranched, scapose or zigzag and leafy. Flowers small, perfect (in ours) and regular, their parts in threes.

Flowering stems scapose, the inflorescence an elongate bractless more or less compact raceme; leaves all basal; sepals and petals concave, early deciduous; carpels united until maturity of fruit ....1. *Triglochin*
Flowering stems leafy, the inflorescence a few-flowered open bracted raceme; leaves folded longitudinally, their sheathing bases much enlarged; sepals and petals oblong, narrow petals persistent; carpels divergent ....2. *Scheuchzeria*

1. TRIGLOCHIN L.   Arrow-grass

Bog or marsh plants of fresh or brackish waters of all continents. Leaves tufted at base of slender scape, somewhat fleshy and rush-like.

Carpels 6 (rarely 3), united until ripening into an oblong or ovoid-prismatic fruit appearing 12-ribbed (each carpel 2-angled on back), and with star-like cap formed of recurving stigmas ....1. *T. maritima*
Carpels 3, united until maturity into a linear or club-shaped fruit blunt at apex and narrow at base, upon ripening separating from base upward and remaining attached to axis near summit..2. *T. palustris*

1. **Triglochin maritima L.**

This, as usually defined, is a species of very wide distribution, essentially circumboreal. However, cytotaxonomic studies in the genus *Triglochin* indicate that *T. maritima* as commonly interpreted is an aggregate species with chromosome numbers ranging "from the diploid number 2n = 12 to the 24-ploid number 2n = 144." The chromosome number of the typical Linnaean species of Europe is 2n = 48; the rather widespread eastern American plant has 2n = 144 chromosomes. Morphological differences between European and eastern American specimens were recognized by Nuttall, who described our plants as *T. elatum* in 1818. This was later reduced to varietal status, then to synonymy with *T. maritimum* [sic] and for over half a century has been completely ignored. The recent chromosome studies give evidence that Nuttall's name should be reinstated, that *T. elatum* Nutt. is our representative of the heterogeneous aggregate, *T. maritima*, and that *T. maritima* L. is a European species (Löve and Löve, 1958). Inconspicuous grass-like plants, the scapes usually 1.5–7 dm tall, bearing a more or less open spike-like raceme. The persistent, radiating stigmas form a star-like cap on the ovoid-oblong fruit, which contains only one ovule. Marly bogs and springs, and lake shores.

2.   **Triglochin palustris** L.

Similar in general appearance to the preceding, but stem and leaves more slender. Readily distinguished by its linear-clavate fruit whose carpels at maturity separate from the stem-end upward, their lower ends slender-pointed and divergent. Marly bogs or swamps; wide-ranging in the North, but rare in Ohio; also in Eurasia and southern South America.

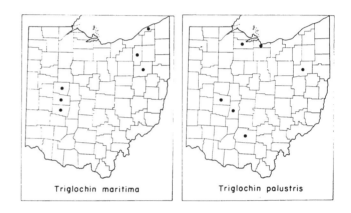

Triglochin maritima   Triglochin palustris

2.   SCHEUCHZERIA L.

A genus of only one species which is separable into an American and a European variety.

1.   **Scheuchzeria palustris** L. var. **americana** Fern.

A low bog plant of northern range, leaves grass-like with greatly expanded sheathing bases; fruit of 3 divergent follicles joined toward base along a central axis.

Scheuchzeria palustris
var. americana

ALISMATACEAE (ALISMACEAE).   Water-plantain Family

Swamp or aquatic herbs widely distributed in tropical and temperate latitudes; about 75 species, variously classified into about 8 to 14 genera. Flowers hypogynous, perfect or unisexual, with 3 green sepals, 3 white or pinkish petals, numerous stamens, and few to many pistils; inflorescence with one to many whorls of pedicels, or of simple or forking branches. Leaves all basal, narrow or broad, with prominent transverse veinlets, aerial, or in some species submersed.

a.   Inflorescence paniculately branched, each branch or pedicel from axil of whorled bracts; carpels flattened, arranged in a single ring on a flat receptacle............1. *Alisma*
aa. Inflorescence not branched (except rarely at lowest node), composed of whorls of bracteate pedicellate flowers; carpels crowded on convex receptacle in dense heads.
   b.   Flowers all perfect, mostly in whorls of more than 3 (to 9 or 10); fruiting heads globular, subtended (in ours) by the reflexed sepals; achenes turgid, strongly ridged, beaked; leaf-blades cordate or subcordate (or lanceolate), with 5–7 principal veins ....2. *Echinodorus*
   bb. Upper flowers staminate, mostly in whorls of 3 (sometimes 1 or 2); achenes flattened; leaf-blades mostly with more than 7 principal veins.
      c.   Sepals broad, concave, partly surrounding fruiting heads; stamens 9–15..........
          3.   *Lophotocarpus*
      cc. Sepals not as above, but spreading or reflexed; stamens usually more numerous
          4.   *Sagittaria*

1. ALISMA L.   WATER-PLANTAIN

Plants of muddy shores, shallow water, or occasionally submersed. Interpretation of species varies greatly, some authors segregating aquatic forms as species.

Fruiting heads 3–4 mm in diam.; flowers 3–3.5 mm broad; achenes 1.5–2.5 mm long; bracts lanceolate ....1. *A. subcordatum*
Fruiting heads 4–7 mm in diam.; flowers 7–13 mm broad; achenes 2.5–3 mm long; bracts ovate to ovate-lanceolate ....2. *A. triviale*

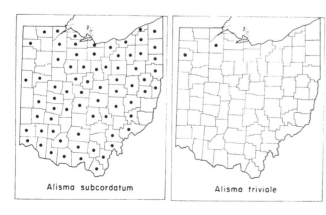

Alisma subcordatum   Alisma triviale

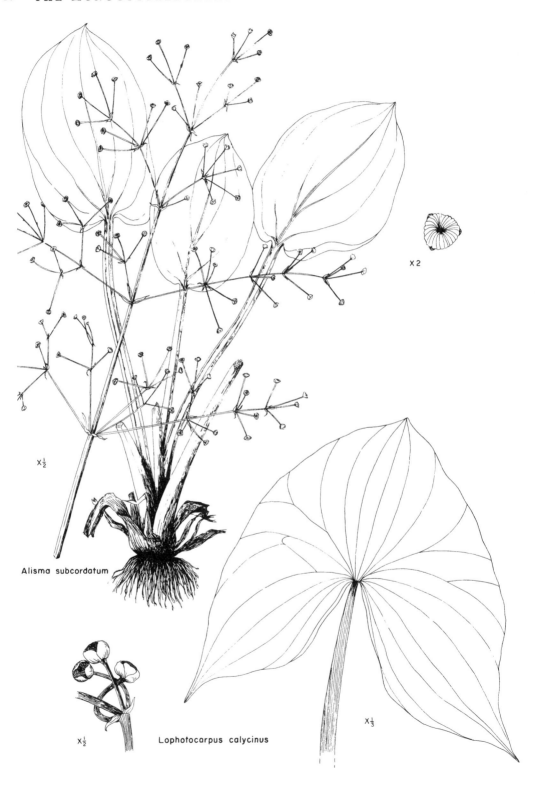

X 2

X½

Alisma subcordatum

X½   Lophotocarpus calycinus

X⅓

1. **Alisma subcordatum** Raf.   Water-plantain
   A. parviflorum Pursh; *A. plantago-aquatica* var. *parviflorum* (Pursh) Farw.;
A. plantago-aquatica of many Amer. auth.
   A plant of muddy shores and shallow water of the eastern half of the
Deciduous Forest. Leaves large, sometimes 2 dm long, broadly ovate or elliptic,
petioled, the petiole-length affected by depth of water, leaves in deeper water
often floating, usually emergent. Inflorescence much branched, flowers small,
petals 1–2 (–2.5) mm long. Distinguished from *A. triviale* Pursh (*A. plantago-
aquatica* of ed. 7 in part) by the much smaller flowers.

2. **Alisma triviale** Pursh
   A. plantago-aquatica L. var. *americanum* Roem & Schult.; *A. brevipes*
Greene
   The American representative of the circumboreal *A. plantago-aquatica;*
differs from the preceding chiefly in the shape of bracts and in the size-characters
given in key.

2. ECHINODORUS Richard.   Bur-head

   Both generic and common name refer to the rough or bur-like appearance
of globose head of plump or turgid achenes which spread apart in fruit and
are tipped with prominent persistent style. A genus of about 18 species, only one
of which occurs in Ohio.

1. **Echinodorus rostratus** (Nutt.) Engelm.
   E. cordifolius of ed. 7 and of Gleason
   A tropical and southern species extending north into our area. Somewhat
similar in appearance to *Sagittaria,* but distinguished by the spiny appearance
of head of plump achenes, cordate or subcordate leaves in the typical form (from
Hamilton County), or in forma *lanceolata* (Engelm.) Fern., lanceolate leaves,
and larger number of flowers in a whorl. The lance-leaved form is smaller than
the typical, often only a few cm tall, and more northern in range.

Echinodorus rostratus

3.   LOPHOTOCARPUS Th. Durand

Strongly resembling *Sagittaria* and included in that genus by Gleason (1952); however, because the idiogram of chromosomes of *Lophotocarpus* differs from that of *Sagittaria,* it seems best to maintain this as a separate genus (Baldwin & Speece, 1955). Distinguished macroscopically by the sepals, which are enlarged in fruit, concave, and appressed to the fruit.

1.   **Lopotocarpus calycinus** (Engelm.) J. G. Sm.
Sagittaria montevidensis Cham. & Schlecht.
Fruiting pedicels thick, some of the lower spreading or recurved; sepals covering about two-thirds of the fruiting head, beaks of the flattened achenes appressed, giving a smooth appearance to head. Leaves vary greatly in shape, sometimes resembling those of *Sagittaria latifolia,* sometimes without lobes. Apparently rare in Ohio.

Lophotocarpus calycinus

4.   SAGITTARIA L.   Arrowhead

Swamp or aquatic stoloniferous perennials, widely distributed in tropical and temperate regions. Petioles with abundant air-spaces, especially near sheathing bases, some (especially the earlier ones) without blade. Flowers showy, the upper usually staminate, or staminate and perfect flowers on different plants, in whorls of threes; sepals 3, spreading or reflexed in fruit (not appressed to fruit as in *Lophotocarpus*), petals 3, white, 1 cm or more long, stamens numerous (mostly more than 12), pistils numerous, the achenes flattened, winged, and crowded into compact head. For identification, mature plants with well-grown leaves and fully developed achenes are necessary. Illustrations of achenes (except of No. 3) are adapted from Smith, 1895. Leaves of all species vary greatly; submersed leaves are bladeless or are narrow and entire; emersed may be narrow or broad. In late summer and fall, tubers are produced at the ends of horizontal rhizomes. These are an important source of food for muskrats (as are also the growing plants in summer) and are stored for winter use. They are also a palatable human food, and were much used by the Indians, who sometimes raided the muskrat's cache.

a. Leaves usually emersed, except early in season; all or most of them with prominent basal lobes; achene with wide dorsal keel.
  b. Bracts obtuse, rarely acute, ovate, less than 1 cm long; beak of achene horizontal, pointing inward ....1. *S. latifolia*
  bb. Bracts acuminate to long attenuate, rarely only acute; lanceolate, more than 1 cm long (1–4 cm); beak of achene erect or bent outward toward prominent dorsal keel.
    c. Beak of achene very short (less than 0.5 mm), erect; achene small (2–2.5 mm), without keels on faces; petioles often curved ....2. *S. cuneata*
    cc. Beak of achene longer (0.5–2 mm), broad-based, ascending or turned back toward dorsal keel; achene larger (2.5–3.5 mm), both faces with keels.
      d. Pedicels of pistillate flowers 1–2 cm long; fruiting heads 2–3 cm in diam.; sinus between dorsal keel and beak of achene wide and rounded
                                                                      3. *S. brevirostra*
      dd. Pedicels of pistillate flowers usually less than 1 cm long, of staminate, longer; fruiting heads 1–1.5 cm in diam.; sinus between dorsal keel and beak of achene deep and narrow ....4. *S. australis*
aa. Leaves submersed or emersed, all or most of them without basal lobes, these when present shorter than terminal lobe; achene with narrow or obscure dorsal keel.
  b. Flower-stem weak, flexuous, plainly bent near lower part of inflorescence; leaves linear to oval, sometimes with basal lobes; lowest (pistillate) flowers subsessile
                                                                      5. *S. rigida*
  bb. Flower-stem slender, erect, or lax on emersed plants; leaves bladeless and flat, or with lanceolate to ovate-elliptic blades, very rarely with basal lobes; flowers all on slender pedicels ....6. *S. graminea*

1. **Sagittaria latifolia** Willd.   COMMON ARROWHEAD
   S. variabilis Engelm.
   An exceedingly variable species, the leaves ranging in shape from bladeless petioles (phyllòdia) when submerged, to broad sagittate or hastate. Basal lobes are about as long as the body of the leaf, wide or narrow, incurved (sagittate) or rarely outcurved (hastate); apex of leaf may be acute or obtuse, as are also the lobes. Varietal or form names have been given to some of the more pronounced variants: var. *obtusa* (Muhl.) Wieg., with leaf-blades obtuse and about as broad as long; forma *hastata* (Pursh) Robins., with body of leaf narrow-triangular; forma *gracilis* (Pursh) Robins., with body of leaf more or less linear, var. *pubescens* (Muhl.) J. G. Sm., with rachis, bracts, and calyx pubescent (specimens from Harrison and Jefferson counties). Distinguished from other similar species by the broad, ovate, obtuse (rarely acute) bracts in the inflorescence, and by the almost horizontal beak of the achene which points away from the dorsal keel. A common and widespread species, growing in shallow water of ponds and slow streams, and in muddy or swampy places.

2. **Sagittaria cuneata** Sheldon
   Equally as variable as the last, with several named forms; aquatic forms with long, simple or branched scape and ribbon-like phyllodia; terrestrial forms usually with sagittate leaves, the scapes and petioles shorter than those of submersed forms, the latter frequently curving outward. Bracts narrowly ovate or

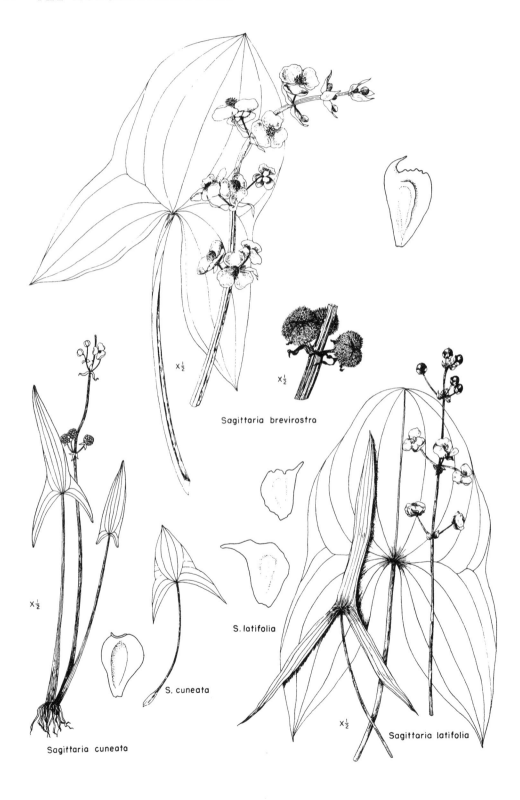

Sagittaria brevirostra

S. latifolia

S. cuneata

Sagittaria cuneata

Sagittaria latifolia

lanceolate, acute to attenuate, and often joined at base; achene with minute beak appearing terminal. A northern species, rarely found in Ohio.

3. **Sagittaria brevirostra** Mackenz. & Bush

This and the next species are very similar, and cannot always be distinguished; in fact, both are included in *S. engelmanniana* J. G. Smith by Gleason (1952). Characters used to distinguish them overlap in Ohio material (especially in Jackson County specimens), some plants with large fruiting heads have short pedicels, others have long pedicels. Curvature of the beak varies, even on a single plant. Ribs or keels on the faces of the achenes are indistinct on even slightly immature achenes. *S. brevirostra* as defined by Fernald is more western in range than *S. australis* and is a stouter, more robust plant. Deam (1940) considers it to be the most common species in Indiana, while *australis* is confined to a single southern county.

The long-acuminate or attenuate bracts of the inflorescence and ascending or recurving beak of the achene distinguish these two species from the similar *S. latifolia*.

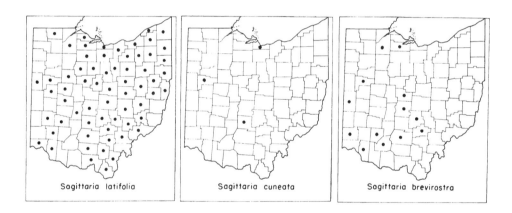

Sagittaria latifolia   Sagittaria cuneata   Sagittaria brevirostra

4. **Sagittaria australis** (J. G. Smith) Small

S. longirostra J. G. Smith in part

A southern or southeastern species, similar to the last, and included in *S. engelmanniana* in Gleason, as is also *S. brevirostra*. Fruiting pedicels short; staminate longer, filiform, equal to or shorter than attenuate bracts.

5. **Sagittaria rigida** Pursh

S. heterophylla Pursh

A northern species found in deep or shallow water, muddy shores, and swamps. Most easily recognized by its bent or flexuous scape, very long in deep water, short in terrestrial plants. Leaves highly variable, long and linear or represented by phyllodia in submersed forms; sometimes broad oval, with one or two small divergent basal lobes or appendages, more often lanceolate or elliptic and without basal lobes. The lower, pistillate flowers sessile or on very short pedicels,

X½

X½

Sagittaria rigida

X½

X½

X½

Sagittaria australis

the staminate on long pedicels; bracts ovate or roundish, obtuse, connate; achene with very narrow keel and slender awl-shaped or hooked beak.

6. **Sagittaria graminea** Michx. GRASS-LEAF ARROWHEAD

A wide-ranging species, but rare and local in Ohio. Submersed leaves thin, flat, bladeless, emersed leaves firmer, narrowly to broadly lanceolate; scapes slender, all the flowers on slender to thread-like pedicels; bracts ovate, connate; achene with narrow dorsal keel and small subulate beak arising below summit of achene.

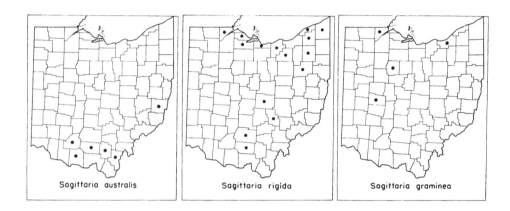

Sagittaria australis   Sagittaria rigida   Sagittaria graminea

BUTOMACEAE. Flowering Rush Family

Aquatic or swamp plants with rather showy, bright-colored flowers; none native to North America. Besides the following, one genus occurs in Africa, one in Asia, two in tropical South America, from whence comes the showy water-poppy, *Hydrocleis nymphoides* Buchen., with floating leaves and light yellow flowers 5–6 cm across; this is sometimes grown in aquaria.

BUTOMUS L. FLOWERING RUSH

A monotypic genus of Eurasia.

1. BUTOMUS UMBELLATUS L.

Thoroughly established on shores of the St. Lawrence Valley and spreading westward along the Great Lakes, where it has reached Michigan. Our first Ohio specimens were collected in 1936 in Lucas and Ottawa counties, and in 1939 in the Erie Islands (Core, 1948). Plant with thick fleshy rootstock bearing, late in the season, numerous easily detached grain-like tubers which facilitate spread of the species. Flowers about 2 cm across, rose-colored (sepals partly green), 25–30 in umbel topping long scape; leaves sheathing at base, 3-angled, often 1 m long.

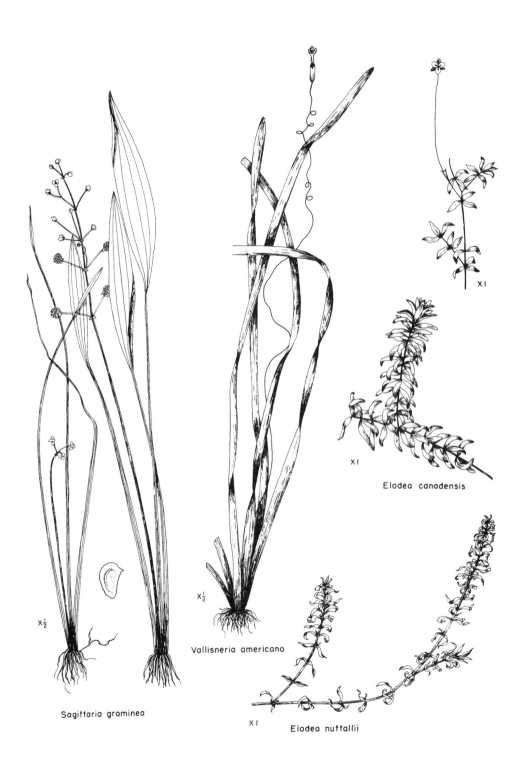

X½

Sagittaria graminea

X½

Vallisneria americana

X 1

Elodea nuttallii

X 1

Elodea canadensis

X 1

HYDROCHARITACEAE. Frogbit Family

Aquatic plants (ours submersed), mostly of warm-country distribution, only a few genera and species in more northern latitudes. Members of the several genera are so unlike in appearance that relationship is not suggested by the vegetative plants. Plants monoecious, dioecious, or sometimes with perfect and imperfect flowers on the same plant (polygamous); flowers solitary, their parts in threes (or multiples of 3), the 3 carpels united into one inferior ovary, sometimes apparently several celled by inward extension of parietal placentae.

Stems leafy, internodes generally shorter than leaves; leaves narrow, whorled or opposite, 1–3 cm long ....1. *Elodea*
Stems short, except the horizontal stolons, leaves essentially basal, ribbon-like, up to 2 m long ....2. *Vallisneria*

1. ELODEA Michx. WATERWEED

Anacharis Richard, *Philotria* Raf.

Familiar aquatic plants of aquaria, and commonly used in the laboratory to demonstrate streaming of protoplasm, which can be observed in the very thin leaves (one layer of cells). Dioecious (in our species), the white flowers from sessile axillary spathes, the staminate inflated and in some species breaking away from plant, the pistillate elevated to water-surface by extreme elongation of hypanthium (perianth-tube), which simulates a thread-like pedicel. Two native species; a third, *E. densa* (Planch.) Caspary, from South America (Argentina), is larger (leaves 2–3 cm long), and commonly grown in aquaria.

Leaves about 2(1–4) mm wide, 6–13 mm long, obtuse, imbricate above ....
    1. *E. canadensis*
Leaves narrower, about 1.3(0.7–1.8) mm wide, acute, spreading ....2. *E. nuttallii*

1. **Elodea canadensis** Michx.
    Anacharis canadensis (Michx.) Rich., *Philotria canadensis* (Michx.) Britt.
    The commoner of our two species. In this both staminate and pistillate

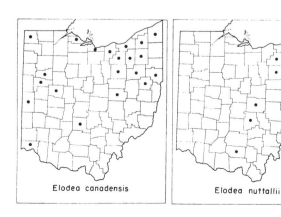

Elodea canadensis   Elodea nuttallii

flowers are elevated on greatly elongated pedicel-like hypanthium, both remaining attached. The usually larger size of the plants, wider and more obtuse leaves, appressed toward tip of stem, aid in recognition.

2. **Elodea nuttallii** (Planch.) St. John

E. minor (Engelm.) Farw., *E. occidentalis* (Pursh) Vict., *Anacharis nuttallii* Planch.

Less wide-ranging than the preceding, and less frequently reported in Ohio. The pistillate flower, by great elongation of the hypanthium, which then resembles a pedicel, reaches the surface of the water; the staminate flower, sessile in a somewhat inflated spathe, breaks away at maturity and floats to the surface, there discharging its pollen. The narrower more acute leaves, not appressed toward tip of stem, distinguish the vegetative plant from the preceding.

2. VALLISNERIA L.   EEL-GRASS. TAPE-GRASS

Two similar species, one American, one Old World. Dioecious; staminate flowers small, crowded in a spike borne on a short club-shaped scape subtended by a 3-valved spathe; pistillate flowers solitary and sessile in a narrow spathe, the slender thread-like scape elongating until the bud just reaches the water-surface. Just before anthesis, the staminate flowers become detached and rise to the surface, where they open and float about, their stamens projecting over the edge of the little "boat" formed by expanded perianth, some finally coming in contact with the opened pistillate flowers with their 3 broad stigmas. After pollination, the scape of the pistillate flower coils like a watch-spring, pulling it well below the water-surface.

1. **Vallisneria americana** Michx.

V. spiralis of Amer. auth., not L.

Distinguished from all other of our aquatic plants by the long, thin, wholly submersed ribbon-like leaves (their ends floating if water is shallow), sometimes 2 m or more in length, shorter in shallow water, and about 1 cm wide. Vegetatively reproducing by off-sets on slender stolons.

Vallisneria americana

GRAMINEAE. Grass Family

By Clara G. Weishaupt

One of the largest families of the flowering plants, cosmopolitan in distribution, the number of species estimated at from 4,000 to more than 6,000. Contains rice, wheat, corn, rye, oats, barley, sugar cane, sorghum, and bamboo, as well as the many pasture, hay, turf, and range grasses of the world; hence it is the most important or one of the few most important families economically. It is estimated that ⅕ to ⅓ of the total land area of the earth was occupied originally by areas of vegetation of which grasses were the dominant plants; these grasslands, found in Europe, Asia, Africa, Australia, and the Americas, are called by such names as prairie, plain, steppe, pampa, savanna, and veldt. However, grasses are not limited to such areas, but grow in forests, swamps, and even deserts. Parts of these grasslands now under cultivation are some of the most fertile cropland of the world; much of the wheat belt and the corn belt of the United States was originally not forest but prairie and plain.

Some grasses, usually those with rhizomes or stolons, form a dense ground cover or sod, with underground parts of individual plants densely intermingled. Others grow in tufts, new shoots from the base growing upright in the basal sheaths. Under certain conditions, especially where rainfall is deficient, the plants grow in tufts more or less widely separated and are known as bunch grasses.

Culms herbaceous or rarely woody, hollow or rarely solid between nodes, nodes hard and solid; rhizomes present or absent; leaves alternate, 2-ranked, linear to lanceolate, parallel-veined, consisting of sheath and blade, with nearly always at the junction of sheath and blade an extension of inner surface of sheath called ligule (see *Milium*, p. 126), and sometimes at base of blade two appendages, one on either side, called auricles (see *Agropyron*, p. 94). Unit of the inflorescence a spikelet, the spikelets aggregated in spikes, racemes, or panicles. Spikelet consists of an axis (rachilla) bearing 2 rows of scales, the 2 lowest of which are empty, the lower called first glume, the other called second glume; scales above glumes called lemmas; in axil of lemma another scale, palea, with back to the rachilla; between lemma and palea, subtended by palea, the flower; lemma, palea, and flower parts constitute a floret; the usual spikelet thus made up of rachilla, 2 glumes, and 1-many florets. Hard downward extension below body of lemma called callus. When spikelet contains more than 1 floret, the rachilla is usually jointed or articulated below each floret and breaks there at maturity, individual florets then falling away leaving the glumes on the pedicel; if articulation is in pedicel below glumes, the whole spikelet falls away as a unit. Flowers hypogynous, usually bisporangiate (bisexual, perfect), with sometimes some monosporangiate (unisexual) flowers present too, or rarely all monosporangiate, the two kinds of flowers then on the same or on separate plants. Perianth of 2 (rarely 3) small scales called lodicules; stamens usually 3 (1–6); carpels united; styles usually 2 (1–3); stigmas usually 2, usually plumose; fruit a grain (one-seeded, pericarp adnate to seed-coat).

Unless otherwise stated, awns are not included in measurements of lemmas and glumes. Width of blades of most species is in range of 3–9 mm; exact width is not stated in descriptions unless it is a distinguishing character or blade is unusually wide or narrow. Height of culms is indicated approximately as follows: short, 1–4 dm; moderately tall, 5–8 dm; tall, 9–12 dm; very tall, more than 12 dm. Stated measurements of parts are the usual ones; departures in either direction may occur.

The serious student will wish to refer to *Manual of Grasses of the United States* (Hitchcock, second Ed.; revised by Chase, 1951) from which most of our illustrations of grass-spikelets have been redrawn.

a.   Culms woody, hollow between nodes, 1–2 cm thick and 2–3 m tall, or sometimes thicker and taller farther south; seldom flowering .... TRIBE I. BAMBUSEAE, p. 65
aa. Culms herbaceous.
    b.   Spikelets enclosed in prickly burs ..... *Cenchrus,* TRIBE X. PANICEAE, p. 141
    bb. Spikelets not enclosed in prickly burs.
        c.   Spikelets of two kinds together in clusters in a spike-like or capitate inflorescence, one kind fertile, with a few perfect florets, the other kind sterile, with several empty lemmas ....*Cynosurus,* TRIBE II. FESTUCEAE, p. 66
        cc. Spikelets not as above.
            d.   Florets bisporangiate (bisexual, perfect); some monosporangiate (unisexual) and/or vestigial florets may be present, also.
                e.   Spikelets attached in definite rows on a rachis; inflorescence a single spike or raceme or a cluster of spikes or racemes.
                    f.   Spikelets attached in two rows, the rows on opposite sides of a single rachis; inflorescence a single spike, not one sided ....
                              TRIBE III. HORDEAE, p. 92
                ff. Spikelets attached in rows, the rows usually on one side of rachis; inflorescence a single one-sided spike or raceme or a cluster of usually one-sided spikes or racemes.
                    g.   Glumes absent or small; lemma and palea folded and keeled, spikelet laterally compressed, its edges formed by keel of lemma and of palea; lemma boat-shaped, awnless, edges firmly holding edges of palea; articulation below glumes when they are present ....
                              TRIBE VIII. ORYZEAE, p. 138
                gg. At least one glume well developed; spikelet not as above.
                    h.   Articulation above glumes (rarely whole spike falling entire), or spikelet laterally compressed; both glumes present, not minute; staminate or vestigial florets, when present, above perfect florets ....
                              TRIBE VI. CHLORIDEAE, p. 131
                hh. Articulation below glumes; spikelets not laterally compressed, sometimes dorsally compressed; one glume sometimes absent or minute; staminate or vestigial floret below perfect floret; sometimes whole spikelet without a perfect floret.
                    i.   Lemma and palea of perfect floret firmer than glumes; if spikelets are in pairs, then first glume minute or absent; below perfect floret a membranous lemma resembling a second or third glume ....TRIBE X. PANICEAE, p. 141
                    ii. Both glumes present, not minute, firmer than lemmas; spikelets in pairs, 1 of pair sessile and perfect, the other pediceled and staminate or vestigial or only pedicel present; rarely both spikelets pediceled and/or perfect ....
                       TRIBE XI. ANDROPOGONEAE, p. 166

ee. Spikelets not attached in definite rows; inflorescence an open or
narrow panicle or dense and spike-like.
  f. Spikelets with 2 or more perfect florets.
    g. At least 1 glume overtopping body of lemma just above it,
        or lower lemma awned from back, or both characters
        present ....TRIBE IV. AVENEAE, p. 101
  gg. Each glume overtopped by body of lemma just above it.
    h. Articulation above glumes (below, in *Melica*).
      i. Inflorescence dense and spike-like, axis and
          branches densely short-hairy; lemma awnless or
          with short awn from just below tip ....
          Koeleria, TRIBE IV. AVENEAE, p. 101
      ii. Inflorescence not dense and spike-like or, if rarely
          so, then axis and branches not densely short-hairy;
          lemma awnless or awned from tip or from between
          or just below terminal teeth ....
          TRIBE II. FESTUCEAE, p. 66
    hh. Articulation below glumes, sometimes in pedicel some
        distance below; upper floret often falling away early
        from lower floret and glumes ....
        TRIBE IV. AVENEAE, p. 101
ff. Spikelets with 1 perfect floret; staminate and vestigial florets
present or absent.
  g. Staminate and vestigial florets absent.
    h. Glumes absent or small; lemma and palea folded and
        keeled, spikelet laterally compressed, edges formed by
        keel of lemma and of palea; articulation below glumes
        when they are present; lemma boat-shaped, awnless
        TRIBE VIII. ORYZEAE, p. 138
    hh. Glumes usually well-developed; articulation usually
        above glumes; if glumes are minute or if articulation
        is below glumes, then lemma is awned ....
        TRIBE V. AGROSTIDEAE, p. 109
  gg. One or two staminate or vestigial florets below, or rarely
      above, perfect floret, the vestigial ones sometimes only
      scales below perfect floret.
    h. Spikelets in pairs, 1 sessile and perfect, the other a
        pedicel only or a pedicel bearing a staminate spikelet
        or a rudiment, rarely both spikelets pediceled and or
        perfect; glumes firmer than lemmas ....
        TRIBE XI. ANDROPOGONEAE, p. 166
    hh. Spikelets not in pairs; glumes not firmer than lemma of
        perfect floret.
      i. Articulation above glumes.
        j. Florets 2, lower staminate, upper perfect ....
            Arrhenatherum, TRIBE IV. AVENEAE, p. 101
        jj. Florets 3, upper perfect, the two lower stam-
            inate or vestigial, each sometimes a minute
            scale....TRIBE VII. PHALARIDEAE, p. 135
      ii. Articulation below glumes (1 glume sometimes
          absent)
        j. Upper floret staminate, its lemma with short
            hooked awn, lower floret perfect; glumes
            longer than florets ....
            Holcus, TRIBE IV. AVENEAE, p. 101
        jj. Upper floret perfect, lower floret staminate or
            vestigial; lemma and palea of perfect floret
            of firmer texture than glumes and lower
            lemma; margins of lemma usually inrolled
            around palea except sometimes at tip ....
            TRIBE X. PANICEAE, p. 141

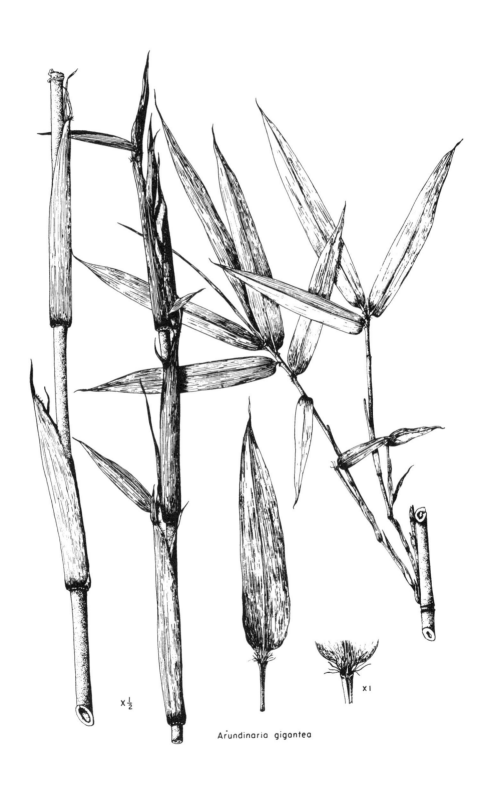

$\times \frac{1}{2}$

$\times 1$

Arundinaria gigantea

dd. Florets monosporangiate (unisexual), the staminate and carpellate (pistillate) in different spikelets on same plant.
  e. The two kinds of spikelets in different parts of same inflorescence.
    f. Inflorescence a terminal panicle, upper branches bearing carpellate (pistillate) spikelets, lower branches bearing staminate spikelets; carpellate spikelets not covered by bead-like involucre
                                TRIBE IX. ZIZANIEAE, p. 140
    ff. Inflorescences axillary; carpellate (pistillate) spikelets surrounded by bead-like involucre from top of which staminate part of inflorescence protrudes ....
             Coix, TRIBE XII. TRIPSACEAE (MAYDEAE), p. 174
  ee. The two kinds of spikelets in different inflorescences, staminate in terminal panicle, carpellate (pistillate) in rows on axillary branch (cob), spikelets and cob forming an "ear" ....
            Zea, TRIBE XII. TRIPSACEAE (MAYDEAE), p. 174

TRIBE I. BAMBUSEAE

Culms usually woody, sometimes climbing, usually hollow between nodes, in some species short, in other species to 30 m or more high and 30 cm thick; spikelets 2-several-flowered, in panicles, racemes, close clusters, or heads; stamens 3–6 or more. Native species found on all continents except Europe; many, both wild and cultivated, of economic importance, their combined uses varied and almost innumerable. One genus in the United States.

1. ARUNDINARIA Michx. BAMBOO. CANE

About 25 species.

1. **Arundinaria gigantea** (Walt.) Chapm.

A plant of the eastern half of southern United States as far north as southern Ohio, Indiana, and Illinois. Forming colonies (canebrakes) by spreading of extensive rhizomes in moist ground and swampy places. It is possible that some of the colonies in Ohio are native, or perhaps all have spread from culti-

Arundinaria gigantea

vation. Culms 2–3 m high and 1–2 cm thick or 2–3 times as tall in the South. Blades of main culm and primary branches sometimes to 4 cm wide, rounded at base to a short petiole and jointed with sheath, those of ultimate branches smaller and often in fan-like clusters. Seldom flowering; spikelets large, 8–12-flowered, lemma 1.5–2 cm long, rachilla breaking above glumes and between florets.

TRIBE II. FESTUCEAE

Spikelets with 2-many florets, rachilla usually articulated above glumes and between florets; each glume shorter than floret immediately above it; lemma awnless or awned from tip or from between or just below terminal teeth; inflorescence an open panicle or rarely narrow or spike-like.

a.   Hairs of rachilla absent or much shorter than body of lemma; callus or body of lemma may be pubescent.
   b.   Spikelets uniform, all containing some fertile florets.
      c.   Spikelets in dense one-sided clusters at ends of panicle-branches; panicle-branches few; spikelets nearly sessile ....10.   *Dactylis*
     cc.   Spikelets not in dense one-sided clusters at ends of panicle-branches.
        d.   Lemma 5-many-nerved (nerves obscure in *Festuca*)
          e.   Lemmas folded very flat lengthwise, many-nerved, lower 1–4 empty; spikelets flat, oval, 2–4 cm long; blades to 2 cm wide ...... 9.   *Uniola*
        ee.   Lemmas and spikelets not as above.
          f.   Rachilla prolonged beyond uppermost palea bearing 2–3 small empty lemmas each enfolded by one below; florets spreading at maturity; lemmas several-nerved; glumes and lemmas with hyaline margins ....13.   *Melica*
        ff.   Rachilla, if prolonged, not bearing a mass of 2–3 small enfolded empty lemmas.
          g.   Lemma 2-toothed at apex, 7 mm long or more, awned from base of teeth or just below or sometimes awnless; sheaths closed.
            h.   Callus of lemma bearded ....14.   *Schizachne*
           hh.   Callus of lemma not bearded but body of lemma may be pubescent .... 2.   *Bromus*
         gg.   Lemma not 2-toothed but sometimes erose at apex; if awned, then awned from tip; sheaths open or closed.
          h.   Lemma keeled, awnless, usually with tuft of long soft hairs at base (web); when, rarely, web is absent, both keel and marginal nerves are pubescent; leaf-blades boat-shaped at tip .... 6.   *Poa*
         hh.   Lemma rounded on back or keeled only at tip; web absent.
            i.   Lemma awned or acute at tip, the 5 nerves obscure; sheaths of awnless species open ....
                3.   *Festuca*
            ii.   Lemma not awned, usually not acute at tip but if rarely so then nerves conspicuous and more than 5.
              j.   Lemma 7–9-nerved; sheaths closed; plants of moist, swamp, or aquatic habitats ....
                5.   *Glyceria*
             jj.   All or most of the lemmas 5-nerved; sheaths open.
                k.   Panicle-branches ascending; plants of moist, swamp, or aquatic habitats ....
                  5.   *Glyceria pallida*

          kk. Lower panicle-branches reflexed; plants of dry or moist, sometimes saline, habitats ....4. *Puccinellia distans*

dd. Lemma with 1–3 obvious nerves.
  e. Nerves of lemma glabrous or scabrous; no tuft of hairs present at base of lemma.
    f. Lemma 3 mm long or less; low to moderately tall plants of various habitats, blades much narrower than 1 cm....7. *Eragrostis*
    ff. Lemma 6–10 mm long, the 3 strong nerves converging in a sharp point at tip; tall woodland plants with blades 1–2 cm wide; lemma and palea separated at maturity by turgid beaked grain ....8. *Diarrhena*
  ee. Lateral nerves or midnerve of lemma, or both, pubescent; or, if rarely all nerves glabrous, then lemma with tuft of long soft hairs at base (web).
    f. Lemma not toothed at apex, not awned, sometimes with tuft of long soft hairs at base (web); blades with boat-shaped tip ....
                                          6. *Poa*
    ff. Lemma toothed or emarginate at apex, midnerve or all three nerves excurrent as short awns.
      g. Panicle open, large, branches often drooping; blades elongate; sheaths not enlarged, not containing cleistogamous panicles; nerves of lemma excurrent as 3 short points ....
                                       15. *Triodia*
      gg. Panicle small; blades and internodes of culm short and many; sheaths enlarged and usually containing cleistogamous panicles; midnerve of lemma excurrent as short awn ....16. *Triplasis*

bb. Spikelets of two kinds together in clusters in a spike-like or capitate inflorescence, one kind fertile, with a few perfect florets, the other kind sterile, with several empty lemmas ....11. *Cynosurus*

aa. Hairs of rachilla equaling or exceeding the long-acuminate lemmas; panicle dense and hairy; tall stout grass of wet places; blades to 4 cm wide ....12. *Phragmites*

2. BROMUS L.   BROME GRASS

    About 100 species; moderately tall to tall annuals or perennials of temperate regions, with closed sheaths. Inflorescence a panicle or, sometimes in certain species and in depauperate specimens of other species, a raceme; spikelets large, 1–4 cm long, of several to many florets; lemma large, 7–20 mm long, usually awned, awn arising below the two-toothed tip; stigmas arising below tip of ovary. Some species are hay and pasture grasses; some of the introduced species are troublesome weeds.

    In most species attachment of awn is below base of notch between lemma-teeth. Since base of notch is covered by the awn, length of actual teeth is difficult to ascertain; also, notch may be deepened by splitting of hyaline tip as lemma is handled. Consequently, "tooth" is used in this key to mean that part of half the lemma from its tip to level of attachment of awn. Before deciding on number of nerves of glumes it is well to look at several spikelets; occasional glumes may have more or fewer nerves than the majority.

a. Length of lemma usually 2½ times width or more; teeth of lemma not more than 0.6 mm long, or very narrow and as much as 2 mm long or more; first glume 1-nerved, second glume 3-nerved (except in No. 1.)

b. First glume 3-nerved or rarely 5-nerved, second glume 5-nerved or rarely 7–9-nerved; panicle drooping, branches slender and flexuous; glumes and lemmas pubescent ....1. *B. kalmii*

bb. First glume 1-nerved, second glume 3-nerved.

    c. Teeth of lemma not more than 0.6 mm long; awn of lemma shorter than body or lemma awnless.

        d. Panicle-branches erect or ascending at maturity or rarely spreading; foliage essentially glabrous except sometimes lowermost sheaths; lemma awnless or with awn to 3 mm, glabrous, rarely villous; roadsides and fields ....2. *B. inermis*

       dd. Panicle-branches spreading and usually drooping at maturity, spikelets near ends of branches only; sheaths or blades or both usually pubescent; lemmas usually pubescent, at least near base and at margin; native woodland plants.

           e. Sheaths, except sometimes the upper, covering nodes; base of blade with prominent flange which ends in auricles; nodes 10 or more ....

                              3. *B. latiglumis*

          ee. Sheaths, except sometimes the lower, not covering nodes; blade without basal flange and auricles; nodes usually fewer than 10.

            f. Lemma more or less evenly pubescent all over or rarely glabrous

                            4. *B. purgans*

           ff. Lemma with band of long hairs on each margin, at least below middle, remainder of surface glabrous; nodes of culm usually pubescent ....5. *B. ciliatus*

    cc. Teeth of lemma to 2 mm long or more, slender and pointed; lemma narrowly lanceolate, awn equaling or longer than body; weedy grasses of open places.

        d. Inflorescence a panicle, branches of axis or most of them branched; awn of lemma 1–1.5 cm long ....6. *B. tectorum*

       dd. Inflorescence racemose, branches of axis or most of them not branched; awn of lemma 2–3 cm long ....7. *B. sterilis*

aa. Length of lemma usually not more than twice width; teeth of lemma 0.5–2 mm long, triangular, from somewhat shorter than to somewhat longer than wide (or teeth and awn sometimes absent or nearly so in No. 13); first glume 3–5-nerved, second glume 5–9-nerved.

    b. Lemma pubescent, strongly nerved; branches of panicle mostly shorter than spikelets, panicle compact, ellipsoid ....8. *B. mollis*

   bb. Lemma glabrous or scabrous; branches of panicle not so short, panicle not compact but branches may be erect.

        c. Sheaths glabrous or lower ones slightly hairy; tip of palea and hairy tip of grain usually projecting slightly beyond tip of lemma at maturity; lemma strongly inrolled when mature, rachilla thus visible; awn usually short, sometimes crimped; anthers 2 mm long or less ....9. *B. secalinus*

       cc. Sheaths, except sometimes the upper, pubescent; palea not exceeding lemma.

           d. Lemma obviously longer than broad, awn well developed.

               e. Lower sheaths soft-villous, hairs somewhat entangled; awn at maturity usually divaricate; palea 1–2 mm shorter than lemma; anther about 1 mm long; panicle diffuse, branches divergent ....10. *B. japonicus*

              ee. Hairs of lower sheaths stiffer, less dense, not entangled, or lower sheaths short-downy to subglabrous; awn not divaricate at maturity; palea little shorter than lemma.

                  f. Teeth of lemma longer than wide, acute; anther 3–4 mm long; awn about equaling body of lemma; panicle usually diffuse; lower sheaths short-downy to subglabrous; spikelets usually somewhat purple at maturity ....11. *B. arvensis*

                 ff. Teeth of lemma broad and obtuse; anther 2 mm long or less; awn usually shorter than body of lemma; panicle usually not diffuse; hairs of lower sheaths rather stiff, sometimes sparse ....

                                12. *B. commutatus*

          dd. Lemma almost as broad as long, inflated, obtuse; awn very short or absent; spikelets flattened, about 1 cm wide ....13. *B. brizaeformis*

1. **Bromus kalmii** Gray

Perennial; distinguished from other native species by the 3-nerved first glume. Open ground or open woods, northeastern fourth of the United States.

2. BROMUS INERMIS Leyss.   SMOOTH BROME.   HUNGARIAN BROME

Perennial; introduced from Europe, cultivated in parts of the United States, and naturalized in the northern half. Common in Ohio; waste places and roadsides.

3. **Bromus latiglumis** (Shear) Hitchc.

Tall perennial of moist woods, especially along streams, in southern Canada and northeastern fourth of the United States; flowering in August and September. Typical form has sheaths pilose at summit, glabrous or somewhat pilose below; forma *incanus* (Shear) Fern. has sheaths densely canescent.

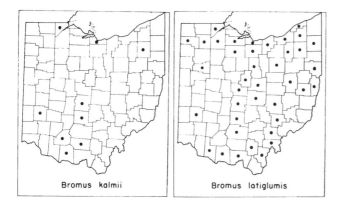

Bromus kalmii   Bromus latiglumis

4. **Bromus purgans** L.

Resembles *B. latiglumis* but differs in the usually more slender stem, less numerous nodes most or all of which are not covered by the sheaths, absence of flange and auricles at base of blade, and earlier flowering time, June and July. Blades usually sparsely villous; lemmas and sheaths usually pubescent; plants with glabrous lemmas have been segregated as var. *laeviglumis* (Scribn.) Swallen or as forma *glabriflorus* Wieg., and those with glabrous sheaths as forma *laevivaginatus* Wieg. Eastern half of the United States.

5. **Bromus ciliatus** L.   FRINGED BROME

Resembles *B. purgans* but differs in having lemma glabrous except for a band of soft hairs on each margin. Canada, south to Pennsylvania, Tennessee, Iowa, Texas, and California.

6. BROMUS TECTORUM L.   DOWNY BROME

Short to moderately tall annual or winter annual with softly pubescent sheaths and blades and drooping panicles that are silvery green when young and often red-tinged with age. Naturalized from Europe in most of the United States; abundant in Ohio; roadsides and waste places.

Bromus kalmii

X½

X½

Bromus purgans

X½

Bromus latiglumis

7. BROMUS STERILIS L.

Annual of medium height with inflorescence-branches long and somewhat nodding but rigidly so. Introduced from Europe. Waste places and roadsides, most of the United States; recorded from 10 Ohio counties.

8. BROMUS MOLLIS L.   SOFT CHESS

B. hordeaceus of authors, not L.

Native of Europe, introduced in northeastern United States and in the West; recorded from Lake and Lorain counties.

9. BROMUS SECALINUS L.   CHEAT. CHESS

Naturalized from Europe, an annual weed in waste places and in wheat and other grain fields throughout the United States. Distinguished by the short, often crimped, awn, distant lemmas strongly inrolled at maturity, exposed tip of palea, spikelets less fragile than in other species, and glabrous or nearly glabrous sheaths. Common in Ohio.

10. BROMUS JAPONICUS Thunb.   JAPANESE BROME

Short to moderately tall annual; naturalized from the Old World in most of the United States, a weed in waste places and along roadsides. General in Ohio.

11. BROMUS ARVENSIS L.

Resembles *B. japonicus;* distinguished from it by the less turgid purplish spikelets, straight awn, and palea only slightly shorter than its lemma. From the Old World; recorded from Van Wert County.

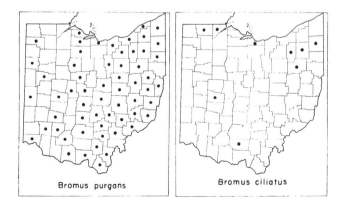

Bromus purgans

Bromus ciliatus

12. BROMUS COMMUTATUS Schrad.

Much like *B. secalinus* but sheaths pubescent, awn usually longer, palea shorter than lemma, and spikelets more overlapping. Cultivated and waste ground, especially in northern half of the United States; general in Ohio.

Bromus racemosus L., usually considered a separate species, is closely related to *B. commutatus* and is distinguished from it by such arbitrary characters as denser panicles and shorter lemmas; no Ohio records.

13. BROMUS BRIZAEFORMIS Fisch. and Mey. RATTLESNAKE CHESS. QUAKE GRASS
Introduced from Europe, cultivated as an ornamental, and occasionally escaped; recorded from Erie County.

3. FESTUCA L. FESCUE

About 100 species of short to moderately tall grasses of temperate and colder regions. Spikelets few-several-flowered, the uppermost floret reduced; glumes narrow, usually unequal, the first 1-nerved, the second 3-nerved; lemma sometimes firm, sometimes awned from tip, 5-nerved, nerves sometimes obscure. In the United States the species more numerous in the West where several are forage grasses.

a. Blades flat when fresh, more than 3 mm wide; lemma awnless or with awns up to 2 mm long; perennial.
    b. Auricles present; spikelets about 1 cm long or more; panicle-branches with spike-lets nearly to base; fields, meadows, roadsides .... 1. *F. elatior*
    bb. Auricles absent; spikelets about 5 mm long; panicle-branches naked for ¾ their length, a few spikelets toward tips; woods .... 2. *F. obtusa*
aa. Blades rolled, folded, setaceous, or flat and less than 3 mm wide.
    b. Perennial, with a dense mass of culms and old leaf-bases; lemma awnless or with awn much shorter than body; stamens 3.
        c. Culms usually decumbent at base; leaf-sheaths closed but soon splitting; basal sheaths red or becoming brown, soon disintegrating into fibers; rhizomes some-times present .... 3. *F. rubra*
        cc. Culms erect at base; leaf-sheaths drab or light in color, not readily disinte-grating into fibers; without rhizomes .... 4. *F. ovina*
    bb. Annual slender plants; stamens usually 1; flowers often cleistogamous (stigmas and anthers not extruded.)
        c. First glume not more than 2 mm long, ½–¼ as long as second glume .... 5. *F. myuros*
        cc. First glume more than 2 mm long, more than half as long as second glume.
            d. Spikelets usually 6–10-flowered; glumes about 2–4 and 3–5 mm long; body of lemma 2.5–5 mm long, awn 1–5 mm long, rarely longer; sheaths and/or culms sometimes puberulent .... 6. *F. octoflora*
            dd. Spikelets usually 4–6-flowered; glumes about 4 and 6–7 mm long; body of lemma 7–8 mm long, awn 10–14 mm long; sheaths and culms glabrous 7. *F. dertonensis*

1. FESTUCA ELATIOR L. MEADOW FESCUE
Native of Europe, a common grass of meadows and roadsides, sometimes cultivated for hay and pasture. Somewhat more robust than the typical variety is var. *arundinacea* (Schreb.) Wimm. (*F. arundinacea* Schreb.), Alta Fescue. This variety and especially a variant of it, Kentucky 31 Fescue, widely planted along highways in Ohio.

2. **Festuca obtusa** Biehler NODDING FESCUE
    F. nutans Willd.
Culms slender, blades soft and lax, panicle nodding, branches erect at first

but later often reflexed; spikelets 3–5-flowered; lemma firm, not awned, obscurely nerved. Woods, eastern half of the United States.

3.   **Festuca rubra** L.   RED FESCUE
Canada and Alaska, ranging into the United States in the East to Georgia and Missouri and in the West in the mountains. Apparently native in parts of the United States, but doubtfully native in Ohio. Recorded from about 15 widely scattered counties. Typical variety decumbent at base, sometimes with creeping rhizomes; var. *commutata* Gaud., Chewings Fescue, more erect without creeping rhizomes; both are used alone or in mixtures as turf grasses. Shade Fescue, var. *heterophylla* (Lam.) Mut., densely tufted, with basal blades filiform, used in shaded lawns.

4.   FESTUCA OVINA L.   SHEEP FESCUE
Introduced in Ohio. Hard Fescue, var. *duriuscula* (L.) W. D. J. Koch, has smooth wider and firmer blades than var. *ovina;* and Hair Fescue, var. *capillata* (Lam.) Alef. (*F. capillata* Lam.), more slender and shorter than var. *ovina,* has capillary flexuous blades and awnless lemmas; all three of these occasionally used in lawn mixtures. Blue Fescue, var. *glauca* (Lam.) Koch, with elongate glaucous blades, cultivated as an ornamental.

5.   FESTUCA MYUROS L.
Vulpia myuros (L.) K. C. Gmel.
Introduced from Europe; open ground along coasts and locally inland. One specimen recorded from Lake County.

6.   **Festuca octoflora** Walt.   SIX-WEEKS FESCUE
Vulpia octoflora (Walt.) Rydb.
Occurs throughout the United States. All or nearly all Ohio specimens referred to var. *tenella* (Willd.) Fern., which is smaller than var. *octoflora,* with culms very slender, inflorescence unbranched or nearly so, and spikelets in lower range of measurements given in key.

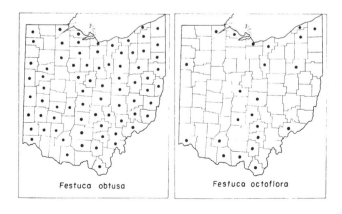

Festuca obtusa   Festuca octoflora

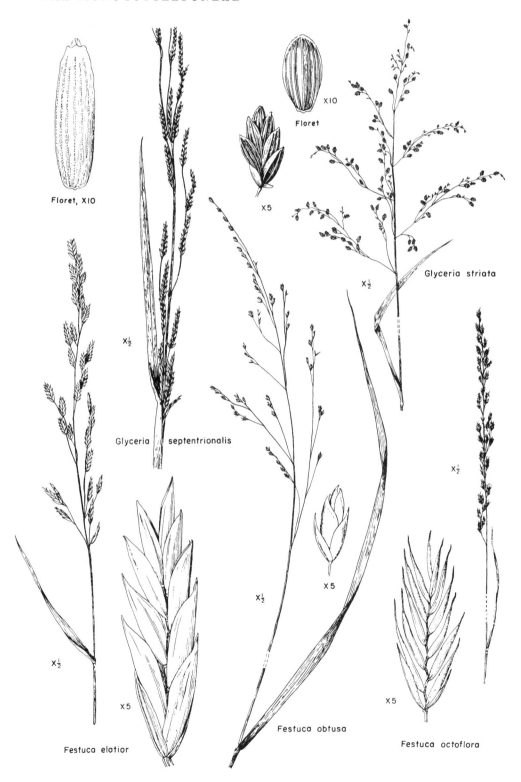

Floret, X10

Floret X10

X5

Glyceria striata

Glyceria septentrionalis

Festuca elatior

X5

Festuca obtusa

X5

Festuca octoflora

7. FESTUCA DERTONENSIS (All.) Aschers. and Graebn.

Introduced from Europe in the West; found rarely as a waif in the Eastern States. Several specimens recorded from Jackson, Pike, and Ross counties.

4. PUCCINELLIA Parl. ALKALI GRASS

About 30 species of temperate and cold regions, chiefly of saline habitats. Spikelets much like those of *Glyceria;* leaf sheaths open.

1. PUCCINELLIA DISTANS (L.) Parl.

Culms short to moderately tall, tufted, blades narrow, sometimes rolled. Panicle open, to 1.5 dm long, branches whorled, the lower spreading or reflexed, the longer with spikelets only above the middle; spikelets 4–6-flowered, about 5 mm long; glumes 1 and 2 mm long; lemma 2 mm long, obtuse. Naturalized from Eurasia; along the coast, Delaware to New Brunswick, locally inland, and in the West. One Ohio record, Butler County.

5. GLYCERIA R.Br. MANNA GRASS

Panicularia Heist.

About 35 species, perennial, usually of aquatic or wet land habitats; culms unbranched, often rooting from lower nodes; sheaths usually closed, at least below; spikelets few-many-flowered, glumes unequal, lemma with 5–9 prominent parallel nerves.

a. Spikelets linear, 1–4 cm long; sheaths compressed; panicle 1.5–5 dm long, narrow, branches appressed or finally somewhat spreading.
  b. Palea tipped by 2 acuminate teeth, projecting 1.5–3 mm beyond tip of lemma; lemma acute at apex, 6 mm long or more ....1. *G. acutiflora*
  bb. Palea projecting beyond lemma less than 1 mm or not at all; lemma obtuse at apex.
    c. Lemma firm, 4–5 mm long, scabrous between nerves, usually exceeded by palea; second glume 3–5 mm long; spikelets 1–2 cm long or more, pedicels short, upwardly-thickened ....2. *G. septentrionalis*
    cc. Lemma thin, 3–4 mm long, glabrous or minutely scabrous between nerves, slightly longer than palea; second glume 2–3 mm long; spikelets mostly 1–1.5 cm long, pedicels slender ....3. *G. borealis*
aa. Spikelets lanceolate, ovate, or oblong, less than 1 cm long; sheaths rounded or nearly so.
  b. Panicle long and slender, the slender branches erect, the whole nodding at tip; glumes 1-nerved; lemma 7-nerved, 2–2.5 mm long; ligule less than 1 mm long ......
                                                        4. *G. melicaria*
  bb. Panicle open, branches spreading or loosely ascending; ligule longer.
    c. Sheaths open; second glume 3-nerved or rarely 5-nerved; lemma 5-nerved or rarely 7-nerved ....5. *G. pallida*
    cc. Sheaths closed, at least below summit; second glume 1-nerved; lemma 7-nerved.
      d. Spikelets not more than 2.5 mm wide; lemma 2–2.5 mm long; palea more than twice as long as wide.
        e. Culms slender; panicle 1–2 dm long; glumes 0.5 (–1) and 1 (–1.4) mm long; lemma about 2 mm long; palea rounded at tip ....
                                                        6. *G. striata*
        ee. Culms stout; panicle 2–4 dm long; glumes about 1.5 and 2 mm long; lemma about 2.5 mm long; palea narrow at tip ....7. *G. grandis*

dd. Spikelet more than 2.5 mm wide; lemma 3–4 mm long, longer than palea; palea less than twice as long as wide, the keels projecting beyond sides of lemma ....8. *G. canadensis*

1. **Glyceria acutiflora** Torr.
Panicularia acutiflora (Torr.) Kuntze
Resembles next two species but easily distinguished by the longer narrower lemma and long projection of palea beyond tip of lemma. Northeastern fourth of the United States.

2. **Glyceria septentrionalis** Hitchc.   EASTERN MANNA GRASS
Panicularia septentrionalis (Hitchc.) Bickn.
Tall; distinguished by the long slender panicle (2–5 dm) of narrow, often many-flowered, spikelets; stem soft and usually thick. Eastern half of the United States; wet places and shallow water.

3. **Glyceria borealis** (Nash) Batchelder   NORTHERN MANNA GRASS
Panicularia borealis Nash
Culms more slender, blades narrower, and spikelets, glumes, and lemmas smaller than in the preceding. A northern species with range as far south as Pennsylvania and Illinois in the East, farther south in the West; wet places and shallow water.

4. **Glyceria melicaria** (Michx.) F. T. Hubbard
Glyceria torreyana (Spreng.) Hitchc.; *Panicularia torreyana* (Spreng.) Merr.
Distinguished from the next four species by the slender panicle and short ligule. Wet woods and swamps, New England to Ohio, Tennessee, and North Carolina.

5. **Glyceria pallida** (Torr.) Trin.
Panicularia pallida (Torr.) Kuntze; *Puccinellia pallida* (Torr.) Clausen
Differs from other Ohio species in having open sheaths; panicle pale green, to 1.5 dm long. Shallow water, northeastern fourth of the United States.

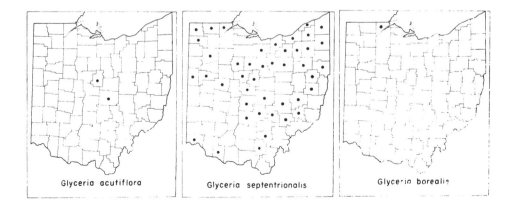

Glyceria acutiflora   Glyceria septentrionalis   Glyceria borealis

6. **Glyceria striata** (Lam.) Hitchc. Fowl Manna Grass
Panicularia nervata (Willd.) Kuntze
Tufted, slender, moderately tall; panicle ovoid or triangular-ovoid, branches ascending, tips drooping; spikelets small, green or purplish. Wet woods, ditches, swamps, almost throughout the United States. Abundant in Ohio.

7. **Glyceria grandis** S. Wats. Tall Manna Grass
Panicularia grandis (Wats.) Nash
Culms generally stouter and taller, and blades wider, than in the other species with short spikelets. A beautiful grass with large open much-branched panicle, the lemmas purple. Range approximately that of *G. borealis.*

8. **Glyceria canadensis** (Michx.) Trin. Rattlesnake Grass
Panicularia canadensis (Michx.) Kuntze
Distinctive because of the diffuse panicle with long drooping branches naked at base and short broad spikelets with closely imbricated florets; culms usually tall. Bogs and wet places, northeastern fourth of the United States.

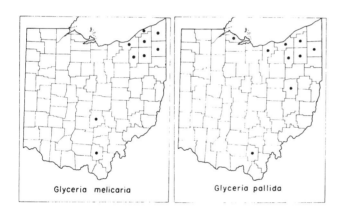

Glyceria melicaria   Glyceria pallida

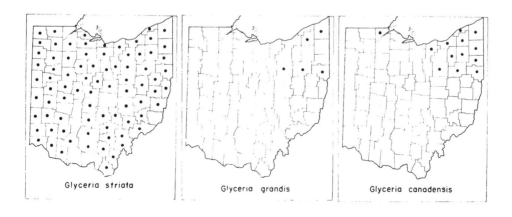

Glyceria striata   Glyceria grandis   Glyceria canadensis

6. POA L. BLUEGRASS.   MEADOW GRASS

About 200 species of temperate and colder regions. Spikelets few-several-flowered, first glume with 1 nerve (rarely 3), second glume with 3 nerves (rarely 5); lemma 2–5 mm long, 5-nerved, sometimes only midnerve and marginal nerves distinct, usually with a tuft of cobwebby hairs (web) at base. Leaf-blades ending in a boat-shaped tip.

a. Spikelets consisting of florets of usual structure.
   b. Low annuals (or winter annuals) of open ground with culms usually no more than 2 dm high; inflorescence ovoid, usually 2–5 cm long.
      c. Lemma without web, the 5 distinct nerves usually pubescent, at least toward base; anthers 0.5–1 mm long ....1. *P. annua*
     cc. Lemma with web, pubescent on keel and marginal nerves, intermediate nerves obscure; anthers about 0.2 mm long ....2. *P. chapmaniana*
  bb. Perennials, usually taller; panicles usually longer.
     c. Lower panicle branches 1–3 at a node.
       d. Panicle narrow, the short branches spikelet-bearing to base; culms flattened, 2-edged, not tufted; lemma with 3 distinct nerves, often bronzed at tip; at least lowest lemma in spikelet usually sparsely webbed ....
                                                    3. *P. compressa*
      dd. Panicle-branches long, spikelet-bearing toward tips only; culms not 2-edged.
        e. Lemma without basal web, pubescent on keel and marginal nerves and between nerves; without creeping rhizomes ....4. *P. autumnalis*
       ee. Lemma, at least the lowest in the spikelet, with basal web.
         f. Marginal nerves and keel of lemma glabrous; lemma 2–3 mm long, obtuse at tip; without creeping rhizomes ..... 5. *P. languida*
        ff. Marginal nerves and keel of lemma pubescent.
          g. Lemma 2.5–3.5 mm long, intermediate nerves obscure ....
                                              6. *P. paludigena*
         gg. Lemma 3.5–5 mm long, intermediate nerves distinct.
           h. Panicle nodding, branches ascending; without creeping rhizomes; anthers 1.5 mm long or less ....7. *P. wolfii*
          hh. Panicle usually erect, lower branches spreading or reflexed; with slender rhizomes; anthers 2–3 mm long; web scant; surface of uppermost lemma sometimes pubescent; basal blades very long, culm blades very short ....8. *P. cuspidata*
   cc. Lower panicle-branches 4 or more at a node; lemma with pubescent keel and basal web.
      d. Marginal nerves of lemma glabrous (except sometimes in No. 9).
        e. Panicle long-exsert, ovoid or oblong, branches spreading-ascending, some of them with spikelets nearly to base; ligule of upper leaves 4–10 mm long; distinct nerves of lemma 5 .... 9. *P. trivialis*
       ee. Panicle included at base, at least at first, branches later spreading; spikelets few, at tips of branches only; ligule of upper leaves 1–2 mm long; distinct nerves of lemma 3 ....10. *P. alsodes*
      dd. Marginal nerves of lemma pubescent.
        e. Lemma with 5 distinct nerves; ligule of upper leaves 1–2 mm long.
         f. Lemma pubescent on lower ½–⅔ of keel, glabrous between keel and marginal nerves; panicle ovoid, exserted, branches spreading or ascending, with spikelets to base or nearly to base of some of them, lowest branches usually in a whorl of 5; lawns and open places ....11. *P. pratensis*
         ff. Lemma pubescent on whole length of keel, somewhat pubescent between keel and marginal nerves; panicle oblong; spikelets few, on upper half of the spreading to reflexed branches; woodlands....
                                                     12. *P. sylvestris*

ee. Lemma with 3 distinct nerves, intermediate nerves obscure; glumes
    acuminate.
    f.   Ligule 2–5 m long; panicle 1–3 dm long ....13. *P. palustris*
    ff.   Ligule 1 ı. long or less; panicle 4–10 cm long; stems slender....
                              14. *P. nemoralis*
aa. All or most spikelets consisting of bulblets the bracts of which are sometimes as much
    as 2 cm long ....15. *P. bulbosa*

1. POA ANNUA L.   ANNUAL BLUEGRASS

Culms erect to spreading, or prostrate and sometimes rooting at nodes; blades soft, bright green; panicle branches solitary or in 2's. Introduced from Europe, occurring almost throughout the United States in moist, usually open, areas; abundant in lawns in Ohio, where it is usually considered a weed.

2. Poa chapmaniana Scribn.

Resembling *P. annua*, the culms usually erect, the sheaths closer. Delaware to Iowa and south to the Gulf, occasionally introduced northward. Recorded from Van Wert County from fallow fields and waste ground with *Alopecurus carolinianus*; doubtfully native in Ohio.

3. POA COMPRESSA L.   CANADA BLUEGRASS

Rhizomatose, not tufted, blue- or gray-green in color. Naturalized from Europe, occurring throughout most of the United States; throughout Ohio in drier, somewhat poorer soils, where it is sometimes a pasture grass.

4. Poa autumnalis Muhl.

A short to moderately tall woodland grass, with long soft basal leaves; panicles 10–20 cm long and broad, spikelets few, at tips of branches only. To be expected in Ohio, but no specimens seen.

5. Poa languida Hitchc.
Poa debilis Torr.

Culms short to tall; panicle 5–10 cm long with a few ascending branches bearing spikelets only near tips. Woods, northeastern fourth of the United States.

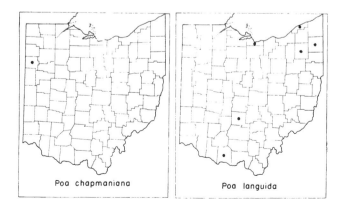

Poa chapmaniana   Poa languida

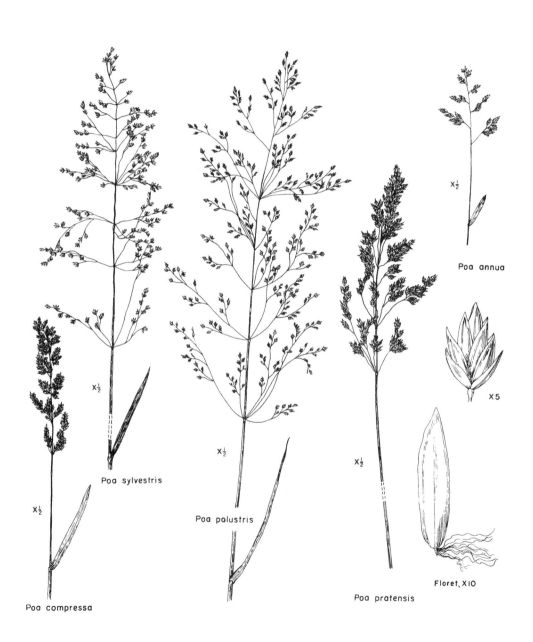

Poa compressa

X½

Poa sylvestris

X½

Poa palustris

X½

Poa pratensis

X½

Poa annua

X½

X5

Floret, X10

6. **Poa paludigena** Fern. & Wieg.

Culms slender, blades narrow, ligules short; panicles mostly 5–10 cm long, branches few, a few spikelets near tips; New York and Pennsylvania to Illinois and Wisconsin.

7. **Poa wolfii** Scribn.

Culms moderately tall to tall; blades mostly near base; panicle 8–15 cm long, branches few, ascending, spikelets near tips only. Open woods and prairie patches, Ohio to Minnesota and Nebraska.

8. **Poa cuspidata** Nutt.

Culms short to moderately tall, tufted; blades abruptly sharp-pointed, basal ones sometimes as long as flowering culms; panicle open, to 12 cm long, branches few, a few spikelets near tips. Woods, early spring; New Jersey to Ohio, south to Georgia and Alabama.

9. Poa trivialis L. Rough Bluegrass

Culms tall or moderately tall, erect from decumbent base, rough below panicle; panicle 6–15 cm long; glumes eventually spreading with tips incurved. Native of Europe, sometimes used in mixtures for meadows, shaded lawns, and pastures; naturalized in much of Ohio.

10. **Poa alsodes** Gray

Culms moderately tall, tufted; basal blades long; panicle 10–25 cm long, lower branches at first included in upper sheath, later spreading. Woods, Maine and Minnesota to Delaware and Tennessee.

11. Poa pratensis L. Kentucky Bluegrass

Culms tufted, short to moderately tall; rhizomes present; lemma about 3 mm long, with a copious web. An important cultivated turf and pasture grass in the North Central, Northeastern, and Northwestern States, certainly the most widely used lawn grass in Ohio. Probably was introduced from Europe in early colonial times and spread rapidly from cultivation; naturalized throughout the United States and northward, or possibly native along our northern border and in Canada.

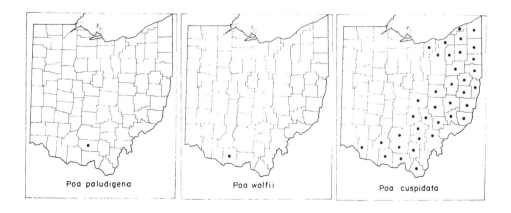

Poa paludigena   Poa wolfii   Poa cuspidata

12. **Poa sylvestris** Gray

Culms erect, tufted, moderately tall; lower blades long; panicle 1–2 dm long, branches slender and flexuous. An early spring woodland grass of eastern United States as far west as Wisconsin, Iowa, and Texas.

13. **Poa palustris** L.   FOWL MEADOW GRASS

P. triflora Gilib.

Culms moderately tall to tall, flattened and decumbent at base; panicle often bronze or purple, branches naked at base, in fascicles of 3–10. Meadows and moist open ground; northern half of eastern United States, farther south in the West.

14. POA NEMORALIS L.   WOOD BLUEGRASS

Culms tufted, moderately tall; blades 1–2 mm wide; lower half of panicle-branches naked. Native of Europe; Minnesota to Virginia and northward. Recorded in Ohio from four northeastern counties and from Fairfield County.

15. POA BULBOSA L.

Culms densely tufted, bulbous at base; panicle ovoid, 6–8 cm long, branches ascending; spikelets consisting mostly of purple-based bulblets and bracts with elongated tips; floret-containing spikelets, when present, about 5-flowered; lemma webbed, keel and marginal nerves pubescent, intermediate nerves obscure. Sparsely introduced from Europe in lawns and fields. Recorded from Franklin County.

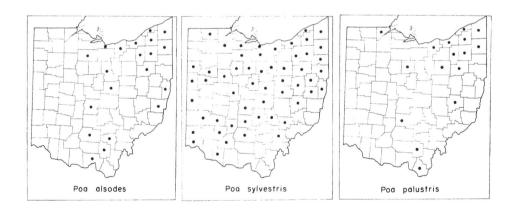

Poa alsodes   Poa sylvestris   Poa palustris

7.   ERAGROSTIS Beauv.   LOVE GRASS

A large genus, many species of which are in the United States. Culms low to moderately tall, usually tufted; ligule a ring of hairs. Spikelets of few to many closely imbricated florets in usually open or diffuse panicles; lemma

3-nerved, awnless, small, in Ohio species 3 mm long or less; palea sometimes persistent after falling of lemma.

a.   Stems creeping, rooting at nodes, short flowering branches erect or ascending; spikelets many-flowered; plants of mud flats and stream banks ....1.   *E. hypnoides*
aa. Stems erect or sometimes bent at base but not creeping.
   b.   Glands present on some or all of the following parts: culms just below nodes, sheaths, margins of blades, pedicels, glumes, lemmas; florets of spikelet usually 10–40, rarely few.
      c.   Spikelet 2.5 mm wide or more; lemma more than 2 mm long ....
                          2.   *E. megastachya*
      cc. Spikelet 2 mm wide or less; lemma less than 2 mm long; glumes and lemmas often almost glandless ....3.   *E. poaeoides*
  bb. Glands not present on the above plant-parts.
      c.   Perennial with hard base; culms unbranched; panicle-branches slender but stiff, at maturity divergent, axils long-pilose; panicle ⅔ to as wide as long, ⅔ height of plant, eventually breaking off and tumbling in the wind; spikelets pink- to red-purple; florets usually 6–12 or rarely few ....4.   *E. spectabilis*
      cc. Annual; culms branched at nodes or at base; panicle-branches slender or capillary, not stiff; spikelets usually lead-color.
         d.   Florets usually not more than 5 per spikelet; panicle much longer than wide, axils usually glabrous.
            e.   Panicle diffuse, to 4 dm long, ⅔ as long as plant; grain with lengthwise furrow; culms branched at base; pedicels much longer than spikelets; plants of dry soil ....5.   *E. capillaris*
            ee. Panicle to 15 cm long, about half length of plant; grain without lengthwise furrow; culms branched above base; pedicels little longer than spikelets; plants of moist soil ....6.   *E. frankii*
        dd. Florets usually more than 5 per spikelet.
            e.   Lateral nerves of lemma conspicuous; panicle 5–15 cm long, not diffuse, axils glabrous or somewhat pubescent; spikelets usually linear, often appressed to panicle-branches, about 1.5 mm wide ....
                          7.   *E. pectinacea*
            ee. Lateral nerves of lemma obscure; panicle diffuse, delicate, branches capillary, axils sparsely pilose; spikelets 1 mm wide, not appressed to branches ....8.   *E. pilosa*

1.   **Eragrostis hypnoides** (Lam.) BSP.

Forming a delicate mat-like growth over mud flats and other wet areas; panicles whitish-green, little branched, 1–6 cm long; spikelets linear, to 1.5 cm long; fruit spherical, much shorter than lemma. Much of the United States.

Eragrostis hypnoides

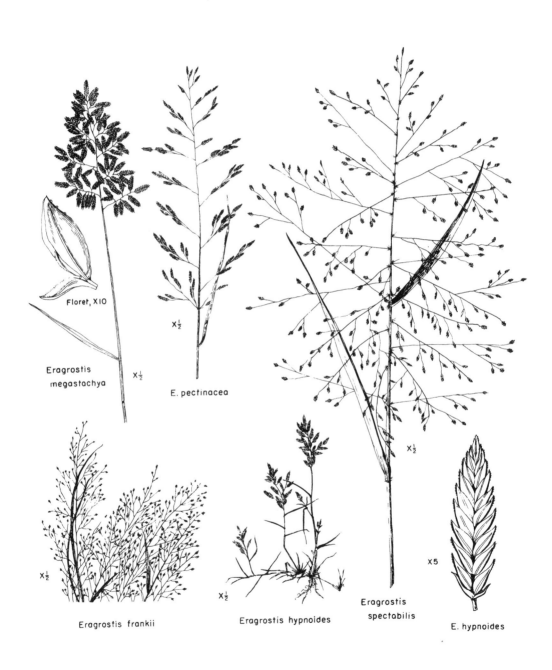

Floret, X10

Eragrostis
megastachya   X½

E. pectinacea   X½

Eragrostis frankii   X½

Eragrostis hypnoides   X½

Eragrostis
spectabilis   X½

E. hypnoides   X5

2. ERAGROSTIS MEGASTACHYA (Koel.) Link   STRONG-SCENTED LOVE GRASS
    E. cilianensis (All.) Lutati; *E. major* Host
Culms short to moderately tall; panicle ovoid, usually dense, 5–15 cm long; glands usually abundant. Plant, when fresh, with somewhat disagreeable odor. Native of Europe, naturalized throughout most of the United States; cultivated ground, roadsides, waste places, sometimes a weed.

3. ERAGROSTIS POAEOIDES Beauv.
    E. minor Host; *E. eragrostis* (L.) Karst.
Resembles the preceding; distinguished from it by shorter culms, more open panicles, more slender dark-gray or purple spikelets, and smaller lemmas; glands often few on glumes and lemmas, more easily found on blades, sheaths, and pedicels. Introduced from Europe; recorded from 16 Ohio counties.

4. **Eragrostis spectabilis** (Pursh) Steud.   PURPLE LOVE GRASS
    E. pectinacea, not (Michx.) Nees
Culms short to moderately tall, tufted, the several panicles of a tuft forming a mass of delicate red-purple color in late summer and autumn along roadsides and in fields and waste places. Eastern United States, west to Minnesota and Arizona.

5. **Eragrostis capillaris** (L.) Nees   CAPILLARY LOVE GRASS
Culms short to moderately tall, erect; panicle diffuse, comprising much of plant; spikelets tiny, long-pediceled, green or lead-color, sometimes very dark. Open woods and fields, eastern half of the United States.

6. **Eragrostis frankii** C. A. Meyer
Resembles *E. capillaris*; sometimes distinguished from it with difficulty when plants are of borderline size, but generally culms shorter, panicles smaller and denser, and pedicels shorter. Absence of furrow in grain is distinguishing but useful only when high magnification is available because grain of both species is very tiny. Moist open ground, eastern half of the United States.

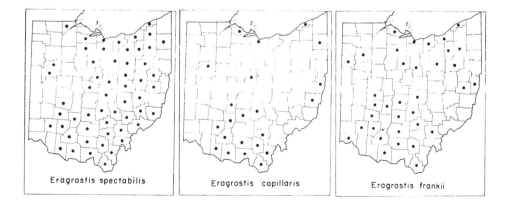

Eragrostis spectabilis   Eragrostis capillaris   Eragrostis frankii

7.   **Eragrostis pectinacea** (Michx.) Nees
    E. purshii Schrad.
    Culms short to moderately tall, slender, tufted, ascending or spreading; spikelets often racemose and appressed along primary panicle-branches. Waste places, gardens, roadsides, throughout the United States; abundant in Ohio, sometimes a weed.

8.   ERAGROSTIS PILOSA (L.) Beauv.   INDIA LOVE GRASS
    Resembles *E. pectinacea* and sometimes not easily distinguished from it; panicle more diffuse and delicate, capillary branches more spreading, spikelets narrower and not appressed to branches, lemmas a little smaller, their lateral nerves obscure. Introduced from Europe; recorded from five scattered counties.

   Eragrostis curvula (Schrad.) Nees, WEEPING LOVE GRASS, is cultivated as an ornamental. A specimen from Portage County may be an escape. Moderately tall to tall; blades elongate, tapering to a long slender point; panicle 2–3 dm long, branches ascending.

Eragrostis pectinacea

8.   DIARRHENA Beauv.

Korycarpus Zea

Two species, one in Japan, one in the United States.

1.   **Diarrhena americana** Beauv.
    D. diandra Wood; *Korycarpus arundinaceus* Zea
    Tall perennial, with rhizomes; blades 1–2 cm wide, elongate, long-tapering to base and tip; panicle 1–3 dm long, exserted, narrow, the few branches appressed; lemma firm and shining, the 3 strong nerves converging in a sharp point; lemma and palea separated at maturity by the plump shining grain, the stout 2-lobed beak of which projects beyond tip of palea. Typical variety has ovate lemma 7–10 mm long, gradually tapering to a cusp; variety *obovata* Gl. has shorter obovate lemma abruptly rounded to shorter cusp, and more turgid

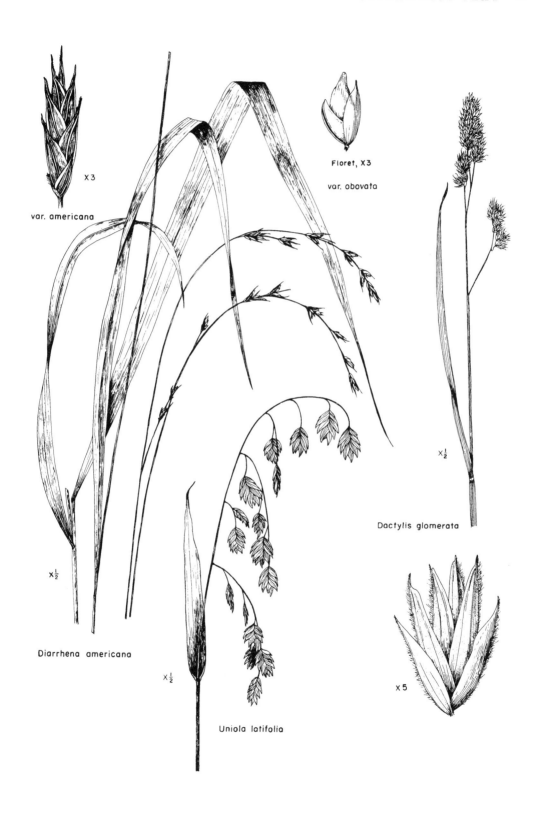

X3

var. americana

Floret, X3

var. obovata

$\times\frac{1}{2}$

Dactylis glomerata

$\times\frac{1}{2}$

Diarrhena americana

$\times\frac{1}{2}$

Uniola latifolia

x5

mature grain projecting farther beyond palea. Most Ohio specimens are var. *americana*; the Lorain and Licking county specimens are var. *obovata*; specimens of both varieties recorded from Franklin and Auglaize counties.

Diarrhena americana

9. UNIOLA L.

About 10 species, all American; spikelets compressed, the lower 1–4 lemmas empty. Panicles of Sea-oats, *U. paniculata* L., a tall grass of coastal dunes and shores of the Southeast, are frequently seen in "arrangements" in florists' windows.

1.   Uniola latifolia Michx.

A handsome and distinctive grass with blades 1–2 cm wide and drooping panicle of large flat many-flowered spikelets on slender pedicels. Lemmas closely imbricated, about 1 cm long; keels of palea winged and projecting beyond sides of lemma. Sometimes cultivated as an ornamental. Woods and steep banks, New Jersey to Kansas and southward.

Uniola latifolia

10. DACTYLIS L.

Two or three species of Eurasia and North Africa.

1. DACTYLIS GLOMERATA L.   ORCHARD GRASS
Culms moderately tall, tufted; foliage green or glaucous, mostly glabrous; panicle 1–2 dm long, the few branches spreading at anthesis but appressed in fruit. Introduced from Europe, naturalized throughout much of the United States, and cultivated as a hay and pasture grass especially in the northeastern fourth.

11. CYNOSURUS L.   DOG'S-TAIL

About 6 Eurasian species. Spikelets short-pediceled or sessile, in clusters, of two kinds, the fertile covered and nearly hidden by the sterile; lemma usually awned.

Inflorescence slender and spike-like; awns inconspicuous, 1 mm long or less ....
1.   *C. cristatus*
Inflorescence ovoid or ellipsoid; awns conspicuous, to 1 cm long ....2.   *C. echinatus*

1. CYNOSURUS CRISTATUS L.   CRESTED DOG'S-TAIL
Perennial; culms short to moderately tall; inflorescence 3–10 cm long; fertile lemma about 3 mm long. From Europe; waste places, lawns, roadsides. Recorded from 3 Ohio counties.

2. CYNOSURUS ECHINATUS L.
Annual; inflorescence compact, bristly because of the awns, somewhat one-sided, 1–5 cm long; fertile lemma 5–7 mm long. Locally introduced from Europe; recorded from Ross County.

12. PHRAGMITES Trin.

Three species, one in South America, one in Asia, the third almost cosmo-politan.

1. **Phragmites communis** Trin.   REED GRASS
P. phragmites (L.) Karst.
Culms stout, 2–4 m tall, with rhizomes and sometimes stolons, usually in colonies; blades usually 2–4 cm wide, tapering to slender apex and to base narrower than summit of sheath; panicle to 4 dm long; spikelets 1 cm long or more, purple and linear when young, spreading at maturity revealing the rachilla-hairs. Banks of streams and lakes, ditches, and swamps, much of the United States; also Canada, Mexico, South America, Eurasia, Africa, and Australia.

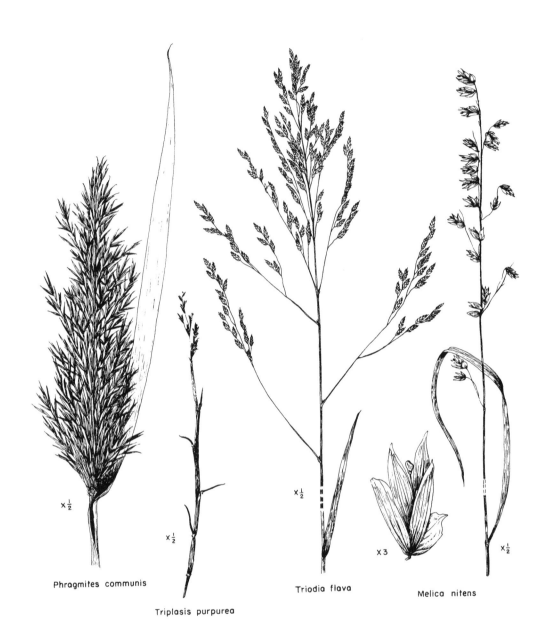

Phragmites communis

Triplasis purpurea

Triodia flava

Melica nitens

13. MELICA L. MELIC GRASS

A large genus of temperate climates. Distinctive in that above uppermost floret are two or three empty lemmas decreasing in size upward, each enclosed by the one just below, the whole forming a small mass at end of rachilla.

1. **Melica nitens** Nutt.   THREE-FLOWERED MELIC
Moderately tall to tall; panicle 1–2 dm long, branches spreading; spikelets usually 3-flowered, about 1 cm long, pendulous; glumes and lemmas loosely spreading at maturity; lemma strongly nerved. Prairies, rocky woods; Pennsylvania and Virginia, west and southwest to Iowa and Texas .

14. SCHIZACHNE Hack.

A monotypic genus of North America and Eastern Asia.

1. **Schizachne purpurascens** (Torr.) Swallen   FALSE MELIC
Avena torreyi Nash; *Melica striata* (Michx.) Hitchc.
Culms slender, moderately tall, sheaths closed, blades narrowed at base; panicle branches few, bearing one or two spikelets each; lemma strongly nerved, about 1 cm long, awned from just below terminal teeth, awn about as long as body. Newfoundland to Alaska, southward in the Rockies to New Mexico and in the East to Kentucky.

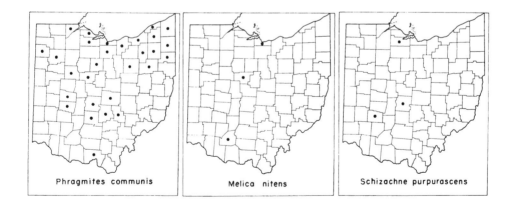

Phragmites communis   Melica nitens   Schizachne purpurascens

15. TRIODIA R. Br.
Tridens Roem. & Schult.

Grasses of temperate climates, with erect culms and terminal panicles of several-flowered spikelets.

1. **Triodia flava (L.) Smyth**   TALL REDTOP. PURPLETOP. GREASE GRASS
    Tridens flavus (L.) Hitchc.

    A tall to moderately tall grass common along roadsides and in old fields, with handsome panicle of purple or rarely yellow spikelets; foliage generally glabrous except for hairs at top of sheaths; panicle-branches drooping, viscid, sometimes covered with adherent dirt particles. Eastern half of the United States.

16.   TRIPLASIS Beauv.

Two species of the United States. Upper sheaths enclosing small panicles of cleistogamous spikelets, lower sheaths sometimes containing spikelets of single florets; culms eventually breaking at nodes.

1. **Triplasis purpurea (Walt.) Chapm.**   PURPLE SAND GRASS

    Tufted annual with short blades; terminal panicle a few cm long, included or exserted, of a few purple spikelets; midnerve of lemma projecting as a short awn to tip of 2 short broad apical teeth; palea soon divergent, hairs on upper part of keels conspicuous. Sandy shores of Atlantic Ocean, Gulf of Mexico, and Great Lakes; locally inland.

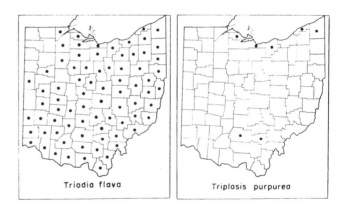

Triodia flava   Triplasis purpurea

TRIBE III.   HORDEAE

Inflorescence a spike with solitary spikelets in two rows, the rows on opposite sides of the rachis, the spikelets alternate; or with groups of 2-several spikelets so attached; spikelets 1–many-flowered, rachilla extending behind uppermost palea as a naked stipe or bearing a vestigial spikelet; glumes and lemmas awned or awnless, glumes sometimes side by side in front of floret or florets. Important grain crops included in this tribe are wheat, rye, and barley.

a.   One spikelet at each node of rachis.
    b.   Each internode of rachis thin at base, thickened at summit, concave on one side,
        a spikelet closely appressed to and fitting into the concavity; rachis jointed between
        spikelets; spikelets in ours cylindric .... 19. *Aegilops*

bb. Rachis and spikelets not as above; spikelets flattened laterally.
 c. Spikelets attached with side to rachis; both glumes present.
  d. Lemma lanceolate; glumes lanceolate or narrower.
   e. Spikelet several-flowered; lemma symmetric, awnless or awned; glumes lanceolate ....17. *Agropyron*
   ee. Spikelet 2-flowered; lemma long-awned, asymmetric, ciliate keel on one side of center; glumes linear ....20. *Secale*
  dd. Lemma and glumes ovate; spikelet 2–5-flowered; lemma awned or awnless, asymmetric ....18. *Triticum*
 cc. Spikelets attached with edge to rachis; glume on edge toward rachis absent except on terminal spikelet ....24. *Lolium*
aa. Two to several spikelets at each node of rachis (sometimes fewer at base or tip of spike).
 b. Two or more spikelets at each node of rachis, all containing perfect florets; if more than 2 spikelets at a node, then spikelets 2–6-flowered.
  c. Both glumes present, about equal, lanceolate to setaceous, not much smaller than lemmas; spikelets ascending at maturity ....21. *Elymus*
  cc. One or both glumes wanting, minute, or reduced to awns much shorter than lemmas; spikelets ascending at first, spreading horizontally at maturity
   22. *Hystrix*
 bb. Three spikelets at each node of rachis, each containing 1 perfect floret and sometimes a second vestigial one, or the 2 lateral spikelets without a perfect floret ....
   23. *Hordeum*

17. AGROPYRON Gaertn. WHEAT GRASS

A large genus of temperate regions, Ohio species perennial. In the United States more of the species occur in the West where some are valuable forage grasses.

a. Spikelets crowded on rachis, divergent, tip of one reaching almost to tip of next above on same side ....1. *A. desertorum*
aa. Spikelets less crowded, appressed or ascending, tip of one sometimes overlapping base of next above on same side but not reaching almost to its tip.
 b. With rhizomes; anthers 3–7 mm long; florets not readily separating from glumes.
  c. Glumes rigid, tapering from near base to sharp point; blades mostly 2–4 mm wide; cartilaginous bands of upper nodes much narrower than their diameter....
   2. *A. smithii*
  cc. Glumes less rigid, tapering from about middle; blades mostly 5–10 mm wide; cartilaginous bands of upper nodes about as wide as their diameter ....
   3. *A. repens*
 bb. Without rhizomes; anthers 1–2.5 mm long; florets readily separating from glumes....
   4. *A. trachycaulum*

1. AGROPYRON DESERTORUM (Fisch.) Schultes
Differs from other species in having crowded overlapping spikelets and short internodes of rachis; spike 5–9 cm long; lemma about 6 mm long; glumes and lemma short-awned. Introduced from Russia, planted in the Great Plains, and adventive farther east; recorded from Pickaway County.

2. AGROPYRON SMITHII Rydb. WESTERN WHEAT GRASS
Native from Iowa and Kansas westward; introduced in Ohio. Differs from *A. repens* in having narrower blades usually rolled in drying, more rigid glumes, and narrower cartilaginous bands of upper nodes. Recorded from 8 counties.

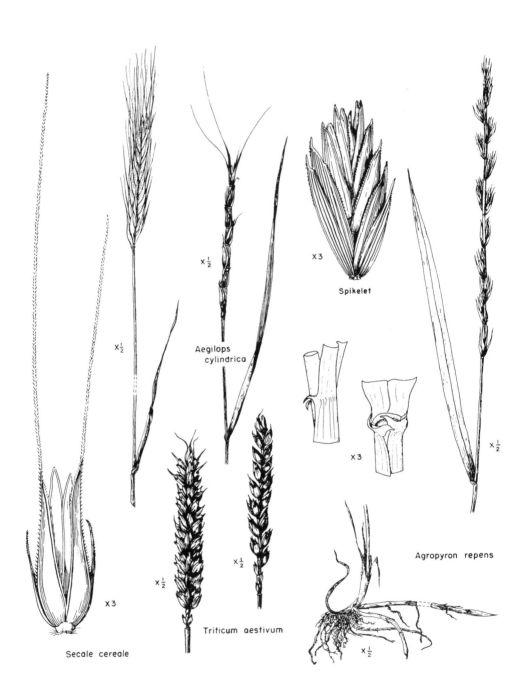

x½

x½

Aegilops
cylindrica

X3

Spikelet

X3

x½

x½

x½

X3

Secale cereale

Triticum aestivum

Agropyron repens

x½

3. AGROPYRON REPENS (L.) Beauv.   QUACK GRASS. WHEAT GRASS

Spreading by long horizontal rhizomes, often a troublesome weed of cultivated fields; lemma awnless or with awn to as long as body; green or glaucous; variable in pubescence. Roadsides, waste places, cultivated fields; abundant in Ohio. Introduced from Eurasia throughout most of the United States and northward, or possibly native in parts of the range.

4. **Agropyron trachycaulum** (Link) Malte

Incl. *A. subsecundum* (Link) Hitchc., *A. pauciflorum* (Schwein.) Hitchc., and *A. caninum* of Am. authors, not (L.) Beauv.

Green or glaucous; spikelets few-flowered, appressed. Treated here (following Fernald, 1950) as an inclusive species in which several varieties are distinguished, two of which are in Ohio. Specimens from Trumbull and Williams counties are referred to var. *trachycaulum* (*A. pauciflorum* (Schwein.) Hitchc., *A. tenerum* Gray), which has distant spikelets, tip of one not reaching base of next above on same side, and lemma awnless or with awn half as long as body or less; remainder of specimens are referred to var. *glaucum* (Pease & Moore) Malte (*A. caninum* of Am. authors, not (L.) Beauv.), which has each spikelet overlapping base of next above on same side, and lemma with awn from nearly as long as to longer than body.

Agropyron trachycaulum

18.   TRITICUM L.   WHEAT

Several species of the Old World, a few wild, the others known only in cultivation, each of which can be placed in one of three groups on basis of chromosome number. Wheat, cultivated since prehistoric times and now worldwide in distribution, is one of the most important food crops. In addition to common wheat, cultivated wheats of the United States include durum, used in manufacture of macaroni and similar products, and club wheat, with short compact spikes, planted mostly in the Northwest. Terms indicating common variants of wheat are: bearded (awned) and beardless; red and white, referring to color of grains; hard and soft, describing texture of grains, the flours differing

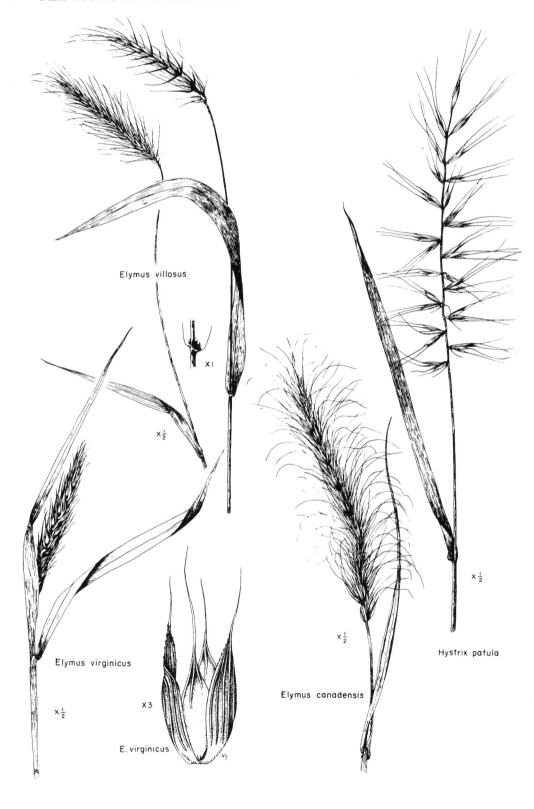

Elymus villosus

X1

X ½

Elymus virginicus

X ½

X 3

E. virginicus

Elymus canadensis

X ½

Hystrix patula

X ½

in certain characteristics and suitable for different uses; winter and spring, winter wheat planted in autumn and harvested the next summer, spring wheat planted in spring and harvested later the same year. Extensive wheat improvement programs of the last seventy-five years have resulted in many races or varieties differing in such ways as morphological features, resistance to disease, milling qualities, and suitability to particular environments.

1. TRITICUM AESTIVUM L.   COMMON WHEAT
Cultivated throughout the world, this is the wheat most extensively planted in the United States.

19. AEGILOPS L.   GOAT GRASS

A small genus of Europe and Western Asia.

1.   AEGILOPS CYLINDRICA Host
Culms short to moderately tall, tufted; spike about 5 mm thick and 5–12 cm long; lemmas, at least those near summit of spike, awned. Introduced from Europe locally throughout the United States; wheat fields, waste places, roadsides. Records from 4 Ohio counties.

20. SECALE L.   RYE

A small genus of annual grasses. One species commonly cultivated in Eurasia and in North America.

1.   SECALE CEREALE L.   RYE
Tall; foliage bluish-green; spike 10–15 cm long. A crop plant sometimes spontaneous but not persisting.

21. ELYMUS L.   WILD-RYE

About 45 species of temperate regions; Ohio species moderately tall to tall perennials, tufted, blades 0.5–2 cm wide, inflorescence a bristly spike. Spikelets 1–several-flowered, 2–several at each node of rachis; glumes somewhat asymmetric, usually stiffly awned; lemma about 1 cm long, usually long-awned. Some species useful for holding sand of dunes and soil of embankments and, in the West, as forage.

a.   Awns outwardly curved when dry; base of glumes not terete, not bowed out; palea of lowest floret of a spikelet from middle of spike mostly 10 mm long or more; often more than 2 spikelets at a node; lemma hirsute (rarely scabrous or glabrous) ....
1. *E. canadensis*
aa. Awns straight or rarely absent; base of glumes hard and terete or subterete; palea of lowest floret of a spikelet from middle of spike mostly 6–8.5 mm long; usually 2 spikelets at a node.
b.   Glumes noticeably widened above base, somewhat spiral, rarely awnless, base yellowish and bowed out; base of spike often included in inflated upper sheath; florets not falling away from glumes ....2. *E. virginicus*

   bb. Glumes scarcely widened above, awned, base little or not at all bowed out; spike
        exserted; florets readily falling away from glumes.
        c. Blades short-villous on upper surface; spikelets 1–2-flowered; lemma and
            glumes usually villous; palea about equaling body of lemma; spike 5–12 cm
            long; summit of sheath with broad flange ....3. *E. villosus*
       cc. Blades glabrous; spikelets 2–4-flowered; lemma and glumes hispidulous to
            glabrous; palea of lowest floret of spikelet 2–5 mm shorter than body of lemma;
            spike to 2 dm long; summit of sheath without broad flange ....4. *E. riparius*

1. Elymus canadensis L.

Tall; foliage sometimes glaucous; blades 1–2 cm wide; spike usually dense, somewhat nodding, to 2.5 dm long, wider than in other species; glumes narrow, up to 3 cm long including awns, less hispid than lemmas. Prairies and sandy soil, much of the United States.

2. Elymus virginicus L.   VIRGINIA WILD-RYE

Moderately tall to tall; distinguished from other species by glumes which have subterete hard yellowish bowed-out bases. Woods, thickets, stream banks; throughout the United States except in the far West. Interpreted, in the treatment here used (following Fernald, 1950), as an inclusive species in which a number of varieties and forms are distinguished. Three of these varieties in Ohio:

   Glumes and lemmas smaller and awned; upper sheath inflated; spike included in upper
      sheath or slightly exserted; glumes and lemmas scabrous on edge or glabrous or, in
      forma *hirsutiglumis* (Scribn.) Fern. (var. *intermedius* (Vasey) Bush, *E. hirsutiglumis*
      Scribn.), glumes and lemmas hirsute (specimens from 84 counties) ....var. *virginicus*
   Glumes and lemmas smaller and awnless or nearly so (specimens from Pickaway and
      Ross counties) ..,...var. *submuticus* Hook.
   Glumes and lemmas larger and awned; upper sheaths not inflated;
      spikelets glabrous or scabrous or, in forma *australis* (Scribn. & Ball) Fern. (var.
      australis (Scribn. & Ball) Hitchc., *E. australis* Scribn.), spikelets hirsute (specimens
      from several counties) ....var. *glabriflorus* (Vasey) Bush

3. Elymus villosus Muhl.

E. striatus of Am. authors, not Willd.

Top of peduncle bent or arching; internodes of rachis short; lemma 7–9 mm long, awn longer than body. In typical form, glumes and lemma villous; in forma *arkansanus* (Scribn. & Ball) Fern. (*E. arkansanus* Scribn. & Ball), glumes and lemmas glabrous or scabrous. Woods, east of the Rocky Mountains.

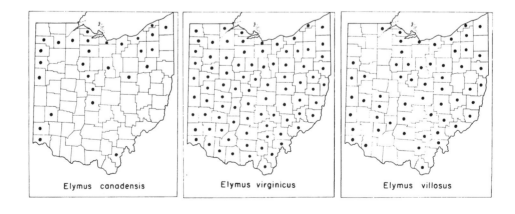

Elymus canadensis   Elymus virginicus   Elymus villosus

4. **Elymus riparius** Wieg.

Spikes nodding; foliage green or glaucous. Banks of streams, woods; eastern half of the United States except the extreme South.

22. HYSTRIX Moench

Four species, one in Asia, one in New Zealand, two in the United States.

1. **Hystrix patula** Moench   BOTTLE-BRUSH GRASS
   H. hystrix (L.) Millsp.

Common name descriptive of distinctive mature inflorescence, the spikelets extending horizontally; internodes of rachis 5–10 mm long; spikelets falling early; tall to moderately tall. Common in woodlands, eastern half of the United States.

23. HORDEUM L.   BARLEY

Three spikelets at each node of rachis, the central, and sometimes all three, containing a perfect floret; rachilla prolonged as a naked stipe or rarely bearing a rudimentary floret; rachis breaking at maturity (except in *H. vulgare*), segment and group of spikelets falling together; glumes of central spikelet located in front of floret.

a.   Awns spreading; whole inflorescence, including awns, as wide as long; lateral spikelets pediceled and vestigial .... 1. *H. jubatum*
aa. Awns erect, when present; inflorescence much longer than wide.
    b.   Spikelets sessile, each containing a fertile floret; lemma of central spikelet with awn several times as long as body, or awnless and 3-lobed at apex; auricles present ....
                                                           2. *H. vulgare*
    bb. Lateral spikelets pediceled, vestigial, awnless; lemma of central spikelet with awn no longer than body; auricles absent ....3. *H. pusillum*

1. HORDEUM JUBATUM L.   SQUIRREL-TAIL BARLEY

Attractive short tufted perennial of fields and roadsides, with soft silvery long-awned inflorescences; lateral spikelets and glumes of fertile spikelets awn-like; lemma 6–8 mm long, long-awned. Most of the United States except the Southeast; probably introduced in Ohio. Recorded from 60 counties.

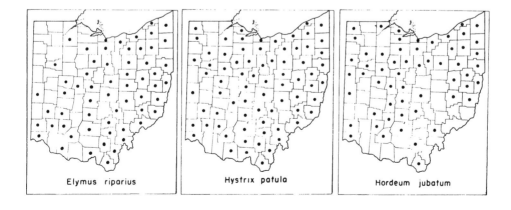

Elymus riparius   Hystrix patula   Hordeum jubatum

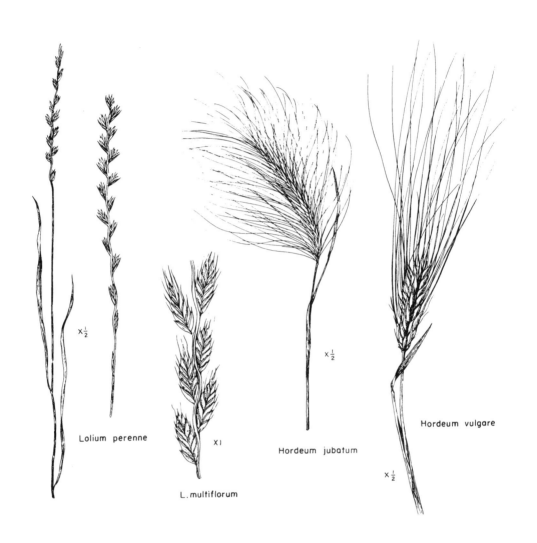

Lolium perenne

x½

L. multiflorum

X1

Hordeum jubatum

x½

Hordeum vulgare

x½

2.   HORDEUM VULGARE L.   BARLEY

Moderately tall annual; spike, exclusive of awns, 3–10 cm long; lemma about
1 cm long, bearing a long coarse flattened awn or, in Beardless Barley, awnless
but prolonged into a hooded appendage with lobe at each side of base. Cultivated; sometimes spontaneous but not persisting.

3.   **Hordeum pusillum** Nutt.   LITTLE BARLEY

Low annual with spikes 2–7 cm long; glumes of fertile spikelet and first
glume of lateral spikelets widened above base. Open ground, much of the
United States.

Hordeum pusillum

24. LOLIUM L.

About 8 species native in the Old World.

a. Body of glume overtopped by uppermost lemma of spikelet.
  b. Perennial; blades folded in bud; lemma awnless or nearly so, 5–7 mm long; in upper spikelets glume usually exceeding lemma immediately above it; spikelets mostly 6–10-flowered ....1. *L. perenne*
  bb. Annual; blades rolled in bud; lemma 7–8 mm long, at least in upper spikelets usually awned; in upper spikelets glume usually not exceeding lemma immediately above it; spikelets mostly 10–20-flowered ....2. *L. multiflorum*
aa. Body of glume overtopping uppermost lemma of spikelet; lemma awned or awnless ....
3. *L. temulentum*

1. LOLIUM PERENNE L. PERENNIAL RYE GRASS. ENGLISH RYE GRASS
Culms short to moderately tall, bases usually reddish. Long cultivated in Europe and important there as forage grass; introduced and less extensively planted in the United States for hay and pasture and in lawn mixtures. Escaped, fields and roadsides.

2. LOLIUM MULTIFLORUM Lam. ITALIAN RYE GRASS
Much like *L. perenne*, but usually taller, with more florets per spikelet, and with awned longer lemmas. Cultivated in Europe and in America as forage and in lawn mixtures; escaped, fields and roadsides.

3. LOLIUM TEMULENTUM L. DARNEL
Introduced from Europe; Pacific States and occasionally eastern United States. One specimen recorded from Ohio. Reputed to contain a poison, the poisonous properties perhaps due to an associated fungus.

TRIBE IV. AVENEAE

Spikelets 2–several-flowered, in open or narrow panicles or rarely in racemes; one or both of the glumes usually equaling or overtopping floret immediately above it or even uppermost floret of spikelet; lemma usually awned from back or from below 2-toothed apex; rachilla usually prolonged behind uppermost floret as a slender stalk, sometimes bearing a reduced floret.

a. Glumes about 2 cm long or more, 7–11-nerved, both usually exceeding upper lemma; lemma awnless or awned from back; inflorescence an open panicle ....30. *Avena*
aa. Glumes shorter, with fewer nerves.
    b. Lower lemma awnless, awned at or near tip, or awned from base of or just below base of terminal teeth.
        c. At least one glume shorter than floret just above it; second glume broadest above middle; spikelet usually 2-flowered; second glume 2–5 mm long.
            d. Panicle dense and spike-like, axis and very short branches densely short-pubescent; articulation above glumes; glumes not differing greatly in width; lemma usually awnless ....25. *Koeleria*
           dd. Panicle sometimes slender but not dense and spike-like, axis and branches glabrous or scabrous; articulation below glumes.
                e. Lemma usually awnless; second glume much broader than first ....
                      26. *Sphenopholis*
                ee. Lower lemma awnless, the second bearing an outwardly-arched awn from below 2-toothed tip; glumes not differing greatly in width ......
                      27. *Trisetum*
        cc. Each glume overtopping floret just above it and usually overtopping body of uppermost floret of spikelet, but not always the awn (when present).
            d. Spikelet 2-flowered, 4–5 mm long; upper floret staminate, lemma with short awn that becomes hooked at tip; lower floret perfect, lemma awnless; panicle usually dense, of many spikelets; whole plant soft-pubescent, gray-green .... 32. *Holcus*
           dd. Spikelet several-flowered, 10–14 mm long; florets perfect, lemmas awned from between bases of terminal teeth, awn bent, spiral at base; inflorescence with few branches, sometimes a raceme, spikelets few ....
                  33. *Danthonia*
  bb. Lower lemma awned from middle of back or below; spikelets 2-flowered; articulation above glumes.
        c. Panicle open, spreading, bases of branches naked; both florets of spikelet perfect; longer glume not more than 5 mm long.
            d. Glumes sometimes unequal, the shorter usually not overtopping uppermost floret, sometimes equaling it; rachilla prolonged; lemma short-toothed or erose at tip .... 28. *Deschampsia*
           dd. Glumes ovate, equal, both exceeding body of uppermost floret but not always the awn; rachilla not prolonged; lemma with 2 setaceous apical teeth ....29. *Aira*
        cc. Panicle narrow, the whorled branches usually bearing spikelets to base; lower floret staminate, lemma with bent awn about 1 cm long; upper floret perfect, lemma usually with short straight awn but rarely with awn similar to that of lower floret; glumes unequal, the longer 7–9 mm long ...31. *Arrhenatherum*

25. KOELERIA Pers.

About 20 species of temperate regions.

1. **Koeleria cristata** (L.) Pers.
    Perennial, short to moderately tall, with narrow blades and slender shining panicle; lemma awnless or rarely short-awned from just below apex; rachilla prolonged as a naked pedicel. Variable and widely distributed in temperate regions of both hemispheres; prairies, open woods, sandy soil.

Koeleria cristata

26. SPHENOPHOLIS Scribn.

A few species, all North American. Slender, perennial, of moderate height, with soft flat blades and slender panicles; lemma almost nerveless; palea hyaline, separating from lemma; articulation below upper lemma and below glumes, upper floret falling away separately, lower floret and glumes falling with a short segment of pedicel.

a. Second glume about as wide as long, firm, rounded at summit, with firm broad papery margin, rarely mucronate; first glume linear-oblong, rounded at tip; internodes of rachilla about 0.5 mm long ....1. *S. obtusata*
aa. Second glume 2/5–2/3 as wide as long, texture less firm; internodes of rachilla about 1 mm long.
    b. Second lemma, and sometimes also the first, scabrous; glumes about equal in length, the first lanceolate, the second broadly rounded at summit or sometimes abruptly pointed; lower sheaths and blades nearly always pubescent ...2. *S. nitida*
    bb. Lemmas glabrous, acute; first glume linear, narrow, usually a little shorter than the acute or subacute second glume; sheaths and blades usually glabrous ....
                                                        3. *S. intermedia*

1. **Sphenopholis obtusata** (Michx.) Scribn.
Distinguished from the other two species by the broader and somewhat cucullate second glume and by the shorter rachilla-internodes; panicle usually dense and crowded, branches short and appressed. Open woods, prairies; throughout the United States.

2. **Sphenopholis nitida** (Biehler) Scribn.
Distinguished by the roughened upper lemma and by the broader first glume; at least the lower blades and sheaths of all but a few Ohio specimens pubescent. Woods, eastern half of the United States.

3. **Sphenopholis intermedia** Rydb.
Sphenopholis pallens (Spreng.) Scribn. of earlier manuals
Distinguished by the very narrow first glume; blades and sheaths of all but a few Ohio specimens glabrous; panicle narrow but somewhat more open than

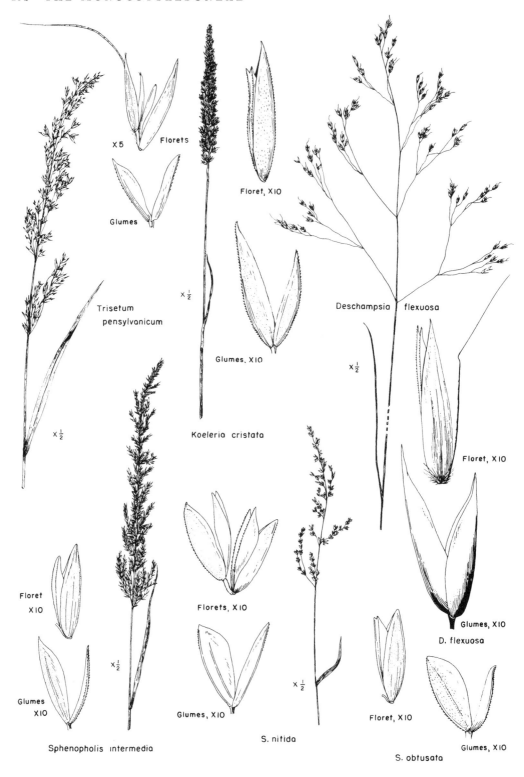

X5   Florets

Glumes

Trisetum
pensylvanicum

X½

Floret, XIO

Glumes, XIO

Koeleria cristata

Deschampsia flexuosa

X½

Floret, XIO

Glumes, XIO

D. flexuosa

Floret
XIO

Glumes
XIO

X½

Florets, XIO

Glumes, XIO

S. nitida

X½

Floret, XIO

Glumes, XIO

S. obtusata

Sphenopholis intermedia

in the other species, branches usually densely flowered to base. Moist places, often in woods; much of the United States.

27. TRISETUM L.

About 65 species of temperate or cooler regions. Spikelets usually 2-flowered, sometimes to 5-flowered, rachilla prolonged behind upper floret; articulation above or below glumes.

1. **Trisetum pensylvanicum** (L). Beauv.   SWAMP-OATS

Resembles species of *Sphenopholis* but distinguished from them by the outwardly curving awn of upper floret; palea shorter than lemma and separating from it; pedicel breaking some distance below glumes, spikelet and a length of pedicel falling together. Swamps and other wet places; eastern United States to Ohio and Louisiana.

Sphenopholis obtusata

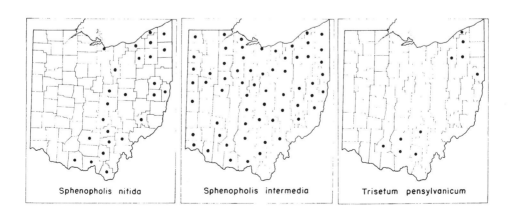

Sphenopholis nitida   Sphenopholis intermedia   Trisetum pensylvanicum

28. DESCHAMPSIA Beauv. HAIR GRASS

About 35 species; spikelets 2-flowered, in panicles; lemma pilose at base.

Blades filiform; ligule of culm leaves rounded at tip, to 3.5 mm long; rachilla between florets about 0.6 mm long, extension behind upper floret short and inconspicuous; awn twisted in lower half, bent, exserted 1–3 mm ....1. *D. flexuosa*
Blades narrow but not filiform; ligule of culm leaves pointed at tip, 5–10 mm long; rachilla between florets about 1.5 mm long, extension behind upper floret conspicuous; awn nearly straight, included or little exserted ....2. *D. caespitosa*

1. **Deschampsia flexuosa** (L.) Trin. WAVY HAIR GRASS
Culms moderately tall, slender, tufted; basal blades much shorter than culms; panicles shining, a few spikelets near tips of capillary branches; glumes hyaline. Transcontinental in the North, southward in mountains, and local elsewhere; also in Eurasia.

2. **Deschampsia caespitosa** (L.) Beauv. TUFTED HAIR GRASS
Densely tufted, slender, but less delicate than No. 1, glumes narrower and less hyaline. Transcontinental in the North, ranging south into our area.

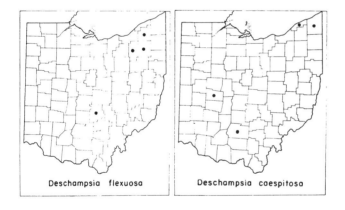

Deschampsia flexuosa   Deschampsia caespitosa

29. AIRA L.

Aspris Adans.

Delicate small annuals with narrow blades.

1. AIRA CARYOPHYLLEA L. SILVER HAIR GRASS
Aspris caryophyllea (L). Nash
Introduced from Europe. One specimen, Lake County.

30. AVENA L. OATS.

Many species, most of which are native in the Old World. Spikelets in ours 2- or 3-flowered, in open panicles.

Lemma glabrous, awnless or with an almost straight awn; spikelets usually 2-flowered

1. A. sativa

Lemma hirsute at base, glabrous or hirsute on back, with bent twisted awn exserted from glumes; spikelets usually 3-flowered ....2. <i>A. fatua</i>

1. AVENA SATIVA L.   COMMON OATS

Foliage gray-green, culms moderately tall. Cultivated extensively in the United States; sometimes escaped but not persistent.

2. AVENA FATUA L.   WILD OATS

Differs from <i>A. sativa</i> in that spikelets are usually 3-flowered and florets readily fall from glumes. Native of Europe, occasionally found in Ohio in waste places and fields.

31. ARRHENATHERUM Beauv.   OAT GRASS

A few species of Europe, West Asia, and North Africa.

1. ARRHENATHERUM ELATIUS (L.) Mert. & Koch   TALL OAT GRASS

Tall handsome perennial with narrow shining panicle 15–30 cm long. Foliage glabrous and upper lemma bearing a short straight awn in most Ohio specimens; but, in specimens from a few counties, foliage pubescent and upper lemma bearing a longer bent twisted awn similar to that of lower lemma. Sometimes cultivated for hay and pasture in the Central and Northern States and in the Pacific Northwest. Escaped, common in Ohio; meadows, railroad rights-of-way, roadsides.

32. HOLCUS L.

Nothoholcus Nash

Small genus of Europe and Africa.

1. HOLCUS LANATUS L.   VELVET GRASS

Nothoholcus lanatus (L.) Nash

Distinguished by its gray-green color, soft pubescence, and ellipsoid dense panicle about 8 cm long; tufted, moderately tall, sheaths somewhat inflated, culm blades short; second glume broad-ovate, wider than first. Introduced from Europe; roadsides, waste places, eastern United States to North Dakota and Louisiana and in the West. Contains a substance that yields hydrocyanic acid; may occasionally cause poisoning when eaten.

33. DANTHONIA Lam. & DC.   WILD OAT GRASS

A large genus most abundant in the south temperate regions of the eastern hemisphere. Tufted low or moderately tall perennials with narrow blades and stiff narrow panicles of a few spikelets; glumes lanceolate; lemma ending in 2 pointed teeth; awn of lemma twisted and flat at base, geniculate.

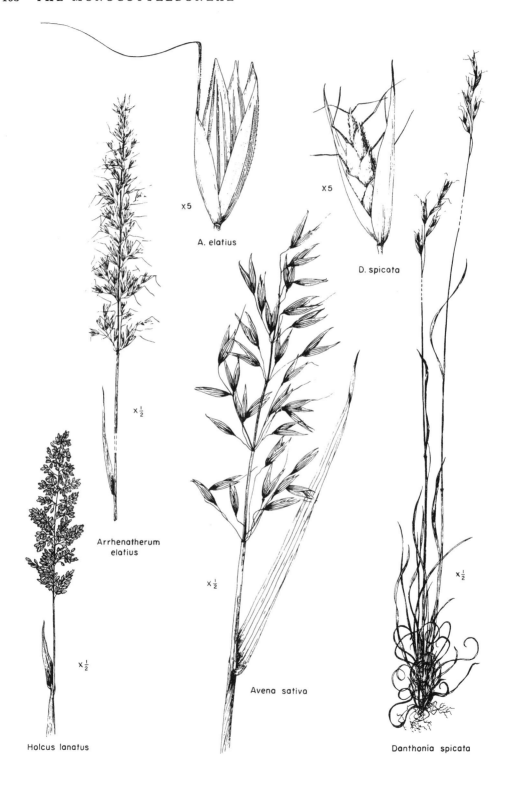

x5

A. elatius

x5

D. spicata

x ½

Arrhenatherum
elatius

x ½

Avena sativa

x ½

x ½

Holcus lanatus

Danthonia spicata

Teeth of lemma acute to subsetaceous, less than 2 mm long; panicle contracted, racemose, or a lower branch or two bearing more than one spikelet ....

1. *D. spicata*

Teeth of lemma aristate, 2–3 mm long; panicle less contracted, primary branches often bearing 2 or 3 spikelets .... 2. *D. compressa*

1. **Danthonia spicata** (L.) Beauv.   POVERTY OAT GRASS

Basal leaves clustered and numerous, blades usually curled; culms terete; sheaths with tuft of hair on each side at summit; inflorescence usually not more than 5 cm long, at maturity much exserted beyond tip of uppermost leaf. Poor soil, open woods; eastern half of the United States and locally westward.

2. **Danthonia compressa** Aust.

Distinguished from the preceding by the longer aristate lemma-teeth, generally taller and stouter culms with lower internodes compressed, and more compound inflorescence less exserted beyond tip of upper blade; however, a number of specimens appear to be intermediate. Maine to Georgia, west to Ohio.

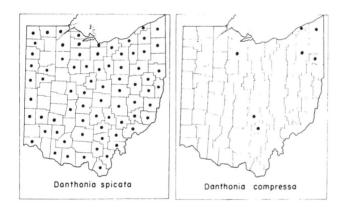

Danthonia spicata   Danthonia compressa

TRIBE V.   AGROSTIDEAE

Spikelet of 1, very rarely 2, usually perfect florets in open, narrow, or even spike-like panicles, but not in definite rows in spikes or racemes.

a. Hairs of callus of lemma half length of lemma or more; lemma with fragile awn from back; extension of rachilla behind palea long-hairy .... 34. *Calamagrostis*
aa. Hairs of callus of lemma, if present, less than half length of lemma.
   b. Inflorescence dense and spike-like, cylindric or subcylindric.
      c. Lemma 9 mm long or more, with tuft of hair at base; inflorescence 1–2.5 cm thick and 1.5–3 dm long .... 35. *Ammophila*
      cc. Lemma much less than 9 mm long; inflorescence usually less than 1 cm thick and 1.5 dm long.
         d. Glumes awned.
            e. Glumes truncate at tip, equal, keels pectinate-ciliate, body longer than lemma; awn about 1 mm long .... 39. *Phleum*
            ee. Glumes tapering to tip, body not longer than lemma, awns varying in length, as much as 4 mm long; inflorescence usually interrupted near base .... 40. *Muhlenbergia*

dd. Glumes not awned.
 e. Glumes equal, edges connate below; lemma equaling glumes, 5-nerved, awned from back ....38. *Alopecurus*
 ee. Glumes unequal in width, edges not connate; lemma exceeding glumes, 1-nerved, awnless ....42. *Heleochloa*
bb. Panicle open or contracted but not dense and spike-like.
 c. One or both glumes minute; lemma about 1 cm long, 5-nerved, with terminal awn to 3 cm long ....43. *Brachyelytrum*
 cc. Glumes not minute or, if rarely so, then lemma not more than 3 mm long.
  d. Lemma not of firmer texture than glumes.
   e. Articulation below glumes; lemma with minute straight awn from just below tip ....37. *Cinna*
   ee. Articulation above glumes.
    f. Lemma 1-nerved; palea often but not always longer than lemma
     41. *Sporobolus*
    ff. Lemma 3–5-nerved; palea not longer than lemma.
     g. Lemma shorter and usually thinner than glumes; palea wanting or not more than ⅔ as long as lemma ... 36. *Agrostis*
     gg. Lemma not shorter than glumes; palea about as long as lemma.
      h. Rachilla prolonged behind palea as a bristle; panicle open ....36. *Agrostis*
      hh. Rachilla not prolonged behind palea; panicle usually contracted, rarely open ....40. *Muhlenbergia*
  dd. Lemma of obviously firmer texture than glumes.
   e. Lemma awnless, edges wrapped around palea at maturity; glumes ovate; ligule membranous, to 6 mm long or more ....44. *Milium*
   ee. Lemma awned, awn deciduous in No. 45.
    f. Awn to 25 mm long, straight or flexuous, not sharply bent, not branched; glumes lanceolate or narrower; lemma narrow, terete, with short blunt callus ....45. *Oryzopsis*
    ff. Either awn much longer than 25 mm and twice bent, or awn branched; glumes elliptic; lemma ellipsoid, dark at maturity, with hard sharp callus.
     g. Awn simple, several times length of body of lemma, twice bent, strongly twisted below ....46. *Stipa*
     gg. Awn 3-parted, lateral branches sometimes short ....
     47. *Aristida*

34. CALAMAGROSTIS Adans.   REED GRASS

Over 100 species of tall or moderately tall plants of temperate and colder regions. Panicle terminal, open or contracted; glumes about equal, acute or acuminate, usually longer than floret; lemma usually 5-nerved, awned from back, with long hairs on callus; rachilla articulated above glumes. Some species useful as range and wild hay grasses.

a. Awn of lemma straight, attached near or above middle.
 b. Rachilla-extension hairy throughout; awn attached near middle of lemma.
  c. Panicle usually open, somewhat nodding; tip of·lemma translucent; callus-hairs about as long as lemma, abundant; awn delicate; culms sometimes branched ....1. *C. canadensis*
  cc. Panicle contracted, branches erect; lemma firm, scarcely translucent because scabrous; callus-hairs ½–¾ as long as lemma ....2. *C. inexpansa*
 bb. Rachilla-extension with hairs in tuft at tip only, the hairs almost reaching tip of lemma; awn attached ¼ distance from tip of lemma; glumes long-acuminate ....
  3. *C. cinnoides*

aa. Awn of lemma bent, twisted below, attached about ¼ distance above base; rachilla-extension hairy throughout, the hairs and some of those of callus reaching ½–¾ distance to tip of lemma ....4.   *C. insperata*

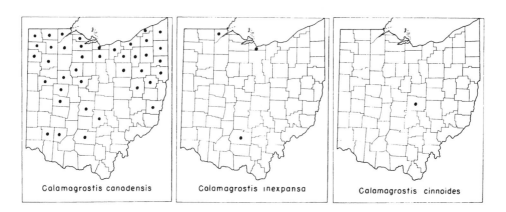

Calamagrostis canadensis   Calamagrostis inexpansa   Calamagrostis cinnoides

Calamagrostis insperata

1.   **Calamagrostis canadensis** (Michx.) Beauv.   BLUEJOINT
Blades elongate, tapering to slender points; panicle 1–2.5 dm long; glumes 3–4 mm long; rachilla extension delicate. Bogs, wet meadows, swamps, sometimes woods; throughout the United States except the southeastern fourth.

2.   **Calamagrostis inexpansa** Gray   NORTHERN REED GRASS
Resembles the preceding in range and habitat; differs in having firmer lemma, narrower panicle, and more rigid rougher blades that are often rolled.

3.   **Calamagrostis cinnoides** (Muhl.) Barton
Distinguished from other Ohio species by the rachilla-extension which is without hairs except for terminal tuft. Bogs and other moist ground, eastern United States as far west as Ohio and Louisiana.

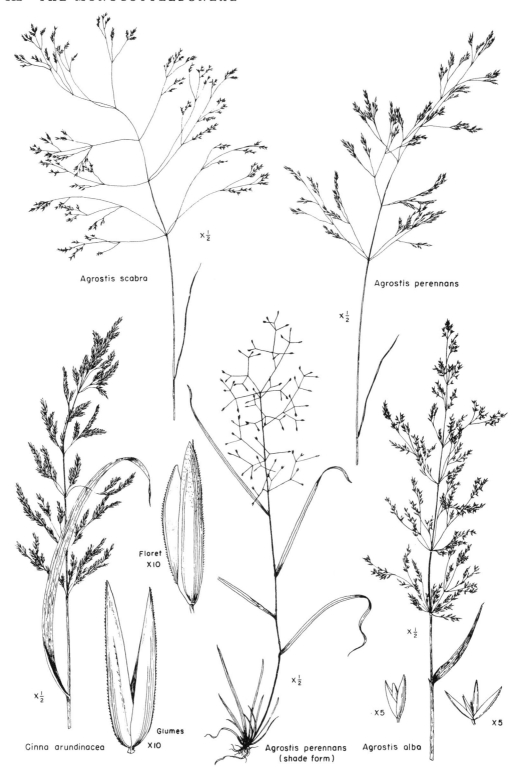

Agrostis scabra

$\times \frac{1}{2}$

Agrostis perennans

$\times \frac{1}{2}$

Floret
X10

Glumes
X10

Cinna arundinacea

$\times \frac{1}{2}$

Agrostis perennans
(shade form)

$\times \frac{1}{2}$

Agrostis alba

$\times \frac{1}{2}$

X5

X5

4. **Calamagrostis insperata** Swallen
Distinguished from other Ohio species by the bent awn attached about ¼ the distance above base of lemma. Discovered in Jackson County by Floyd Bartley and Leslie L. Pontius in 1934 and later collected by Bartley in Vinton County.

35. AMMOPHILA Host   BEACH GRASS

Two or three species of coarse moderately tall erect perennials with extensive rhizomes; both the European species, *A. arenaria* (L.) Link, and the American species are effective soil-binders, planted for the purpose of controlling dunes. Panicle dense and spike-like; rachilla articulated above glumes and prolonged behind palea as a short bristle.

1. **Ammophila breviligulata** Fern.   AMERICAN BEACH GRASS
A. arenaria of authors, not Link
Distinctive because of the pale nearly cylindric panicle sometimes included or partly included in upper sheath, and the curved elongate rolled blades. Sand dunes, Great Lakes and Atlantic Coast as far south as North Carolina.

Ammophila breviligulata

36. AGROSTIS L.   BENT GRASS

Incl. *Apera* Adans.

About 100 species of moderately tall to short plants of temperate and colder regions, often with slender culms and delicate panicles. Several introduced species cultivated for lawn, turf, meadow, and pasture; some native species are forage plants in parts of the West. Glumes about equal, 1-nerved; lemma usually 3-nerved, awned or awnless.

a. Palea almost as long as lemma; rachilla prolonged behind palea as a bristle; lemma firm, with delicate long awn from just below tip, about equaling glumes ....
1. *A. spica-venti*

aa. Palea not more than ⅔ as long as lemma, or wanting; rachilla not prolonged behind palea; lemma shorter and usually thinner than glumes.
   b. Palea ½–⅔ as long as lemma; perennial.
       c. Panicle with several branches at each node, some short and bearing spikelets to base, branches spreading in anthesis, ascending in fruit; ligule 2–6 mm long
                                                                           2. *A. alba*
      cc. Panicle delicate, open, not densely flowered, branches and branchlets spreading; ligule 1–2 mm long; lemma sometimes awned from near base, awn bent, exceeding glumes ....3. *A. tenuis*
  bb. Palea minute or absent.
       c. Lemma 1–1.5 mm long, awned from just below tip with delicate flexuous awn 5–10 mm long; annual ....4. *A. elliottiana*
      cc. Lemma awnless or rarely with short awn; perennial.
          d. Panicle-branches capillary, fragile, very rough to touch, branching only above middle, bearing spikelets only near tips; panicle diffuse.
             e. Glumes 1.5–1.7 mm long; lemma 1–1.2 mm long, hardly longer than grain; spikelets aggregated in small terminal clusters; anthers 0.2 mm long; flowering May to early June ....5. *A. hyemalis*
            ee. Glumes 2–2.7 mm long; lemma 1.5–1.7 mm long, longer than grain; spikelets not aggregated in clusters; anthers 0.4–0.5 mm long; flowering middle June to July ....6. *A. scabra*
        dd. Panicle-branches slightly rough to touch, slender, branching below or at middle; glumes mostly 2–3 mm long; panicle open but not diffuse; flowering August to September ....7. *A. perennans*

1. AGROSTIS SPICA-VENTI L.
    Apera spica-venti (L.) Beauv.
    Moderately tall tufted annual with slenderly ovoid or ellipsoid panicle; glumes, lemma, and palea about equal, 2–3 mm long. Introduced from Europe in a few places in the United States; records from Lake, Lorain, and Wayne counties.

2. AGROSTIS ALBA L. REDTOP
    A. stolonifera L. var. *major* (Gaud.) Farw.
    Tall to moderately tall; panicles ovoid, often reddish, 1–2 dm long. Introduced from Europe, cultivated for meadows, pastures, and lawns in much of the United States and widely escaped; common in Ohio.

3. AGROSTIS TENUIS Sibth. COLONIAL BENT
    Moderately tall, tufted, slender, glabrous, with narrow blades and short ligules; glumes 2–2.5 mm long. Introduced for use in lawns and pastures and escaped in northeastern and northwestern United States. Reported from 5 Ohio counties.

4. **Agrostis elliottiana** Schultes
    Slender tufted annual with open or diffuse panicle about half length of entire plant or more, the capillary branches with spikelets near tips only; lemma sharply nerved, the nerves sometimes extending as short awn-tips, usually with awn, rarely awnless; palea absent. Sandy fields, roadsides, waste places; Maine to Kansas and southward.

5. **Agrostis hyemalis** (Walt.) BSP.
    Culms low to moderately tall, slender, tufted; blades filiform or flat, narrow; panicle diffuse, the spreading or drooping fragile branches again branched above

the middle, short-pediceled spikelets in small clusters at tips of ultimate branches. Old fields, roadsides, waste places, eastern half of United States.

6. **Agrostis scabra** Willd.

    A. hyemalis (Walt.) BSP. var. *tenuis* (Tuckerm.) Gl.

    Similar to *A. hyemalis* in appearance and habitat; differing in the usually taller culms, larger and more diffuse panicles, larger and longer-pediceled spikelets more loosely arranged at tips of branches, and later flowering. Much of the United States.

7. **Agrostis perennans** (Walt.) Tuckerm.

    Tufted perennial; panicle ellipsoid, 1–2.5 dm long, slender branches forking near middle. Varying in appearance from forms that have panicle-branches ascending and culms and blades stiff and upright, to forms that have panicle-branches widely divaricate, spikelets on long pedicels, culms weak and decumbent, and blades soft and narrow. Plants of the latter appearance have been segregated as var. *aestivalis* Vasey (*A. schweinitzii* Trin.). Both extremes and intergrades common in Ohio. Woods and open ground, eastern half of the United States.

Agrostis elliottiana

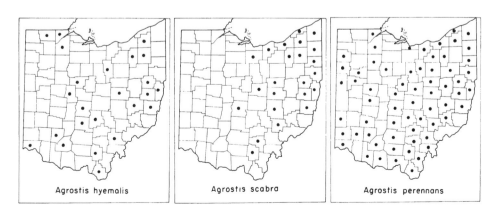

Agrostis hyemalis   Agrostis scabra   Agrostis perennans

37. CINNA L. Wood Reed

Three species of the Americas and Eurasia. Tall or moderately tall perennials; palea 1-keeled.

Second glume 3-nerved, the hyaline margins narrow; panicle silvery-green, densely-flowered, branches ascending ....1. *C. arundinacea*
Second glume 1-nerved or with 2 additional obscure nerves, the hyaline margin about half the width from edge to midvein; panicle green, open, loosely flowered, branches spreading ....2. *C. latifolia*

1. **Cinna arundinacea** L.
Woodland grass with distinctive silvery-green nodding panicle 1.5–3 dm long; blades usually less than 1 cm wide; ligule elongate; spikelets 4–6 mm long. Eastern half of the United States; common in Ohio.

2. **Cinna latifolia** (Trevir.) Griseb.
More northern in range than *C. arundinacea*, usually with wider blades and smaller spikelets. Woods, northern United States, in the mountains to North Carolina and California; Canada and northern Eurasia.

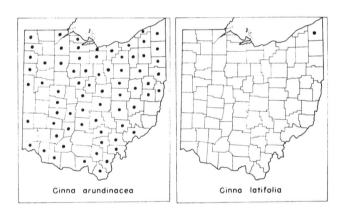

Cinna arundinacea   Cinna latifolia

38. ALOPECURUS L. Foxtail

Low to moderately tall plants of the north temperate zone, with dense spike-like inflorescence resembling that of timothy.

a. Spikelets 4–7 mm long; anthers 2–3 mm long; awn bent.
   b. Panicle usually not more than 5 mm thick; spikelets about 6 mm long; keels of glumes rough or short-ciliate; awn exserted 5 mm or more ....1. *A. myosuroides*
   bb. Panicle 5–10 mm thick; spikelets about 5 mm long; keels of glumes long-ciliate; awn exserted 2–5 mm .... 2. *A. pratensis*
aa. Spikelets less than 3 mm long; anthers 1 mm long or less.
   b. Awn straight, attached about middle of lemma, included or only slightly exserted, scarcely evident when panicle is viewed with unaided eye; perennial ....
                                   3. *A. aequalis*
   bb. Awn bent, attached toward base of lemma, exserted 2–3 mm, evident as short bristle of panicle; tufted annual ....4. *A. carolinianus*

1. ALOPECURUS MYOSUROIDES Huds.

Tufted annual with narrow blades. Adventive from Eurasia; recorded from Franklin County.

2. ALOPECURUS PRATENSIS L.   MEADOW FOXTAIL

Introduced from Eurasia, sometimes cultivated for pasture and meadow in eastern United States and in Pacific Northwest, and occasionally escaped. Reported from 2 Ohio counties.

3. **Alopecurus aequalis** Sobol.

A. aristulatus Michx; *A. geniculatus* L. var. *aristulatus* Torr.

Culms erect or decumbent at base; panicle 2–7 cm long and about 4 mm thick. Throughout the United States except the southeastern fourth, mostly in wet places.

4. **Alopecurus carolinianus** Walt.

A. ramosus Poir.

Much like *A. aequalis* in appearance except that lemma-awns are long enough to be evident. Open ground, fields, and gardens, throughout most of the United States; sometimes a weed.

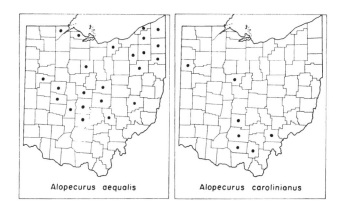

Alopecurus aequalis   Alopecurus carolinianus

39. PHLEUM L.   TIMOTHY

About 10 species of temperate regions, with dense cylindric panicles.

1. PHLEUM PRATENSE L.   TIMOTHY

Moderately tall, culms with swollen bases, blades to 1 cm wide; spike-like panicle usually 5–10 cm long, sometimes longer or very short. Our most important hay grass; native of Eurasia, cultivated in eastern half of the United States north of the cotton states and in parts of the West, and widely naturalized.

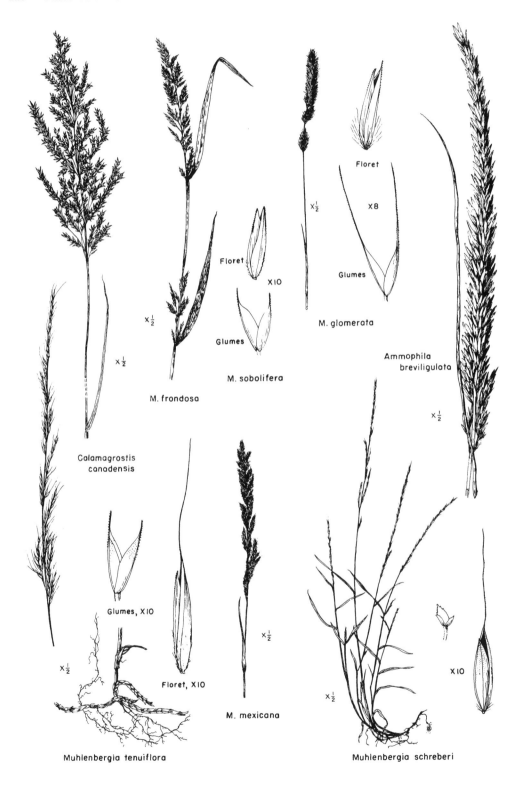

Floret
X10

Glumes

M. sobolifera

Floret

X8

Glumes

M. glomerata

Ammophila
breviligulata

x½

x½

x½

M. frondosa

Calamagrostis
canadensis

Glumes, X10

Floret, X10

x½

x½

M. mexicana

x½

X10

x½

Muhlenbergia tenuiflora

Muhlenbergia schreberi

40.   MUHLENBERGIA Schreb.

Over 100 species, mostly American, most numerous in the arid and semi-arid Southwest where many are important range grasses. Low to moderately tall, mostly perennial with rhizomes; spikelets 1-flowered (rarely 2-flowered), the rachilla breaking above glumes; glumes usually 1-nerved, often acuminate or awned, rarely minute; lemma usually 3-nerved, sometimes awned from tip or just below.

a. Panicle open or diffuse, pedicels capillary.
　　b. Spikelets 1.5–2 mm long, sometimes 2-flowered; lemma glabrous, awnless ....
　　　　　　　　　　　　　　　　　　　　　　　　　　　1. *M. asperifolia*
　　bb. Spikelet 3–4 mm long; lemma minutely puberulent at base, awned ....
　　　　　　　　　　　　　　　　　　　　　　　　　　　2. *M. capillaris*
aa. Panicle narrow, no more than 4 cm wide, branches erect or ascending; or rarely panicle dense and spike-like.
　　b. Lemma not pilose at base, surface minutely pubescent, cuspidate at tip; blades erect or ascending, 1–2 mm wide; without rhizomes; with hard bulb-like base....
　　　　　　　　　　　　　　　　　　　　　　　　　　　3. *M. cuspidata*
　　bb. Lemma pilose at base.
　　　　c. Culms weak, decumbent at base, somewhat reclining; one or both glumes minute or less than half as long as lemma; lemma awned.
　　　　　　d. Glumes minute, the second no more than 0.5 mm long, the first sometimes wanting; lemma about 2 mm long, its awn once or twice length of body ....4. *M. schreberi*
　　　　　　dd. Glumes varying from minute to larger on same plant; lemma 2.5–3 mm, awn shorter than body ....5. *M. curtisetosa*
　　　　cc. Culms erect or reclining; rhizomes bearing overlapping scales; glumes half length of body of lemma or more, with pointed or awned tips.
　　　　　　d. Glumes lanceolate, tapering from base to tip, at least one nearly as long as to longer than lemma.
　　　　　　　　e. Panicle dense and spike-like, interrupted below; glumes awned, awns exceeding lemma; anthers 1–1.5 mm long ....6. *M. glomerata*
　　　　　　　　ee. Panicle not spike-like, but branches sometimes short and dense; glumes awn-pointed, apex of at least one equaling or nearly equaling body of lemma; scales of rhizome ovate, base cucullate-arched; anthers 0.3–0.6 mm long.
　　　　　　　　　　f. Internodes of culm glabrous; stem much branched, whole plant often bushy or sprawling; panicles many, often somewhat open, the axillary often included; lemma usually not awned, short-pilose at base ....7. *M. frondosa*
　　　　　　　　　　ff. Internodes puberulent, at least toward their summit; lemma long-pilose at base.
　　　　　　　　　　　　g. Panicle stiff, spikelets short-pediceled to subsessile, the short densely-flowered branches 1–3 cm long; lemma usually not awned ....8. *M. mexicana*
　　　　　　　　　　　　gg. Panicle flexuous and slender, the erect slender branches often elongate, not densely-flowered, lemma usually awned
　　　　　　　　　　　　　　　　　　　　　　　　　　　9. *M. sylvatica*
　　　　　　dd. Glumes ovate, abruptly pointed, shorter than lemma; anthers 0.8–1.5 mm long; scales of rhizome narrowly ovate, closely appressed.
　　　　　　　　e. Nodes and internodes glabrous; lemma 2–2.5 mm long, blunt or appearing 3-toothed as result of slight projection of nerves at tip, awnless or with awn shorter than body ....10. *M. sobolifera*
　　　　　　　　ee. Nodes and at least upper portion of internodes pubescent; lemma 3–4 mm long, awned, awn equaling to much longer than body ....
　　　　　　　　　　　　　　　　　　　　　　　　　　11. *M. tenuiflora*

1. **Muhlenbergia asperifolia** (Nees and Mey.) Parodi   SCRATCH GRASS
   Sporobolus asperifolius Nees and Mey.
   Culms short, branching, usually decumbent at base; blades short, narrow; panicle diffuse, at first included at base, finally breaking away from culm; spikelets very small, sometimes 2-flowered. Ohio specimens, all along railroads, may be introduced; recorded from Crawford, Fayette, and Jackson counties.

2. **Muhlenbergia capillaris** (Lam.) Trin.
   Distinctive because of the large diffuse ellipsoid panicle, ⅓–½ height of plant, with elongate pedicels ending in red-purple spikelets; lemmas with awns 5–15 mm long. Tall to moderately tall tufted perennial; Massachusetts to Kansas and southward.

3. **Muhlenbergia cuspidata** (Torr) Rydb.   PLAINS MUHLENBERGIA
   Short, in dense tufts, culms slender; panicle narrow, the short branches erect; lemma about 3 mm long, glumes ½–⅔ as long. Prairies and dry slopes, mostly west of Ohio.

4. **Muhlenbergia schreberi** Gmel.   NIMBLEWILL
   Blades short and narrow; panicles terminal and axillary, slender. Distinguished from other Ohio species by the minute glumes. Throughout eastern half of the United States in lawns and gardens, where often a weed, and in waste places; abundant in Ohio.

5. **Muhlenbergia curtisetosa** (Scribn.) Bush
   Much like *M. schreberi* but with coarser stem, longer glumes, and shorter lemma-awns. Recorded from only a few scattered stations in the United States; taxonomic status doubtful.

6. **Muhlenbergia glomerata** (Willd.) Trin.
   Incl. in *M. racemosa* (Michx.) BSP. in earlier manuals.
   Culms erect, unbranched or with a few basal branches; inflorescence dense and spike-like, interrupted below; lemma about 3 mm long, pointed at tip. Bogs and wet places, northeastern United States and in Canada.

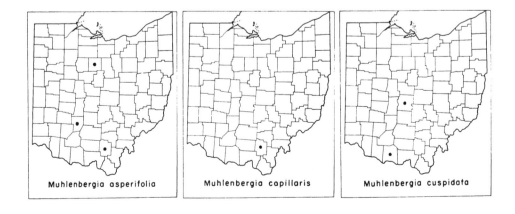

Muhlenbergia asperifolia   Muhlenbergia capillaris   Muhlenbergia cuspidata

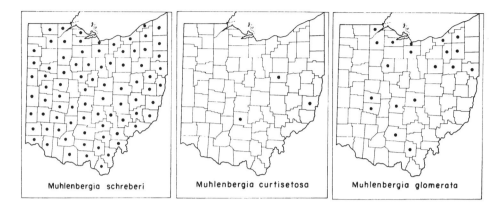

Muhlenbergia schreberi   Muhlenbergia curtisetosa   Muhlenbergia glomerata

7. **Muhlenbergia frondosa** (Poir.) Fern.

M. mexicana of earlier manuals, not (L.) Trin.

Culms moderately tall to tall, much branched, branches again branched, base ascending or decumbent and rooting at nodes, whole plant reclining or sprawling. Glumes variable in length, equal or unequal; lemma 2–3 mm long, typically awnless. Specimens from 8 scattered counties are referred to forma *commutata* (Scribn.) Fern. in which the lemmas have awns 4–10 mm long. Thickets, fence rows, moist soil; common in Ohio, sometimes a weed; eastern half of the United States except extreme southeast.

8. **Muhlenbergia mexicana** (L.) Trin.

M. foliosa (Roem. and Schult.) Trin.

Culms usually unbranched below but with stiff ascending branches from middle nodes; panicles mostly well exserted; lemma 2–3 mm long. A specimen from Ashtabula County has the lemma awned (forma *ambigua* (Torr.) Fern.), and one from Lucas County has the glumes awned (forma *setiglumis* (S. Wats.) Fern.). Thickets, woods, and open ground, throughout the United States except extreme southern states from Texas eastward.

9. **Muhlenbergia sylvatica** Torr.

M. umbrosa Scribn.

Culms erect, sparingly branched from middle nodes, retrorsely rough below nodes; blades ascending or spreading; panicles slender, nodding, branches ascending; lemma 2–3 mm long. Woods, eastern half of the United States except a few southern states.

10. **Muhlenbergia sobolifera** (Muhl.) Trin.

Culms with a few erect branches, glabrous or minutely rough below nodes; blades divergent, tapering to base, 3–8 mm wide; panicle slender, 0.5–1.5 dm long, the branches erect. Woods; range approximately that of *M. sylvatica*.

11. **Muhlenbergia tenuiflora** (Willd.) BSP.

Resembles *M. sobolifera* in panicle, branching of culm, and shape and position of blades; culms coarser, blades slightly wider, lemma longer; nodes, upper portion of or whole internodes, and usually bases of sheaths, retrorsely pubescent. Woods; range approximately that of the 2 preceding species.

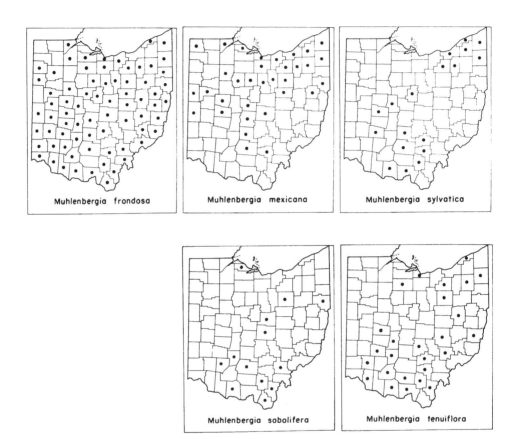

Muhlenbergia frondosa

Muhlenbergia mexicana

Muhlenbergia sylvatica

Muhlenbergia sobolifera

Muhlenbergia tenuiflora

41. SPOROBOLUS R. Br. DROPSEED

About 100 species of tropical and temperate regions, about 40 of which are North American. Blades narrow, flat or rolled; glumes and lemma 1-nerved; lemma awnless, often shorter than palea; pericarp free from seed. Fruit falling early from lemma and palea; hence the common name. In the United States most abundant in the southern Great Plains and in the Southwest where some species are of value as range and revegetation grasses.

a. A conspicuous tuft of long white hairs at top of sheath; terminal panicle usually included at base but open above ....1. *S. cryptandrus*
aa. No conspicuous tuft of hairs at top of sheath but sheath may be pubescent.
    b. Panicle-branches ascending; panicle open, narrowly ellipsoid, exserted; glumes acuminate, second much longer and wider than first, equaling or slightly exceeding lemma and palea; fruit globose, splitting the palea as it enlarges; spikelet dark ....
                          2. *S. heterolepis*
    bb. Panicle-branches appressed; panicle not open, narrow, usually wholly or partly included in sheath.
        c. Sheaths not inflated or only the uppermost inflated; leaf-blades elongate, tapering to fine point; plants tall or moderately tall.
            d. Lemma and palea glabrous, blunt at tip; uppermost sheath inflated ....
                          3. *S. asper*
            dd. Lemma and palea pubescent, tapering to narrow point at tip; uppermost sheath scarcely inflated ....4. *S. clandestinus*
        cc. All the sheaths inflated; panicles terminal and lateral, usually less than 5 cm long; plants usually short.
            d. Lemma glabrous; spikelet 2–3 mm long ....5. *S. neglectus*
            dd. Lemma pubescent; spikelet 3 mm long or more ....6. *S. vaginiflorus*

1. **Sporobolus cryptandrus** (Torr.) Gray SAND DROPSEED
Panicle 1–2 dm long; spikelets usually lead-color, crowded except at base of branches; second glume, lemma, and palea about equal, first glume much shorter. Sandy soil, much of the United States, but more abundant in the Southwest where useful as revegetation and range grass.

2. **Sporobolus heterolepis** Gray PRAIRIE DROPSEED
Perennial, tufted, moderately tall, with long narrow blades; panicle 1–2 dm long; spikelets few, 3–6 mm long. Prairies.

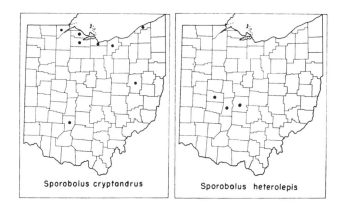

Sporobolus cryptandrus   Sporobolus heterolepis

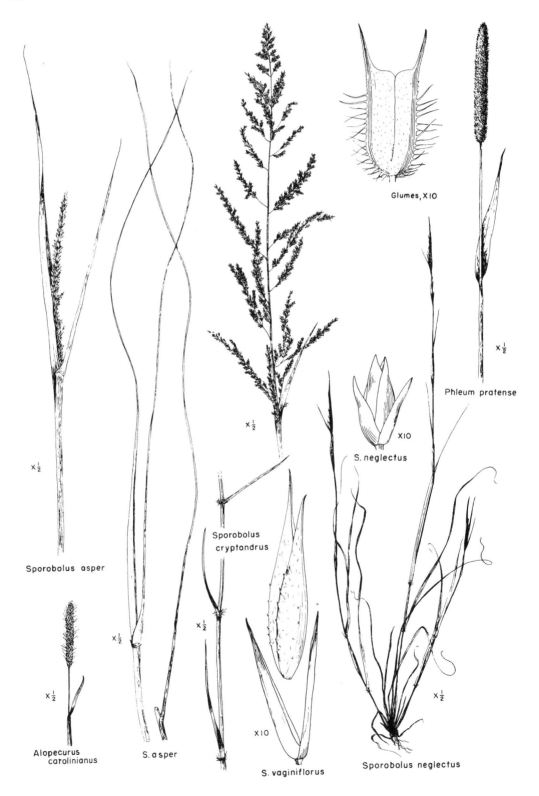

Sporobolus asper

$\times \frac{1}{2}$

Alopecurus
carolinianus

$\times \frac{1}{2}$

S. asper

$\times \frac{1}{2}$

Sporobolus
cryptandrus

$\times \frac{1}{2}$

$\times \frac{1}{2}$

X10

S. vaginiflorus

Glumes, X10

S. neglectus

X10

Phleum pratense

$\times \frac{1}{2}$

Sporobolus neglectus

$\times \frac{1}{2}$

3. **Sporobolus asper** (Michx.) Kunth

Easily recognized because of the long-attenuate often rolled blades and the inflated upper sheath which encloses or partly encloses the panicle; culms tufted; lemma and palea nearly equal, 3.5–6 mm long, longer than the unequal glumes. Prairies and sandy fields, much of the United States; often along roadsides in Ohio.

4. **Sporobolus clandestinus** (Biehler) Hitchc.

Palea longer than lemma; lemma somewhat longer than the slightly unequal lanceolate glumes. Credited to Ohio in Hitchcock and Chase (1951); no specimens seen.

5. **Sporobolus neglectus** Nash

Annual, culms branching from base, spreading or ascending; internodes and culm-blades short; sheaths inflated; axillary and sometimes also terminal panicles included; lemma and palea about equal, longer than the subequal glumes. Sandy or leached soil, much of eastern half of the United States; local westward. Ohio specimens often from roadsides.

6. **Sporobolus vaginiflorus** (Torr.) Wood

Similar to S. *neglectus* in range, habitat, and distribution but distinguished from it by spikelet characters; lemma and palea lanceolate, pubescent, acute or acuminate, equal or palea longer; lemma sometimes mottled with dark spots. Common in Ohio.

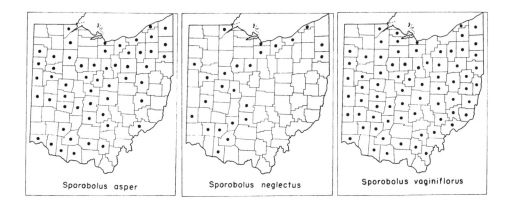

Sporobolus asper   Sporobolus neglectus   Sporobolus vaginiflorus

42. HELEOCHLOA Host

About 8 Eurasian species of tufted annuals with dense short spike-like panicles, short blades, and ligules of hairs.

1. HELEOCHLOA SCHOENOIDES (L.) Host

Culms low, erect to spreading; panicle 8–10 mm thick and no more than 4 cm long, base included in usually inflated sheath; spikelet 3 mm long. Intro-

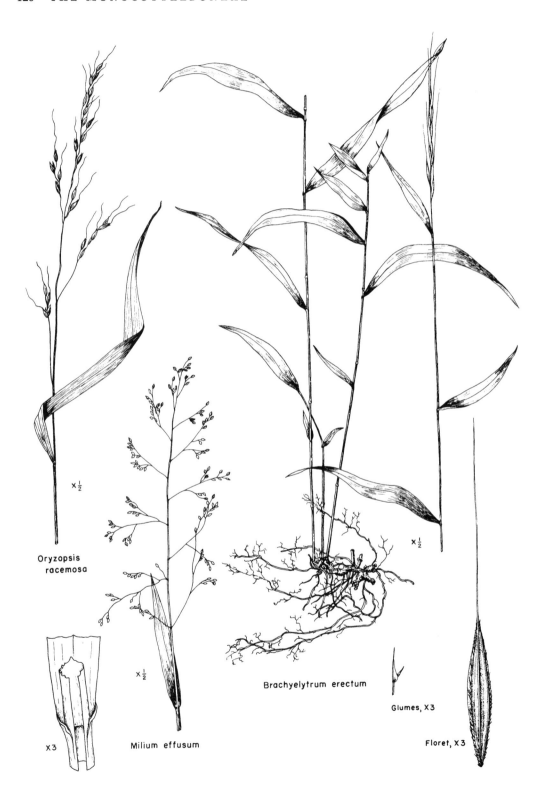

Oryzopsis
racemosa

×½

×½

×3

Milium effusum

Brachyelytrum erectum

×½

Glumes, X3

Floret, X3

duced from Europe at scattered stations in northeastern United States; recorded from Greene County.

43.   BRACHYELYTRUM Beauv.

A monotypic genus.

1.   **Brachyelytrum erectum** (Schreb.) Beauv.

Culms moderately tall to tall; blades spreading, lanceolate, narrowed to base, acuminate at tip, to 16 mm wide; panicle narrow, branches ascending or appressed, of few spikelets; glumes minute or the second to a few mm long; palea shorter than lemma; rachilla-extension a bristle more than half as long as lemma. Woods, eastern half of the United States and adjacent Canada.

44.   MILIUM L.

About 6 Eurasian species, one of which is native to North America also. Glumes rounded on back; lemma and palea of like texture, firm and pale, edges of lemma wrapped about palea, the whole like upper floret of *Panicum*.

1.   **Milium effusum** L.

Moderately tall; blades to 1.5 cm wide; panicle ovoid, open, 1–2 dm long, branches spreading or reflexed; spikelets 3–3.5 mm long. An attractive woodland grass worthy of cultivation as an ornamental. Northeastern fourth of the United States, southeastern Canada, and Eurasia.

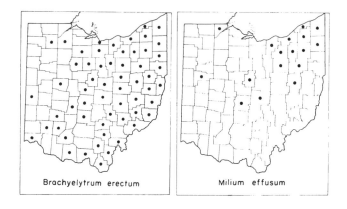

Brachyelytrum erectum   Milium effusum

45.   ORYZOPSIS Michx.   RICE GRASS

About 25 species of north temperate regions. Spikelets in panicles; lemma firm, enclosing a palea of like texture, awned, the awn deciduous; glumes membranous. Indian Rice Grass, *O. hymenoides* (Roem. & Schult.) Ricker, of arid

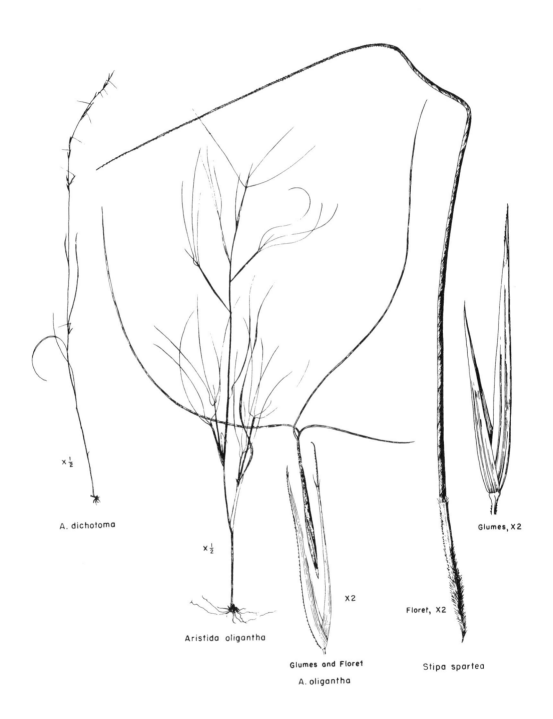

A. dichotoma

x ½

x ½

Aristida oligantha

X2

Glumes and Floret
A. oligantha

Floret, X2

Glumes, X2

Stipa spartea

and semi-arid regions of the West, is useful as forage and its grain has been used by Indians as food.

1. **Oryzopsis racemosa** (Smith) Ricker
Moderately tall, blades to 1.5 cm wide; panicle 1–2 dm long, branches few, spreading; spikelets few, 7–9 mm long. Woods, northeastern fourth of the United States.

46. STIPA L.   Needle Grass

A large genus of tropical and temperate regions, species growing in drier grasslands of all continents. Rachilla disarticulating obliquely above glumes; the sharp-pointed bearded callus sometimes penetrates the skin of grazing animals causing injury; awn of lemma often hygroscopic, unwinding when moist, twisting when dry. Species of the United States mostly western, where they are important range plants if grazed before inflorescence is formed. Commercial fiber is obtained in Spain and North Africa from S. *tenacissima* L.

Glumes 1.5 cm long; lemma about 1 cm long; awn to 6 cm. long ....1.   S. *avenacea*
Glumes 3–4 cm long; lemma 1.6–2.5 cm long; awn to 20 cm long ....2.   S. *spartea*

1. **Stipa avenacea** L.   Blackseed Needle Grass.   Black Oat Grass
Moderately tall to tall; blades narrow, sometimes rolled; lemma rough at summit. Woods, eastern United States to Michigan and Texas.

2. **Stipa spartea** Trin.   Porcupine Grass
Tall; blades sometimes rolled; body of lemma pubescent below, with line of pubescence extending to top. Prairies, Pennsylvania to Montana and New Mexico.

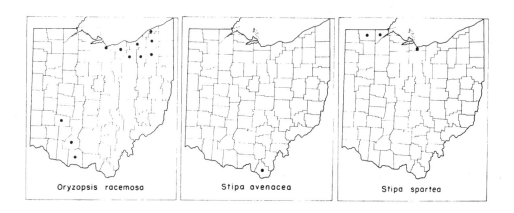

Oryzopsis racemosa   Stipa avenacea   Stipa spartea

47. ARISTIDA L.   Triple-awn Grass.   Three-awn

Numerous species, mostly of warmer and drier regions. Usually slender, with narrow blades, ours with narrow panicles; rachilla breaking obliquely above glumes leaving a hard-pointed, usually bearded, callus; glumes pointed or awned; lemma linear, bearing a trifid awn. The hard callus causes mechanical injury when eaten by sheep and cattle, but some species are useful in the West as range grasses if grazed before inflorescence is formed.

a. Central awn and lateral awns not greatly unequal.
    b. Lemma about 2 cm long; awns 3–7 cm long; annual ....1. *A. oligantha*
    bb. Lemma about 7 mm long; awns 1.5–3 cm long; perennial .... 2. *A. purpurascens*
aa. Central awn twice as long as lateral awns or more.
    b. Central awn coiled ½–1 turn and horizontally bent, 3–10 mm long; lateral awns straight, about 1 mm long; lemma averaging 5–6 mm long, usually appressed-pubescent; panicle usually less than 10 cm long ....3. *A. dichotoma*
    bb. Central awn not coiled, but bent outward at base, 3–20 mm long; lateral awns ⅓–½ as long, rarely 1 mm long, less divergent than central; lemma averaging 4–5 mm long, with short stiff hairs on nerves; panicle 10–20 cm long ....
                                                                    4. *A. longespica*

1. **Aristida oligantha Michx.**
    Short, slender, with branched stem; panicle about half height of plant, conspicuous because of the long-branched awns. Dry open ground, eastern half of the United States, and a few western states.

2. **Aristida purpurascens Poir.**
    Moderately tall, tufted, the purplish panicle half to one-third length of entire plant. Dry sandy soil, eastern half of the United States.

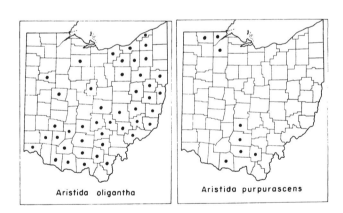

Aristida oligantha   Aristida purpurascens

3. **Aristida dichotoma Michx.**
    Culms low, slender, branched at base; terminal inflorescence short, sometimes a raceme; axillary panicles small, sometimes enclosed in sheaths. Dry or leached open ground, eastern half of the United States.

4. **Aristida longespica Poir.**
    A. gracilis Ell.
    Culms low, slender, branched; inflorescence with few branches or unbranched. Range and habitat of the preceding.

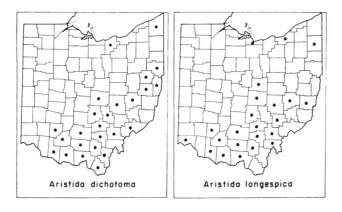

Aristida dichotoma   Aristida longespica

TRIBE VI. CHLORIDEAE

Spikelets 1–several-flowered, in two rows along one side of a rachis forming a one-sided spike or spike-like raceme which constitutes the inflorescence or, more often, a branch of the inflorescence.

a. Inflorescence of 2–8 divergent spikes clustered at tip of peduncle, or some of the spikes so clustered and 1 or 2 attached a short distance below; rarely the inflorescence a single spike.
   b. Spikes about 5 mm wide, 2–8 in number, usually 1 or 2 attached a short distance below the others; plants tufted; spikelets with 3–6 florets ....49. *Eleusine*
   bb. Spikes 2–3 mm wide, usually 4–5 in number, all at tip of peduncle; extensive rhizomes or stolons present; spikelets with 1 floret ....50. *Cynodon*
aa. Inflorescence of several to many spikes or racemes scattered along an elongate axis or its branches.
   b. Spikes horizontal to reflexed, each spike 1–2 cm long, of 4–8 spikelets, eventually falling entire; spikelet of 1 perfect floret with rudiments of 1 or more above ....
            54. *Bouteloua*
   bb. Spikes ascending to appressed, short; or spikes longer and usually bearing more than 8 spikelets.
      c. Spikes erect or ascending; spikelets closely imbricated, with 1 perfect floret and no vestigial floret; articulation below glumes; of wet habitats.
         d. Glumes inflated, equal, a little shorter than lemma; spikelet flattened laterally, suborbicular; spikes short ....51. *Beckmannia*
         dd. Glumes not inflated, first glume shorter than lemma, second glume longer than lemma, awned; tall plants ....52. *Spartina*
      cc. Spikes spreading to almost erect, long and slender; spikelets not closely imbricated; articulation above glumes.
         d. Blades of culm several, 2–7 cm long, about 1 cm wide, spreading, rounded or truncate at base; spikelet 1– or rarely 2–3-flowered, with a rudiment that is sometimes only an awn; panicle about as wide as long, branches spreading at maturity ....53. *Gymnopogon*
         dd. Blades of culm fewer, longer, and narrower; spikelet with 2–several perfect flowers; lateral, or all 3, nerves of lemma pubescent, at least below
            48. *Leptochloa*

48. LEPTOCHLOA Beauv. SPRANGLETOP

About 20 species, mostly of warmer regions; Ohio species annual.

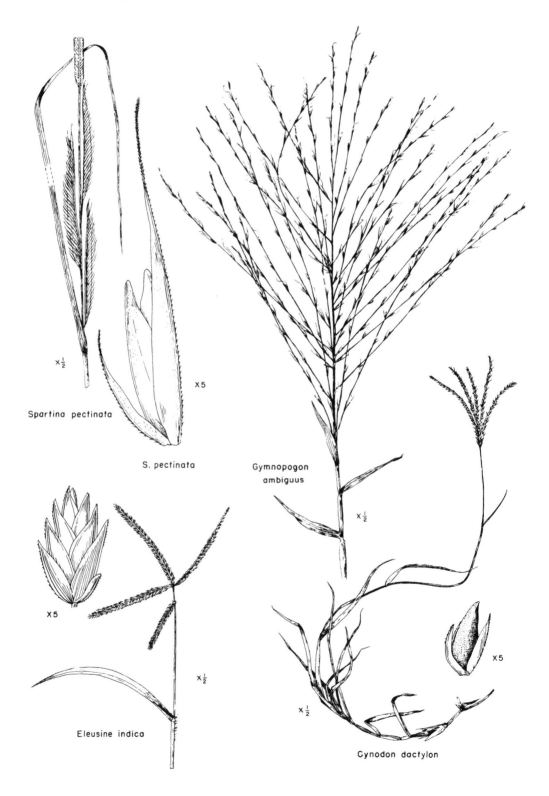

Spartina pectinata

×½

S. pectinata

×5

Gymnopogon
ambiguus

×½

Eleusine indica

×5

×½

Cynodon dactylon

×½

×5

Spikelets about 2 mm long, 2–4-flowered; lemma 1–1.5 mm long, awnless; sheaths
  papillose-pilose, sometimes sparsely so ....1. *L. filiformis*
Spikelets 7–12 mm long, 5–12-flowered; lemma 4–5 mm long, awned; sheaths glabrous
  or scabrous ....2. *L. fascicularis*

1. **Leptochloa filiformis** (Lam.) Beauv. RED SPRANGLETOP
Moderately tall; the beautiful ellipsoid, often red-purple, panicle sometimes
half height of entire plant, the many racemes very slender. A weed of fields
and gardens in southern United States.

2. **Leptochloa fascicularis** (Lam.) Gray
Diplachne fascicularis Beauv.
Culms tufted, branched, moderately tall to tall; panicle-base included in
sheath, the elongate ascending branches less slender than in the preceding.
Marshes along the Atlantic Coast; inland from Ohio westward.

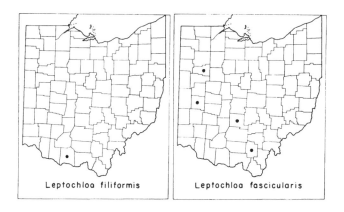

Leptochloa filiformis   Leptochloa fascicularis

49. ELEUSINE Gaertn.

Tufted annuals native in warmer parts of the Old World. Seed loosely
enclosed in pericarp.

1. ELEUSINE INDICA (L.) Gaertn. YARD GRASS. GOOSE GRASS
Easily recognized by the digitate arrangement of the wide (5 mm) spikes;
culms short, compressed, branched at base, spreading or ascending, mostly
glabrous; spikelets 3.5–5 mm long. Common weed of lawns, gardens, paths, and
waste places throughout eastern half of the United States; occasional in the West.

50. CYNODON Richard
Capriola Adans.

Low perennials of warm regions, mostly of South Africa and Australia.

1. CYNODON DACTYLON (L.) Pers.   BERMUDA GRASS
   Capriola dactylon (L.) Kuntze
   Culms erect or ascending from extensively spreading stolons or rhizomes; inflorescence superficially like that of *Digitaria*, the spikes about 4 cm long; spikelets 2–2.5 mm long. Cultivated for lawns and pastures in the South; sometimes a weed. Records from about 25 scattered Ohio counties.

51. BECKMANNIA Host   SLOUGH GRASS

Two species of erect annuals, one Eurasian.

1. **Beckmannia syzigachne** (Steud.) Fern.   AMERICAN SLOUGH GRASS
   Moderately tall grass of swamps and ditches, with panicle of several to many short branches bearing flat orbicular spikelets; acuminate apex of lemma protruding slightly beyond tips of transversely-wrinkled inflated glumes. Formerly referred by some authors to the European species *B. erucaeformis* (L.) Host, which has 2-flowered spikelets. Native westward; our specimen from stockyards of Cuyahoga County may be adventive.

52. SPARTINA Schreb.   CORD GRASS

Coarse perennials of wet places and of shallow water, mostly of temperate regions. When extensive colonies of rhizomatose species grow on mud flats or in marshes along sea coasts or along inland waters, soil particles are held in the mass of rhizomes and land may eventually be built.

1. **Spartina pectinata** Link   PRAIRIE CORD GRASS.   SLOUGH GRASS
   S. michauxiana Hitchc.
   Tall to very tall, with elongate blades tapering to long slender tips and with narrow panicle of several to many stiff coarse ascending spikes; lemma 7–9 mm long, awnless. Wet prairies, swamps, and coastal marshes, throughout much of the United States.

Spartina pectinata

53. GYMNOPOGON Beauv.

About 10 species, mostly American. Spikelets 1–3-flowered, nearly sessile; rachilla prolonged behind uppermost floret bearing a vestige sometimes reduced to a short awn.

1. **Gymnopogon ambiguus** (Michx.) BSP.
Panicle about half length of entire plant, the spikes at maturity stiffly spreading, short awns of lemmas and rudiments the most conspicuous part of spikelets; glumes 4–6 mm long. Woods, New Jersey to Kansas and southward.

54. BOUTELOUA Lag. GRAMA

About 40 species, all American, chiefly of arid and semi-arid regions of the West where many are among the most valuable range and hay grasses.

1. **Bouteloua curtipendula** (Michx.) Torr. SIDE-OATS GRAMA
Atheropogon curtipendulus (Michx.) Fourn.
Easily recognized by the narrow panicle of many spreading to pendulous, often secund, short spikes; culms slender, moderately tall, tufted; lemma about 5 mm long. Prairies and dry hills, in Eastern and Midwestern states; plains, farther west.

Bouteloua gracilis (HBK.) Lag., Blue Grama, and *B. hirsuta* Lag., Hairy Grama, important grasses of the short-grass plains, sometimes adventive eastward, are recorded from Franklin County. Both have inflorescence of 1–4, usually 2, spikes of closely imbricated spikelets; in *B. hirsuta*, rachis of spike extends conspicuously beyond spikelets.

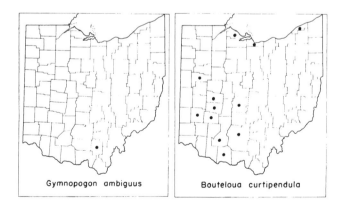

Gymnopogon ambiguus   Bouteloua curtipendula

TRIBE VII.   PHALARIDEAE

Spikelet of 1 perfect floret and, below that, 2 staminate florets or 2 scale-like vestiges of florets; rachilla articulated above glumes.

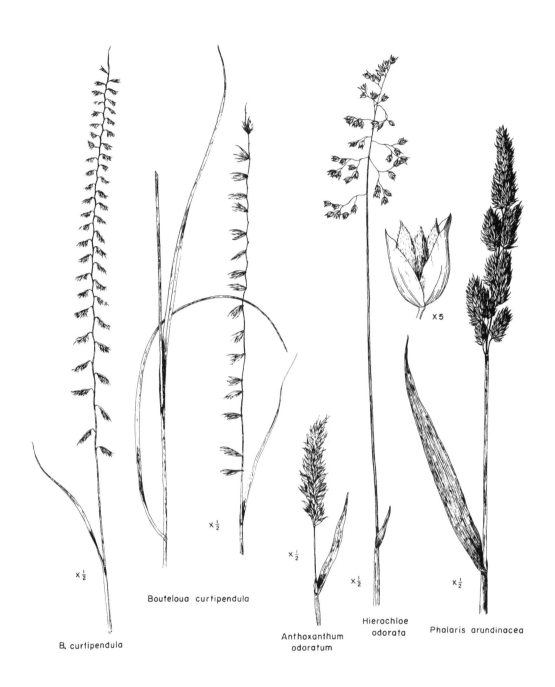

B. curtipendula

Bouteloua curtipendula

Anthoxanthum
odoratum

Hierochloe
odorata

Phalaris arundinacea

a. Glumes rounded on back, hyaline or with broad hyaline edge; lemma brown and shining at maturity; plants fragrant.
  b. Inflorescence ovoid, an open panicle; glumes equal and similar; fertile floret between and equaled by 2 brown staminate florets ....55. *Hierochloe*
  bb. Inflorescence narrow and spike-like; glumes unequal; fertile floret between and exceeded by 2 brown hairy awned scales ....56. *Anthoxanthum*
aa. Glumes boat-shaped, equal, keeled in upper half, keel sometimes winged; fertile lemma shorter than glumes, pale, at its base 2 narrow hyaline scales about half its length ....
57. *Phalaris*

55. HIEROCHLOE R. Br. HOLY GRASS

Torresia Ruiz & Pav.

Slender erect fragrant moderately tall perennials of cold and temperate regions; lemmas and glumes about equal. Name derived from Greek words meaning "sacred" and "grass", a reference to use of *H. odorata* in northern Europe in early times on holy days.

1. **Hierochloe odorata** (L.) Beauv. SWEET GRASS. VANILLA GRASS
Torresia odorata (L). Hitchc.
Tufted, moderately tall; culm blades short; panicle 4–12 cm long; spikelet 4–6 mm long. Used by Indians for basket-making. Swamps, bogs, wet meadows; throughout the United States except the southeastern fourth, and in Eurasia.

56. ANTHOXANTHUM L. VERNAL GRASS

A few species of sweet-scented grasses native in Eurasia. First glume shorter than second, both acute or awn-pointed; fertile lemma folded about palea at maturity; sterile lemmas 2-lobed at apex, awned from back, awn of upper one twisted and bent.

Culm unbranched, erect; perennial ....1. *A. odoratum*
Culm usually branched, often geniculate at base; annual ....2. *A. aristatum*

1. ANTHOXANTHUM ODORATUM L. SWEET VERNAL GRASS
Tufted, slender, low to moderately tall, with long-exserted narrow panicle 3–7 cm long and yellow-brown spikelets 8–10 mm long. Naturalized in eastern half of the United States and in the Pacific States; recorded in about one-fourth the counties of Ohio. Lawns, pastures, roadsides.

2. ANTHOXANTHUM ARISTATUM Boiss.
A. puelii Lecoq & Lamotte
Culms shorter than in the preceding, panicles smaller and less dense, spikelets a little smaller; found locally over the same range. Recorded from 2 Ohio counties.

57. PHALARIS L. CANARY GRASS

About 20 species of temperate regions; panicles narrow and dense; spikelets laterally flattened, with keeled or winged-keeled glumes.

a. Panicle usually 1–2 dm long, branches obvious, appressed at maturity; keels of glumes narrowly winged at summit; fertile lemma sparsely pubescent; scales at base of lemma villous ....1. *P. arundinacea*
aa. Panicle usually not more than 5 cm long, dense and spike-like; fertile lemma densely pubescent.
    b. Spike ovoid or ovoid-oblong; glumes 7–8 mm long, keels broadly winged, a prominent green stripe on each side of keel; scales at base of lemma glabrous ....
                                                2. *P. canariensis*
    bb. Spike narrowly ellipsoid; glumes 5–6 mm long, keel narrowly winged; scales at base of lemma hairy ....3. *P. caroliniana*

1. **Phalaris arundinacea L.**　REED CANARY GRASS

A wet-land pasture and forage grass in the Northern and Northwestern States. Perennial, with rhizomes, growing in colonies; culms tall, erect; blades to 2 cm wide. Moist ground, throughout the United States except the Southeast. Ribbon Grass, forma *variegata* (Parnell) Druce (var. *picta* L.), with white-striped leaves, is cultivated as an ornamental and sometimes escapes.

2. PHALARIS CANARIENSIS L.　CANARY GRASS

Short to moderately tall annual the grain of which is commercial canary seed; infrequent along roadsides and in waste places. Adventive from Europe.

3. **Phalaris caroliniana Walt.**

Annual, with inflorescence somewhat longer and narrower than in the preceding; native south of Ohio. Our one specimen from a garden in Muskingum County probably a waif.

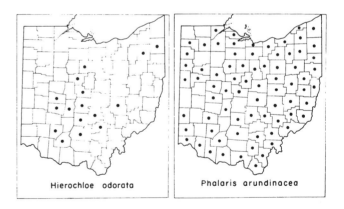

Hierochloe odorata　　　Phalaris arundinacea

TRIBE VIII.　ORYZEAE

Glumes minute or absent; spikelet of one perfect floret, flattened laterally, articulation below glumes when glumes are present. Includes Rice, *Oryza sativa* L., one of the most important food plants of the world. One genus in the United States.

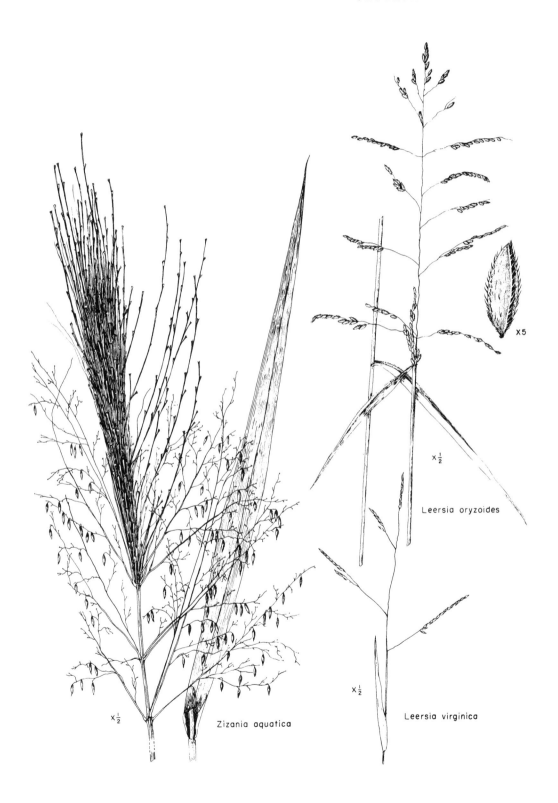

$\times \frac{1}{2}$

Zizania aquatica

Leersia oryzoides

$\times \frac{1}{2}$

X5

$\times \frac{1}{2}$

Leersia virginica

58.   LEERSIA Swartz

Homalocenchrus Mieg

Ten species of perennials of temperate and warmer regions; glumes absent; spikelets in racemes, the racemes in panicles; lemma firm, 5-nerved, marginal nerves clasping margins of narrower 3-nerved palea.

Panicle-branches flexuous, usually 2 or more at lower nodes; edges of blades usually spinulose-ciliate, sharp, cutting; spikelets over 4 mm long; rhizome slender ....
1.   *L. oryzoides*
Panicle-branches solitary at nodes, distant, few; edges of blades slightly rough but not cutting; spikelets usually less than 4 mm long ....2.   *L. virginica*

1.   **Leersia oryzoides** (L.) Swartz   RICE CUTGRASS
Homalocenchrus oryzoides (L.) Poll.
Culms tall, sometimes decumbent at base; spinules present on edges, and sometimes also on surfaces, of blades and on sheaths, or rarely sheaths and blades scarcely rough; terminal panicle ovoid, 1–2 dm long, sometimes included; axillary panicles partly or wholly included in sheaths, bearing cleistogamous spikelets. Ditches, wet places, and edges of streams, ponds, and lakes; often the dominant plant of swampy pond- and lake-borders; much of the United States.

2.   **Leersia virginica** Willd.   WHITE GRASS
Homalocenchrus virginicus (Willd.) Britt.
Culms moderately tall to tall, branching, weak and often sprawling; blades tapering to both ends; panicle usually exserted, branches naked at base, simple or little branched; spikelets appressed. Moist woods and thickets, eastern half of the United States.

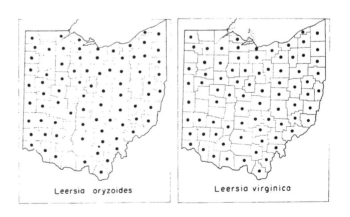

Leersia oryzoides   Leersia virginica

TRIBE IX.   ZIZANIEAE

Spikelets monosporangiate (unisexual), with one floret; glumes wanting or smaller than lemma. One genus in Ohio.

59. ZIZANIA L. WILD-RICE

Three species, two in the United States, one in Asia; upper branches of panicle bearing carpellate (pistillate) spikelets, lower branches bearing staminate spikelets; glumes reduced to small ring-like ridge.

1. **Zizania aquatica** L. ANNUAL WILD-RICE
Annual aquatic; culms tall to very tall, stout; blades elongate, 1–4 cm wide; panicle 3–6 dm long, upper branches erect or ascending, lower branches ascending or spreading; spikelets readily deciduous; lemma of carpellate (pistillate) spikelet about 2 cm long, firm, long-awned; lemma of staminate spikelet membranous, acuminate or short-awned, about 1 cm long; stamens 6. Grain important as food of waterfowl and used to some extent as human food. Usually in shallow water; eastern half of the United States.

Zizania aquatica

TRIBE X. PANICEAE

Spikelet containing a terminal fertile floret and below that a floret consisting of lemma only, of lemma and palea, or of lemma, palea, and staminate flower; fertile lemma and its palea of firmer texture than glumes and lower lemma; articulation below glumes.

a. Spikelets (usually 2) enclosed in a bur-like involucre, the burs in a spike or a spike-like raceme .... 66. *Cenchrus*
aa. Spikelets not enclosed in a bur-like involucre.
    b. Each spikelet subtended by 1 or more bristles longer than spikelet; inflorescence a cylindric spike-like panicle .... 65. *Setaria*
    bb. Spikelet not subtended by bristles.
        c. First glume minute or wanting; below fertile lemma and palea are two similar membranous structures, second glume and lower lemma.
            d. Inflorescence a diffuse panicle; spikelets much shorter than their pedicels; hyaline margins of lemma not inrolled .... 61. *Leptoloma*
            dd. Inflorescence of 1–several one-sided spike-like racemes clustered at top of peduncle or scattered along an axis, each with spikelets attached in 2 rows on one side of rachis but sometimes lying in more than 2 rows.

e. Spikelet plano-convex and rounded or blunt at apex or, if acute and flattened, then rachis is leaf-like and broader than the rows of spikelets it bears; upper lemma pale, hard, margins inrolled ....62. *Paspalum*

ee. Spikelet flattened, acute at apex; rachis not leaf-like, not broader than the rows of spikelets it bears; upper lemma sometimes dark, firm but not hard, margins hyaline, not inrolled ....60. *Digitaria*

cc. First glume usually ⅓–¾ as long as spikelet; below fertile palea are two membranous structures, lower lemma and first glume, and below fertile lemma, one membranous structure, second glume.

d. Lemma not inrolled at tip; spikelets often awned, often hispid, crowded and nearly sessile along branches of panicle; ligule absent ....
64. *Echinochloa*

dd. Lemma inrolled all around palea; spikelets not awned, either pediceled in an open panicle or, if rarely crowded and nearly sessile, then ligule present ....63. *Panicum*

60. DIGITARIA Heist.   Crab Grass.   Finger Grass

Syntherisma Walt.

Ohio species annual, characterized by digitate inflorescence of spike-like racemes, with spikelets attached in 2 rows on one side of rachis. Name derived from Latin word for finger.

a. Rachis winged, wings at least as wide as midrib of rachis; culms often decumbent and spreading at or near base.

b. Blades and lower sheaths pubescent (rarely nearly glabrous); first glume minute but evident, second glume ⅓–⅔ as long as pale upper lemma; spikelet 2.5–3 mm long ....1.   *D. sanguinalis*

bb. Blades and lower sheaths glabrous or nearly so; first glume hyaline, almost invisible; second glume about as long as dark upper lemma; spikelet about 2 mm long ....2.   *D. ischaemum*

aa. Rachis triangular in cross-section, not or very narrowly winged; lower sheaths pubescent; first glume wanting; second glume a little shorter than dark upper lemma; spikelet 1.5–2 mm long ....3.   *D. filiformis*

1. Digitaria sanguinalis (L.) Scop.   Crab Grass
Syntherisma sanguinale (L.) Dulac
Culms branched near base and rooting at lower nodes, sometimes 1 m long, flowering culms ascending; racemes few to several, in a single terminal whorl or in one to three approximate whorls. Naturalized throughout the United States; lawns, gardens, roadsides, waste places, a common and troublesome weed.

2. Digitaria ischaemum (Schreb.) Muhl.   Smooth Crab Grass
Syntherisma ischaemum (Schreb.) Nash
Resembles *D. sanguinalis* but usually shorter, less coarse, and almost glabrous. Naturalized in most of the United States, a common and troublesome weed; lawns, gardens, roadsides, waste places.

3. **Digitaria filiformis** (L.) Koel.
Syntherisma filiforme (L.) Nash
Culms short to moderately tall, very slender, erect, of several lengths in one tuft; racemes 1–6, usually not whorled but scattered along axis; rachis slender

and sometimes flexuous; spikelets often in threes. Open ground, eastern half of the United States.

61. LEPTOLOMA Chase

Four species, 3 in Australia, 1 in the United States and in Mexico. Panicle diffuse, breaking away in age and tumbling in the wind; spikelets long-pediceled; second glume and lower lemma about equal in length.

1. **Leptoloma cognatum** (Schultes) Chase FALL WITCH GRASS
Culms decumbent at base, often much branched; panicle purple, half the height of plant, branches slender and stiff, axils pilose; often mistaken for a species of *Panicum* but can be distinguished by minute first glume; spikelet narrowly elliptic, flattened, acute, 2.5–3 mm long. Eastern half of the United States; westward, in the South, to Arizona.

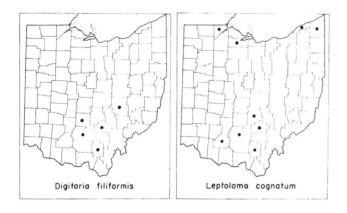

Digitaria filiformis

Leptoloma cognatum

62. PASPALUM L.

Over 200 species, more abundant in warmer regions. Inflorescence of 1-many spike-like divergent to ascending racemes scattered along an axis; spikelets usually plano-convex, attached singly or in pairs in 2 rows along rachis, when in pairs one pedicel of pair usually shorter.

a. Inflorescence of several to many racemes, the rachis with appearance of a narrow blade wider and longer than the rows of spikelets it bears; usually growing in or near water; culms creeping or floating; spikelets acute, 1.8 mm long or less, usually pubescent ....1. *P. fluitans*
aa. Inflorescence of 1–5 (8) racemes, rachis narrower than the rows of spikelets; plants not aquatic.
    b. Spikelets in pairs at most nodes of rachis; one spikelet of pair sometimes vestigial, but pedicel present.
        c. Spikelets more than 2.5 mm long; racemes usually 4 or more; no axillary panicles present.
            d. Culms erect; spikelets about 4 mm long, ovoid, not lying in 4 rows ....
                                                2. *P. floridanum*

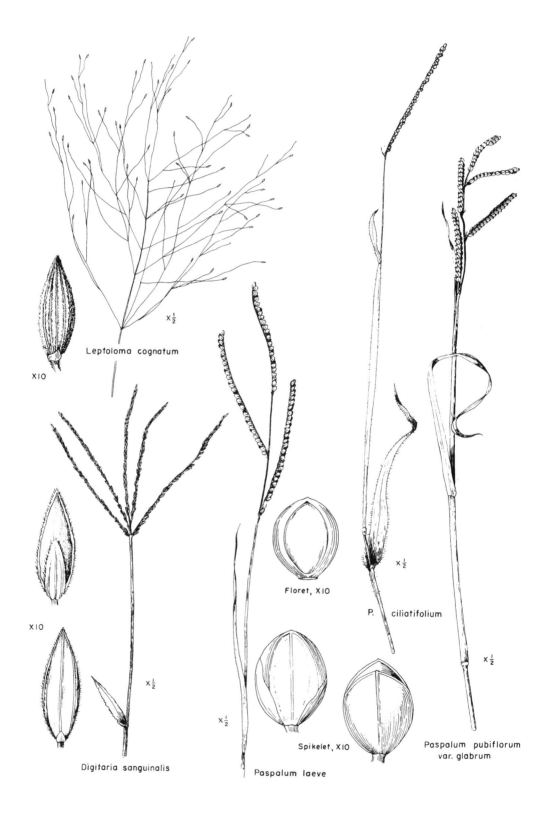

Leptoloma cognatum

X10

X½

Digitaria sanguinalis

X10

X½

X½

Paspalum laeve

Spikelet, X10

Floret, X10

P. ciliatifolium

X½

Paspalum pubiflorum
var. glabrum

X½

dd. Culms decumbent at base, rooting at lower nodes; spikelets about 3
    mm long, obovoid, usually lying in 4 rows ....3.   *P. pubiflorum* var. *glabrum*
  cc. Spikelets 2.3 mm long or less; racemes slender, usually 1–3; axillary racemes
    usually present.
      d. Spikelets elliptic-obovate, 1.4–1.7 mm long....................4.   *P. setaceum*
     dd. Spikelets suborbicular, 1.8–2.3 mm long ....5.   *P. ciliatifolium*
bb. Spikelets solitary at nodes of rachis, broadly oval or orbicular, 2.5–3.2 mm long....
                                    6.   *P. laeve*

1.   **Paspalum fluitans** (Ell.) Kunth
Low annual of shallow water or wet soil with distinctive panicle of many
narrow leaf-like branches. Southeastern fourth of the United States.

2.   **Paspalum floridanum** Michx.
Tall, with coarse panicle of usually 2–5 racemes of glabrous spikelets.
Southeastern fourth of the United States; our record a roadside specimen.

3.   **Paspalum pubiflorum** Rupr. var. **glabrum** Vasey
Conspicuous because of coarse panicle of thick racemes on which the
glabrous spikelets usually lie in 4 rows; blades 1–2 cm wide. Moist soil, southeastern fourth of the United States; our specimens from roadsides. The typical
variety, with pubescent foliage and spikelets, more southern in range.

4.   **Paspalum setaceum** Michx.
Incl. *P. longipedunculatum* Le Conte
Culms short to moderately tall, very slender; terminal peduncle elongate;
blades papillose-ciliate on margin. Typical variety has blades densely villous on
both sides; var. *longipedunculatum* (Le Conte) Wood (*P. longipedunculatum*
Le Conte) has blades glabrous to sparsely strigose. Both varieties recorded from
Jackson County.

5.   **Paspalum ciliatifolium** Michx.
Incl. *P. pubescens* Muhl. (*P. Muhlenbergii* Nash) and *P. stramineum* Nash
Variable in size and pubescence of spikelets and in pubescence of leaves.
Treatment followed here is that of Fernald (1950) who combines under one
species and its varieties the several species of other authors. Most of our

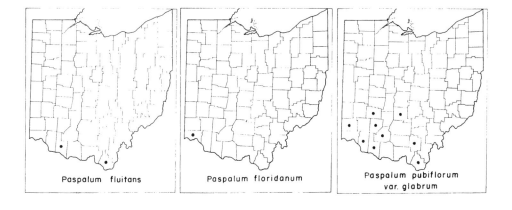

Paspalum fluitans   Paspalum floridanum   Paspalum pubiflorum
                                             var. glabrum

specimens referred to var. *muhlenbergii* (Nash) Fern. (*P. pubescens* Muhl.; *P. muhlenbergii* Nash), which has leaf surfaces long pilose; specimens from Lucas, Sandusky, and Jackson counties referred to var. *stramineum* (Nash) Fern. (*P. stramineum* Nash) which has upper surface of blades puberulent and sparsely pilose and lower surface nearly glabrous; typical variety, with leaf surface glabrous, not recorded from Ohio. Minnesota to Arizona, and eastward; most of our specimens from roadsides and fields.

6.   Paspalum laeve Michx.

Incl. *P. circulare* Nash and *P. longipilum* Nash.

Culms erect or ascending, usually moderately tall; racemes 2–6. Variable in pubescence of sheaths and· blades and in size and shape of spikelets; has been separated into varieties or species by some authors. Massachusetts tc Kansas and southward; most of our specimens from roadsides and fields.

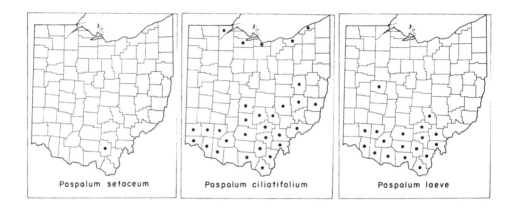

Paspalum setaceum   Paspalum ciliatifolium   Paspalum laeve

63.   PANICUM L.   PANIC GRASS

Largest genus of the family, species estimated at about 500. Spikelet, often small, consisting of upper lemma and palea of firm texture enclosing a perfect flower; below this a membranous lemma enclosing a palea and sometimes a staminate flower; and two membranous glumes, the second similar to and often about the same size as the lower lemma. In many species, each of the culms growing in spring (vernal culms) has at base a rosette of leaves (winter rosette) which grew the previous autumn and lived over winter, the blades shorter than and sometimes wider than those of culm. A few species are constituents of wild hay and a few are cultivated for grain and forage, but most species not of economic importance.

a.   Basal leaves and those of culm similar, elongate; winter rosettes not formed; perennials with rhizomes or with hard rhizomatose bases, or annuals; flowering once per season, grain forming in upper floret; spikelets, in ours, glabrous, acute or subacute at apex; panicles terminal or both terminal and axillary.

b. Second glume and lower lemma covered with wart-like tubercles; slender weak-stemmed annual; first glume minute; spikelet subacute, about 2 mm long ....
1. *P. verrucosum*

bb. Second glume and lower lemma without wart-like tubercles, strongly nerved; spikelets acute at apex.

   c. Annual; glabrous or blades rarely pilose; first glume broad, rounded or obtuse at apex, not more than ¼ as long as spikelet; spikelet 2–3.5 mm long, lanceolate; culms often spreading from geniculate base; ligule a dense ring of hairs 1–2 mm long; anthers orange ....2. *P. dichotomiflorum*

   cc. Perennials or pubescent annuals; first glume acute or acuminate at apex, ⅓ to more than ½ length of spikelet or, if short or blunt, then plants pubescent.

      d. Perennials, culms compressed below, sheaths compressed and often keeled; spikelets short-pediceled, subsecund, crowded and overlapping on branchlets.

         e. Spikelet bent above first glume at angle of about 30 degrees, 3–4 mm long; panicle-branches bearing spikelets almost to base; ligule membranous, 0.5 mm long; rhizomes stout, scaly ....3. *P. anceps*

         ee. Spikelet not bent above first glume, less than 3 mm long; plants tufted, with short caudex.

            f. Panicle narrow, branches slender, erect or ascending; ligule 2–3 mm long, base short, summit fimbriate or ciliate; blades narrow ....4. *P. longifolium*

            ff. Panicle ellipsoid or ovoid; ligule about 1 mm long, membranous.

               g. Lower branches of panicle divaricate; spikelet less than 2.5 mm long; perfect floret essentially sessile above glume and lower lemma ....5. *P. agrostoides*

               gg. Lower branches of panicle ascending; spikelet 2.5 mm long or more, acuminate; perfect floret with stipe 0.2–0.5 mm long above glume and lower lemma ....6. *P. stipitatum*

      dd. Perennials with culms terete, sheaths not compressed and keeled, or annuals; spikelets in open or diffuse panicles, not subsecund, not crowded and overlapping on branchlets (or somewhat crowded in No. 8).

         e. Leaf-sheaths glabrous; perennial; tall, erect, with elongate blades and large diffuse panicle of stiff ascending branches naked at base; spikelet usually gaping, 3.5–5.5 mm long; first glume ½–¾ length of spikelet ....7. *P. virgatum*

         ee. Leaf-sheaths papillose-hispid; annual.

            f. Spikelet 4.5–5.5 mm long, ovoid; panicle nodding, branches ascending; blades to 2 cm wide ....8. *P. miliaceum*

            ff. Spikelet smaller.

               g. Panicle very diffuse, about as wide as long, ½–⅔ height of plant, base often included, branches divaricate at maturity; spikelets acuminate, mostly less than 3 mm long, rarely to 4 mm ....9. *P. capillare*

               gg. Panicle not diffuse, less than half height of plant or, if as much as half, then 2–3 times as long as wide.

                  h. Spikelets 3–3.5 mm long, narrow, acuminate; panicle narrowly ellipsoid, 2–3 times as long as wide, branches ascending ....10. *P. flexile*

                  hh. Spikelets shorter, acute; panicle ovoid or ellipsoid, less than half as long as plant.

                     i. Culms stout, soon decumbent and rooting at nodes; blades 6–10 mm wide; panicles many, terminal and axillary, ellipsoid, base usually included; spikelets turgid, 2–2.5 mm long, 2 or more racemosely attached along branchlets ....
11. *P. gattingeri*

                    ii. Culms slender, branched at base, erect or decumbent at base, blades 3–8 mm wide; terminal panicles exserted, ovoid or rhombic, solitary

branches slender; spikelets few, usually 2 mm long or less, usually in twos, each on a short appressed pedicel near end of branchlet ....
  12. *P. philadelphicum*

aa. Basal blades shorter and often wider than those of culm, in rosettes formed in autumn and living over winter (or, in Nos. 13 and 14, true winter rosettes not formed); in vernal phase, culms unbranched, bearing terminal panicles in which grains usually not formed; in autumnal phase, after or sometimes before maturity of vernal panicles, culms branching, blades usually reduced, panicles axillary, grains forming; spikelets usually rounded at apex, often pubescent.
  b. Blades 20 times as long as wide or more, 5 mm wide or less, erect; true winter rosettes not formed; reduced panicles of autumnal phase hidden or almost hidden among lower sheaths.
    c. Spikelet more than 3 mm long, glabrous or pubescent; second glume and lower lemma extending beyond upper lemma in a slightly twisted pointed beak ....13. *P. depauperatum*
    cc. Spikelet less than 3 mm long, sparsely pubescent; second glume and lower lemma not forming a twisted pointed beak ....14. *P. linearifolium*
  bb. Blades less than 20 times as long as wide (except in No. 16, with ascending blades, uppermost usually longest), usually not more than 10 times as long as wide, erect to spreading.
    c. Spikelets 3 mm long or more.
      d. Some hairs of spikelet as long as half width of spikelet or more, first glume narrow, ½ length of spikelet or more; spikelet 3.4–4 mm long; blades papillose-hispid ....24. *P. leibergii*
    dd. Hairs of spikelet, if present, shorter; first glume broader and usually shorter.
      e. Widest blades of culm (not rosette blades) rarely 1.5 cm wide, not cordate-clasping at base.
        f. Culms spreading-pilose, nodes retrorsely villous; blades and panicle-branches velvety pubescent; spikelets pilose, 2.9–3 mm long ....25. *P. malacophyllum*
        ff. Culms ascending- or appressed-pilose to glabrous; nodes not retrorsely villous; blades and panicle-branches not velvety-pubescent; spikelets sparsely pubescent.
          g. Spikelets ellipsoid, 2.9–3 mm long; culms and sheaths glabrous, nodes sometimes pubescent ....17. *P. calliphyllum*
          gg. Spikelets obovoid, turgid, blunt, 3.0–3.5 mm long; culms and sheaths more or less pubescent ....
            26. *P. oligosanthes* var. *scribnerianum*
      ee. Widest blades of culm (not rosette blades) 1.2–4 cm wide, cordate-clasping at base.
        f. Nodes retrorsely bearded; spikelets 4–4.5 mm long ....
            27. *P. boscii*
        ff. Nodes not retrorsely bearded but sometimes pubescent; spikelets less than 4 mm long.
          g. Vernal sheaths papillose-hispid, or at least papillose, rarely glabrous; autumnal sheaths overlapping, beset with bristle-like hairs and inflated with enclosed panicles; vernal blades 1.2–3 cm wide, long-tapering from widest point to apex; spikelets 2.7–3 mm long ....28. *P. clandestinum*
          gg. Vernal sheaths not papillose-hispid or only slightly so; autumnal sheaths without bristle-like hairs.
            h. Spikelets 3.4–3.7 mm long, rarely shorter; leaf-blades 1.5–4 cm wide ....29. *P. latifolium*
            hh. Spikelets usually less than 3 mm, rarely to 3.2 mm, long; leaf-blades 1–2.5 cm wide ....30. *P. commutatum*
  cc. Spikelets less than 3 mm long.
    d. Some or all blades 10–20 times as long as wide, erect or ascending, uppermost usually longest, glabrous except for ciliate base, 3–8 mm wide; spikelets 2.3–2.8 mm long, long-pediceled, glabrous or pubescent....
        16. *P. bicknellii*

dd. Blades usually no more than 10 times as long as wide, erect to spreading, reduced upward.

    e. Vernal sheaths papillose-hispid or at least papillose, rarely glabrous; autumnal sheaths overlapping, beset with bristle-like hairs and inflated with enclosed panicles; vernal blades 1.2–3 cm wide, long-tapering from widest point to apex; spikelets 2.7–3 mm long ....
                                        28. *P. clandestinum*

  ee. Without the above set of characters.

    f. Culms and sheaths glabrous, except sometimes near base of plant, but nodes may be pubescent and sheaths may be ciliate at margin.

      g. Ligule of hairs 3–5 mm long, clearly visible at base of blade; spikelets 1.3–2.1 mm long, pubescent; blades usually less than 1 cm wide ....33. *P. lanuginosum*

    gg. Ligule not more than 1 mm long.

      h. Nodes densely retrorsely bearded; spikelets 1.5–1.8 mm long, ellipsoid, glabrous or rarely minutely pubescent; blades 8–15 mm wide ....18. *P. microcarpon*

      hh. Nodes not densely retrorsely bearded; sometimes pubescent or the lower slightly bearded.

        i. Spikelets 1.8 mm long or less, pubescent, spherical-obovoid at maturity, ellipsoid when young; rosette blades broad, with white cartilaginous margins; panicle-branches viscid-spotted.

          j. Length of panicle 1–2 times width; blades of culm 7–15 mm wide, usually reduced upward ....22. *P. sphaerocarpon*

          jj. Length of panicle 2–4 times width; upper blades of culm, usually 3, conspicuously larger than those below; blades 15–25 mm wide ....23. *P. polyanthes*

        ii. Spikelets 2 mm long or more, ellipsoid at maturity.

          j. Spikelets glabrous (rarely pubescent in No. 19).

            k. Sheaths not pale-spotted; second glume no longer than and often shorter than upper lemma; spikelet 2–2.2 mm long ....
                            19. *P. dichotomum*

            kk. Sheaths white- or pale-spotted; second glume and lower lemma extending beyond upper lemma in a short slightly twisted point; spikelet 2.3–2.5 mm long
                            20. *P. yadkinense*

          jj. Spikelets pubescent.

            k. Spikelets 2–2.2 mm long, 1 mm wide; nodes usually glabrous; leaf-blades 5–14 mm wide, the upper erect or nearly so
                            21. *P. boreale*

            kk. Spikelets longer and wider; nodes villous or puberulent; leaf-blades ascending or spreading.

              l. Spikelets 2.9–3 mm long; blades 9–15 mm wide, primary nerves differentiated from secondary on lower surface ....17. *P. calliphyllum*

              ll. Spikelets 2.2–3.2 mm long; blades 5–25 mm wide, primary nerves scarcely differentiated from secondary on lower surface ....
                        30. *P. commutatum*

  ff. Culms or sheaths, usually both, pubescent or puberulent; spikelets pubescent.

g. Ligule less than 2 mm long.
   h. Sheaths pubescent; nodes retrorsely bearded.
      i. Plants velvety-pubescent throughout; spikelets 2.9–3 mm long ....25. *P. malacophyllum*
      ii. Sheaths pilose with long, sometimes reflexed, hairs; culms glabrous to pilose; spikelets 1.9–2.3 mm long ....15. *P. laxiflorum*
   hh. Sheaths puberulent, sometimes with long hairs intermixed; nodes not retrorsely bearded.
      i. Spikelets 1.4–2 mm long; sheaths and culms crisp-puberulent, sometimes with long hairs intermixed; blades 3–7 mm wide ....
                  31. *P. columbianum*
      ii. Spikelets 2.2–3.2 mm long; sheaths and culms puberulent; blades cordate at base, 5–25 mm wide ....30. *P. commutatum*
gg. Ligule of dense hairs 2–5 mm long; culms and sheaths villous or papillose-pilose.
   h. Spikelets 2.2–2.5 mm long; hairs of culms and sheaths long and abundant, spreading or appressed ....
                  32. *P. villosissimum*
   hh. Spikelets 1.3–2.1 mm long; hairs of culms and sheaths varying in length and abundance ....
                  33. *P. lanuginosum*

1. **Panicum verrucosum** Muhl.

Panicle diffuse, branches slender, bearing a few spikelets toward the ends. Wet soil, Massachusetts to Florida, west to Michigan and Texas.

2. **Panicum dichotomiflorum** Michx. FALL PANIC GRASS

Short to tall, somewhat succulent; blades 3–20 mm wide, with conspicuous white midrib; panicles terminal and axillary, up to 4 dm long, often included at base, branches ascending. Moist soil, waste places, sometimes a weed in cultivated fields, eastern half of the United States; abundant in Ohio.

3. **Panicum anceps** Michx.

Culms moderately tall to tall; panicle 1.5–4 dm long, often pale green, branches remote; spikelets bent above first glume, crowded on short branchlets. Moist soil, New Jersey to Kansas and southward to Gulf.

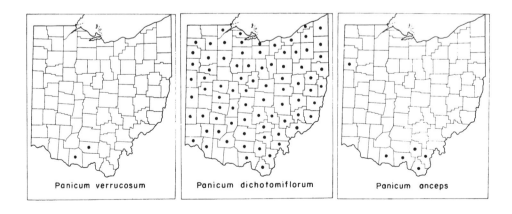

Panicum verrucosum   Panicum dichotomiflorum   Panicum anceps

4. **Panicum longifolium** Torr.

Short to tall, in tufts; basal and cauline blades elongate. Moist soil, Massachusetts to Florida and Texas, north to Indiana and southern Ohio.

5. **Panicum agrostoides** Spreng.

Culms moderately tall to tall, leafy at base; cauline blades erect, elongate; panicles terminal and axillary, the terminal ovoid, lower branches remote and divergent; spikelets 1.7–2.2 mm long, often purple. Wet open places, eastern half of the United States.

6. **Panicum stipitatum** Nash

P. agrostoides Spreng. var. elongatum Scribn.

Resembles *P. agrostoides* and has approximately the same range; can be distinguished from it by longer spikelets, narrower panicle with stiff ascending branches, and stipitate fertile floret.

7. **Panicum virgatum** L.

With scaly rhizomes; usually growing in large clumps; anthers purple. Prairies and open ground throughout most of the United States.

Panicum longifolium

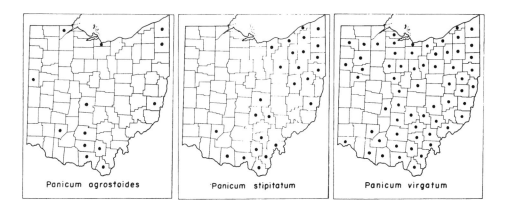

Panicum agrostoides   Panicum stipitatum   Panicum virgatum

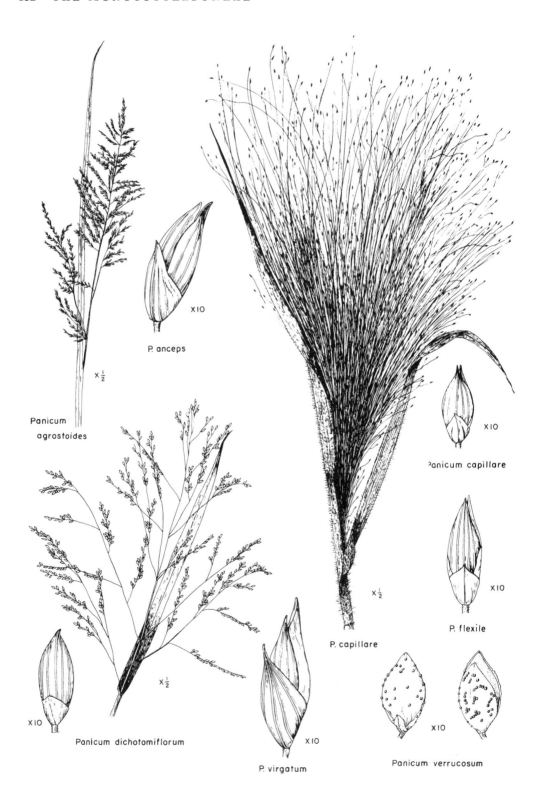

X10

P. anceps

X ½

Panicum
agrostoides

X ½

Panicum dichotomiflorum

X10

X ½

P. virgatum

X10

P. capillare

X ½

Panicum capillare

X10

P. flexile

X10

Panicum verrucosum

X10

8. Panicum miliaceum L. Broom-corn Millet

Stout annual, moderately tall to tall; panicle to 3 dm long, branches spikelet-bearing toward ends. Occasionally cultivated for forage and grain; adventive in Northeastern States and occasionally westward.

9. **Panicum capillare** L. Witch Grass

Recognized by the combination of papillose-hispid sheaths and blades and the large, often purple, panicle that breaks away and becomes a tumbleweed; culms short to moderately tall. Typical variety, with spikelets 2–2.5 mm long, very common in Ohio; var. *occidentale* Rydb., with long-acuminate spikelets 2.5–4 mm long, and with terminal panicles usually exserted, the branches divaricate, recorded from a few scattered counties. Waste places, roadsides, cultivated soil, chiefly east of the Rocky Mountains.

10. **Panicum flexile** (Gatt.) Scribn.

Culms slender, short to moderately tall, branched at base; blades elongate, narrow, erect, soft; panicle 1–2 dm long, spikelets long-pediceled. Eastern half of the United States; mostly damp soil, meadows and open woods; several of our specimens from bogs and swamps.

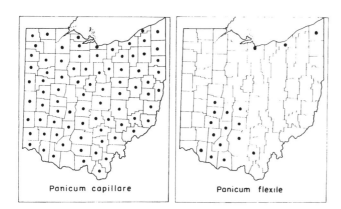

Panicum capillare

Panicum flexile

11. **Panicum gattingeri** Nash

P. capillare L. var. *campestre* Gatt.

Culms sometimes long, decumbent, rooting at lower nodes, hispid; panicles 10–15 cm long, many, ovoid or ellipsoid; spikelets ovoid, turgid and short-pointed. Open ground, waste places, and cultivated soil, New York to North Carolina and westward to Minnesota and Arkansas.

12. **Panicum philadelphicum** Bernh.

Incl. *P. tuckermani* Fern.

Culms slender, short to moderately tall; blades usually erect, yellow-green; panicle open, 1–2 dm long, about as wide. *P. tuckermani* Fern. (*P. philadelphicum* var. *tuckermani* (Fern.) Steyermark & Schmoll), with spikelets more racemosely arranged and with glabrous pulvini, has been segregated from *P. philadelphicum*.

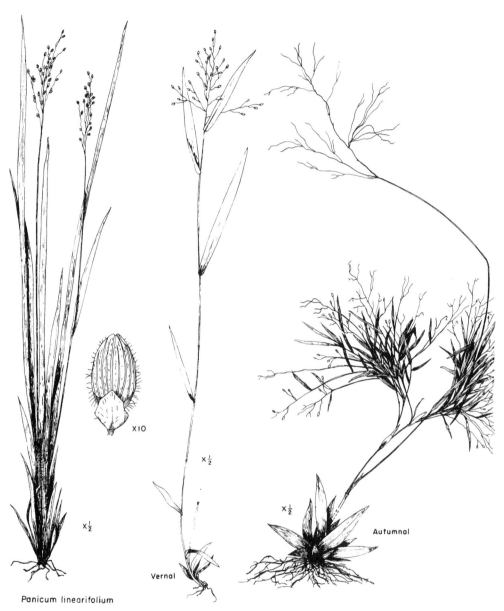

X10

$\times \frac{1}{2}$

$\times \frac{1}{2}$

$\times \frac{1}{2}$

Autumnal

Vernal

Panicum linearifolium

Panicum dichotomum

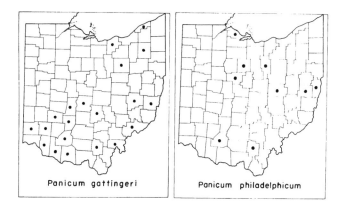

Panicum gattingeri   Panicum philadelphicum

13. Panicum depauperatum Muhl.

Recognized by the densely tufted culms and by the long narrow erect blades, appearing to be mostly basal, little exceeded by the panicle of a few glabrous or sparsely pubescent spikelets mostly 3.2–3.8 mm long; autumnal phase like vernal phase except that panicles are reduced and partly hidden among basal leaves. Woods, eastern half of the United States.

14. Panicum linearifolium Scribn.

Incl. *P. werneri* Scribn.

Resembles *P. depauperatum* in habitat, range, and general appearance, but has long-exserted vernal panicles and pilose or glabrous spikelets 2.2–2.7 mm long. Typical variety has pilose sheaths while variety *werneri* (Scribn.) Fern. (*P. werneri* Scribn.) has glabrous sheaths; both varieties in Ohio.

15. Panicum laxiflorum Lam.

Incl. *P. xalapense* HBK.

Recognized by the combination of very short ligule, small spikelets, and long-pilose nodes and sheaths; hairs to 4 mm long, those of nodes, and sometimes of sheaths, reflexed. Autumnal phase branched near base, panicles small, blades not much reduced. Woods, eastern half of the United States as far north as Virginia and Missouri.

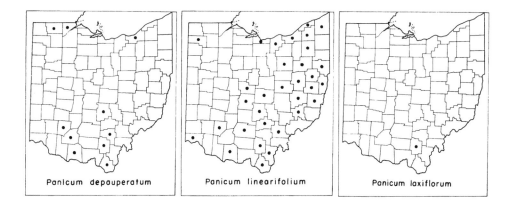

Panicum depauperatum   Panicum linearifolium   Panicum laxiflorum

16. Panicum bicknellii Nash

In appearance intermediate between *P. depauperatum* and *P. dichotomum*. Autumnal phase erect, branched, blades ascending and not much reduced, panicles small and partly concealed. Woods, parts of eastern half of the United States.

17. Panicum calliphyllum Ashe

P. bicknellii Nash var. *calliphyllum* (Ashe) Gl.; incl. in *P. xanthophysum* Gray by some earlier authors

Resembles the preceding but blades broader, less stiffly ascending, bright green; spikelets larger. Local, rare; woods, a few stations in eastern half of the United States.

18. Panicum microcarpon Muhl.

Recognized by the retrorsely bearded nodes, tiny ellipsoid spikelets, and glabrous culms, sheaths, and blades; culms short to tall; sheaths sometimes white-spotted. Autumnal phase a dense mass of many branches from all nodes with reduced blades and small few-flowered panicles. Woods and swamps, eastern half of the United States.

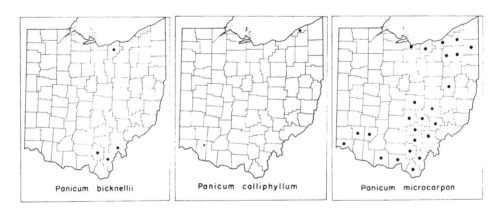

Panicum bicknellii   Panicum calliphyllum   Panicum microcarpon

19. Panicum dichotomum L.

Incl. *P. barbulatum* Michx.

Vernal culms slender and glabrous except that nodes may be pubescent or somewhat bearded; cartilaginous bands at nodes very narrow; blades 3–10 mm wide, glabrous, or sometimes ciliate at base, usually spreading. Autumnal phase erect or reclining at base, much branched at middle nodes; blades many, reduced, sometimes involute; whole mass tree-like. Coarser plants with slightly larger spikelets, wider blades, more strongly bearded nodes, and more diffusely branched reclining autumnal phase, have been segregated as var. *barbulatum* (Michx.) Wood (*P. barbulatum* Michx.).

20. Panicum yadkinense Ashe

Coarser and usually taller than *P. dichotomum*; distinguished from it by the spotted sheaths, pointed somewhat larger spikelets, and autumnal phase less densely branched from middle nodes. Woods, much of eastern half of the United States.

21. **Panicum boreale** Nash

Short to moderately tall, erect, essentially glabrous; blades erect or ascending; spikelets small, pubescent, often purple. Woods or open ground; a northern species.

22. **Panicum sphaerocarpon** Ell.

Recognized by the well-exserted panicle of many tiny globose spikelets, broad white-margined rosette-blades, and essentially glabrous culms, sheaths, and blades; panicle 5–10 cm long; culms low to moderately tall; autumnal phase sparingly branched. Eastern half of the United States.

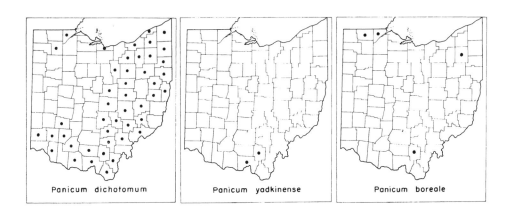

Panicum dichotomum   Panicum yadkinense   Panicum boreale

Panicum sphaerocarpon

23. **Panicum polyanthes** Schultes

Spikelets and rosette-blades like those of *P. sphaerocarpon*, but culms taller, and culm-blades wider and longer; panicle to 25 cm long; autumnal phase erect, usually a few branches from lower nodes. Woods and thickets or openings, eastern half of the United States except extreme North.

24. **Panicum leibergii** (Vasey) Scribn.

Recognized by large spikelets with long papillose-based hairs, long narrow first glume, and papillose-hispid sheaths and blades; culms low to moderately tall; autumnal phase sparingly branched from lower and middle nodes. North-central States, prairies.

25. **Panicum malacophyllum** Nash

Culms, sheaths, and blades velvety-pubescent, nodes retrorsely bearded; culms short to moderately tall; autumnal phase branched from middle and upper nodes. Thin woods and open places, Tennessee to Kansas and Texas; our one specimen from Highland County may be adventive.

26. **Panicum oligosanthes** Schultes var. **scribnerianum** (Nash) Fern.

P. scribnerianum Nash

With large blunt turgid spikelets and relatively narrow blades (6–12 mm); nerves of second glume and lower lemma prominent; branched from middle and upper nodes in autumnal phase. Prairies and open woods, much of the United States except the southeastern states.

Panicum polyanthes

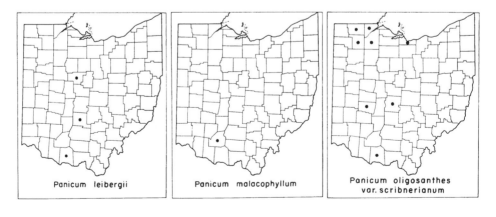

Panicum leibergii

Panicum malacophyllum

Panicum oligosanthes var. scribnerianum

27. **Panicum boscii** Poir.

Distinguished from other species with wide blades and large spikelets by the retrorsely bearded nodes; culms low to moderately tall. Plants with culms and sheaths villous and blades velvety have been separated, as var. *molle* (Vasey) Hitch. & Chase, from typical variety which has culms glabrous or minutely puberulent and sheaths and blades glabrous or nearly so; but several Ohio specimens are intermediate in degree of pubescence. Autumnal phase branched from middle nodes, the small panicles included or partly included in sheaths. Woods and thickets, eastern half of the United States.

28. **Panicum clandestinum** L.

Most easily recognized in the autumnal phase, the sheaths of which are covered with bristle-like hairs and inflated with enclosed or nearly enclosed panicles; often growing in large patches; culms with a few branches from upper nodes. Vernal culms tall to moderately tall, sheaths usually papillose-hispid and ciliate, blades glabrous except ciliate base; or rarely culms and sheaths glabrous. Moist fencerows, thickets, and roadsides, eastern United States to Iowa and Texas; common in Ohio.

29. **Panicum latifolium** L.

Vernal phase differs from the preceding in the usually glabrous sheaths, broader and shorter blades, and narrower panicles with fewer and larger spikelets. Autumnal phase with culms sparsely branched from middle nodes, blades little reduced, panicles small and wholly or partly enclosed in the glabrous sheaths. Woods, eastern half of the United States.

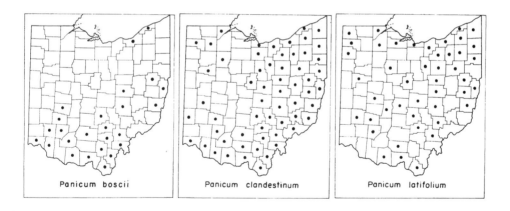

Panicum boscii   Panicum clandestinum   Panicum latifolium

30. **Panicum commutatum** Schultes

Incl. *P. ashei* Pearson

Culms short to moderately tall; variable in width of blades and size of spikelets; blades glabrous or nearly so; autumnal phase divergently branched from upper and middle nodes, panicle and upper portion of vernal culm deciduous. Two intergrading varieties: typical, with culms and sheaths glabrous or nearly so, blades 12–25 mm wide, and spikelets 2.2–3.2 mm long; var. *ashei*

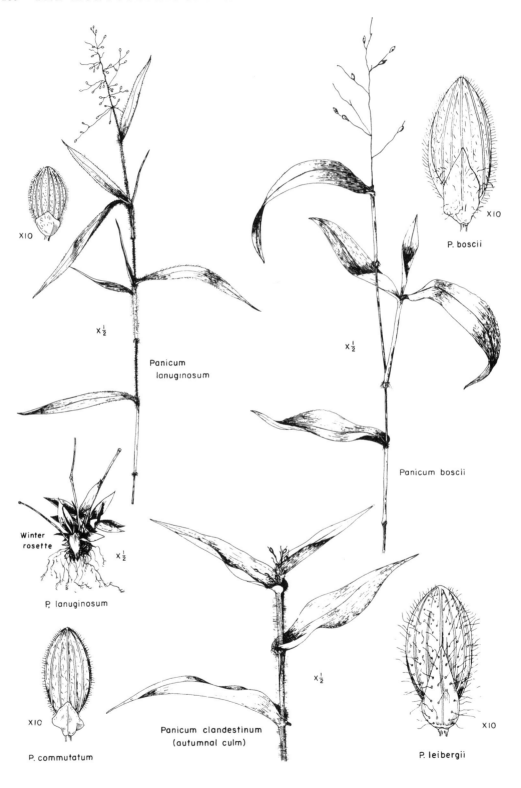

X10

$\times \frac{1}{2}$

Panicum
lanuginosum

P. boscii

X10

P. boscii

$\times \frac{1}{2}$

Panicum boscii

Winter
rosette

$\times \frac{1}{2}$

P. lanuginosum

X10

P. commutatum

Panicum clandestinum
(autumnal culm)

$\times \frac{1}{2}$

X10

P. leibergii

(Pearson) Fern. (*P. ashei* Pearson), often purplish, with sheaths and culms puberulent, blades 5–10 mm wide, and spikelets 2.2–2.7 mm long. Eastern half of the United States.

31. Panicum columbianum Scribn.

Incl. *P. tsugetorum* Nash

Culms, sheaths, and lower surface of blades puberulent, lower internodes of culm sometimes long-pilose, also; spikelets small, blades narrow; autumnal phase spreading or decumbent. Woods and open ground, much of eastern United States.

32. Panicum villosissimum Nash

Incl. *P. pseudopubescens* Nash and *P. scoparioides* (Ashe) Fern.

Conspicuous features are long hairs of culms, sheaths, and blades, and of ligules; vernal culms short to moderately tall; autumnal culms sometimes prostrate, panicles inconspicuous among ascending branches and blades. Two varieties in Ohio: typical, pilose throughout, hairs of culms and sheaths spreading; and var. *pseudopubescens* (Nash) Fern. (*P. pseudopubescens* Nash), center strip of upper leaf surface glabrous to sparsely pubescent and hairs of culms and sheaths ascending. Thin woods and open places, eastern half of the United States.

33. Panicum lanuginosum Ell.

Incl. *P. huachucae* Ashe, *P. tennesseense* Ashe, *P. implicatum* Scribn., *P. lindheimeri* Nash, and *P. languidum* Hitchc. & Chase

Ligule of long hair conspicuous; culms, sheaths, and blades densely long-pubescent to nearly glabrous; culms tufted, short to moderately tall; autumnal phase much branched, becoming spreading or prostrate, blades and panicles reduced. Many habitats, wooded and open.

Five varieties, treated by some authors as species, are distinguished by Fernald (1950). Three of these varieties include the Ohio specimens.

a. Axis of panicle, upper sheaths, and blades glabrous or nearly so; spikelets 1.6 mm long or less (a few Ohio specimens) ....
          var. *lindheimeri* (Nash) Fern. (*P. lindheimeri* Nash)
aa. Axis of panicle spreading-pilose; upper sheaths and lower surface of blades usually pilose (most Ohio specimens)
 b. Spikelets 1.5 mm long or less; blades long-pilose above ....
         var. *implicatum* (Scribn.) Fern. (*P. implicatum* Scribn.)
 bb. Spikelets 1.6 mm long or more; blades short-pilose, sparsely long-pilose, or glabrous above ....
    var. *fasciculatum* Fern. (*P. huachucae* Ashe, *P. tennesseense* Ashe, and
                 P. languidum Hitchc. & Chase)

Many Ohio specimens in the last two varieties can not be separated on bases listed. Fassett (1951), who follows treatment of Shinners (1944) and of Pohl (1947) in using the name *P. implicatum* Scribn. to include the varieties *implicatum* and *fasciculatum* of Fernald, says: "In Wisconsin there is no correlation between spikelet size and pubescence, and the spikelet size ranges from 1.3–2.0 mm, showing a normal curve with a mode at 1.6 mm. Dr. Pohl published almost exactly the same figures for Pennsylvania." While more detailed study is necessary before a valid conclusion can be reached, indications are that Ohio specimens are like those of Wisconsin and Pennsylvania in this respect.

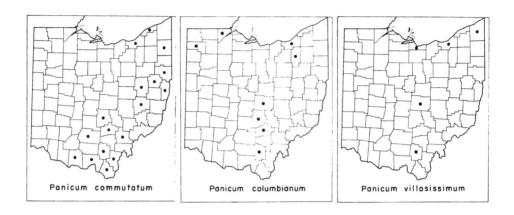

Panicum commutatum   Panicum columbianum   Panicum villosissimum

Panicum lanuginosum

64.   ECHINOCHLOA Beauv.

Ohio species annuals without ligules at base of blades. Panicle of stout densely-flowered branches; first glume ⅓–½ length of spikelet, second glume and lower lemma about equal, pointed at apex, or lower lemma and sometimes also glume awned; fertile lemma acuminate, edges inrolled around sides of palea but not around tip. A few species supply forage and grain.

a.   Sheaths, at least the lower, usually papillose-hispid; second glume and lower lemma both awned; spikelet about 3 times as long as wide; panicle a dense mass of spikelets and long awns ....1.   *E. walteri*
aa.   Sheaths glabrous; second glume awnless; spikelet about twice as long as wide.
   b.   Panicle-branches 1–2 cm long, appressed, simple; spikelets in about 4 rows, short-pubescent to glabrous; lower lemma awnless, short-pointed at apex ... 2.   *E. colonum*

bb. Panicle-branches usually more than 2 cm long, ascending or spreading, sometimes branched; spikelets usually not in rows, hispid on the nerves, usually some or all of the hairs papillose-based; lower lemma awned or awnless ....3. *E. crusgalli*

1. **Echinochloa walteri** (Pursh) Nash
Culms stout, erect, tall; panicle 1–3 dm long, usually reddish-purple, long-awned spikelets crowded on overlapping branches. Wet places, eastern half of the United States.

2. Echinochloa colonum (L.) Link   Jungle-rice
Introduced from Old World; no Ohio specimens recorded but possibly to be found; mostly more southern.

3. **Echinochloa crusgalli** (L.) Beauv.   Barnyard Grass
Culms often coarse, erect or decumbent at base, often branched, moderately tall to tall; blades to 15 mm wide or more; panicles 1–2 dm long, often purple. Open, often wet, waste or cultivated ground. As treated in Hitchcock and Chase (1951), an inclusive species, nearly cosmopolitan, occurring throughout most of the United States. Several authors consider that there are two species, *E. crusgalli*, native of the Old World and naturalized in America, and *E. pungens* (Poir.) Rydb., native in America. The two have been distinguished as follows:

Tip of fertile lemma differing in texture from lustrous body, withering, marked off from body by ring of short stiff hairs; papillose-based hairs of spikelet wanting or confined to margin ....*E. crusgalli*
Tip of fertile lemma merging gradually into lustrous body, firm and sharp, not marked off from body by ring of hairs; papillose-based hairs of spikelet usually present and abundant ....*E. pungens*

Both occur throughout Ohio. *E. pungens* is mapped.
The rarely cultivated *E. crusgalli* var. *frumentacea* (Roxb.) W. F. Wight (*E. frumentacea* (Roxb.) Link), Japanese Millet, with brown-purple awnless spikelets in thick ascending racemes, may be found as an escape; recorded from Ashtabula County.

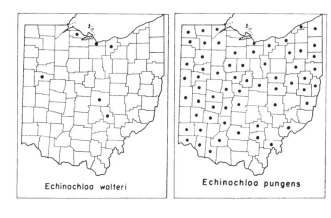

Echinochloa walteri

Echinochloa pungens

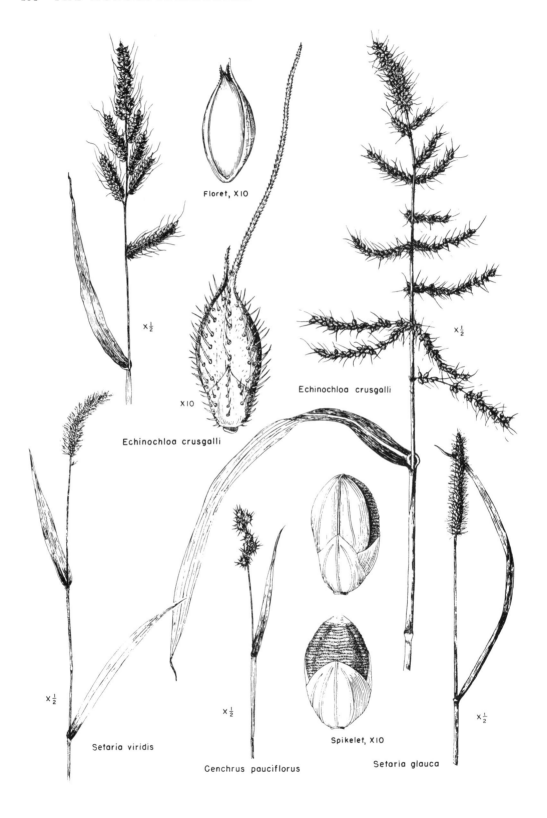

Floret, X10

Echinochloa crusgalli

X10

Echinochloa crusgalli

X½

X½

X½

X½

X½

Setaria viridis

Cenchrus pauciflorus

Spikelet, X10

Setaria glauca

65. SETARIA Beauv. FOXTAIL GRASS
Chaetochloa Scribn.

About 75 species, mostly of warmer regions; Ohio species all annuals introduced from Europe or Asia. Inflorescence bristly, usually dense, spike-like; each spikelet, whether fertile or abortive, subtended by 1–several bristles.

a. Second glume ½–⅔ as long as rugose fertile lemma; spikelet about 3 mm long and 2 mm wide, usually one per branch of main axis of inflorescence; 5–20 bristles per spikelet, yellowish at maturity ....1. *S. glauca*
aa. Second glume ¾ to as long as slightly rugose fertile lemma; usually more than one spikelet per branch of main axis of inflorescence.
    b. Bristles downwardly barbed, those of one inflorescence often entangled with those of another; spikelet about 2 mm long; bristles short, one per spikelet ....
                            2. *S. verticillata*
   bb. Bristles upwardly barbed.
      c. Inflorescence usually about 1 cm thick or less, excluding bristles, usually green, rarely purple; spikelet at maturity falling whole.
         d. Blades glabrous; spikelets 2–2.5 mm long, 1–3 bristles below each ....
                          3. *S. viridis*
        dd. Blades usually pubescent; spikelets 2.5–3 mm long, often 3–6 bristles below each ....4. *S. faberii*
     cc. Inflorescence often more than 1 cm thick, excluding bristles, variously colored, sometimes purple, branches often evident; at maturity, grain and its lemma and palea falling away from remainder of spikelet ....5. *S. italica*

1. SETARIA GLAUCA (L.) Beauv. YELLOW FOXTAIL GRASS
S. lutescens (Weigel) Hubb.; *Chaetochloa glauca* (L.) Scribn.
Distinguished by the yellow mature inflorescence and by the turgid spikelets with rugose lemma only partially covered by second glume; culms short to moderately tall; blades loosely spiral; panicle 5–10 cm long. Naturalized from Europe, common in Ohio; cultivated fields, roadsides, waste places.

2. SETARIA VERTICILLATA (L.) Scribn.
Chaetochloa verticillata (L.) Scribn.
Moderately tall; noticeable because the several panicles of a tuft are usually entangled with one another; bristles 1–3 times as long as spikelets. Naturalized from Europe; locally common in waste and cultivated ground, often in alleys and elsewhere near habitation.

3. SETARIA VIRIDIS (L.) Beauv. GREEN FOXTAIL GRASS
Chaetochloa viridis (L.) Scribn.
Culms short to moderately tall; panicle 2–10 cm long, erect or slightly curved. Naturalized from Europe in much of the United States; common in Ohio, cultivated fields, roadsides, waste places, often a weed.

4. SETARIA FABERII Herrm.
Resembling *S. viridis* and, when in lower range of size, difficult to distinguish from it; usually much taller, as much as 1.5 m, with wider usually pubescent blades, longer and thicker more nodding panicles, larger spikelets, and more rugose lemmas. Introduced from China, it has spread rapidly in Ohio during about the last fifteen years as a weed in cultivated and waste ground.

5.   SETARIA ITALICA (L.) Beauv.   FOXTAIL MILLET
   Chaetochloa italica (L.) Scribn.
   Long cultivated, existing in many variations of size of panicles and color
of bristles and of fertile lemmas; resembles S. *viridis*, but usually coarser, with
panicles thicker and somewhat lobed. Escaped, roadsides and waste places.

66. CENCHRUS L.   SANDBUR GRASS

   About 25 species with spike-like racemes of prickly globose burs composed
of coalescent involucral bristles, each bur surrounding two or more spikelets
and falling with spikelets enclosed; lemma and palea of perfect floret firm,
lemma not inrolled around palea.

1.   **Cenchrus pauciflorus** Benth.   FIELD SANDBUR
   Incl. *C. longispinus* (Hack.) Fern.
   Easily recognized by the burs, about 0.5 cm wide excluding spines, in a
spike-like raceme 3–8 cm long; spines flat, spreading or reflexed; culms short
to moderately tall. Throughout most of the United States, in sandy open ground.

Cenchrus pauciflorus

TRIBE XI.   ANDROPOGONEAE

   Inflorescence of racemes clustered at top of peduncle or in a panicle, each
raceme consisting of a rachis along which spikelets are arranged in pairs, one
of each pair fertile and usually sessile, the other fertile, staminate, or vestigial,
and pediceled; fertile spikelet consisting of terminal perfect floret and below
that a staminate or vestigial floret; lemma of perfect floret hyaline, often awned,
more delicate than glumes; palea small. Rachis usually jointed, breaking at
joints at maturity, pair of spikelets and rachis-segment (illustrated) falling
together.

   a.   Inflorescence terminal, of many racemes; lateral inflorescences usually absent.
      b.   Racemes of many spikelets, obviously hairy, densely aggregated; both spikelets
         of pair fertile, each with tuft of long hairs at base.

    c.   Racemes 1–2 dm long, many of them equaling or longer than axis of panicle, ends spreading, whole panicle fan-like; spikelets of pair unequally pediceled; rachis continuous ....67. *Miscanthus*

    cc. Racemes shorter, long hairs obscuring spikelets, panicle dense, ellipsoid; one spikelet of pair pediceled, one sessile; rachis jointed ....68. *Erianthus*

bb. Racemes short, of few spikelets, the inflorescence appearing a usual panicle instead of a series of racemes; hairs at base of spikelet, if present, much shorter than spikelet; pediceled spikelet staminate, vestigial, or wanting.

    c.   Racemes densely-flowered; spikelets pale or tinged with red or purple; pediceled spikelet usually staminate; stem solid ....71. *Sorghum*

    cc. Racemes loosely-flowered; spikelets tawny when young, brown at maturity; pediceled spikelet wanting, only the pedicel present; stem hollow ....
                    72. *Sorghastrum*

aa. Inflorescence of 1-few racemes clustered at or near end of peduncle, only terminal, or both terminal and lateral, or inflorescence a panicle of such clusters.

    b.   Culms weak, reclining or decumbent; blades lanceolate, 3–8 cm long; pediceled spikelet perfect ....69. *Eulalia*

    bb. Culms erect; blades linear, elongate; pediceled spikelet not perfect ....
                    70. *Andropogon*

67. MISCANTHUS Anderss.

A small genus of tall perennials. Fertile lemma with bent and flexuous awn.

1. MISCANTHUS SINENSIS Anderss. EULALIA

Culms tall and stout, in large clumps, the many leaves mostly on lower half; panicle silky; some forms with variegated leaves. Native of eastern Asia; cultivated as an ornamental and rarely escaped.

68. ERIANTHUS Michx. PLUME GRASS

About 20 species of tall perennials mostly of warm regions. Terminal panicles usually conspicuously silky; lemma of perfect floret awned.

1. **Erianthus alopecuroides** (L.) Ell. SILVER PLUME GRASS
    E. divaricatus (L.) Hitchc.

A tall handsome grass with silvery, tawny, or purplish dense panicle 2–3 dm long, spikelets almost hidden by long silky hairs. Gulf States north to New Jersey and Missouri; woods and open ground.

Erianthus alopecuroides

$\times \frac{1}{2}$

Andropogon gerardi

Pair of spikelets, X5

Andropogon scoparius

$\times \frac{1}{2}$

69. EULALIA Kunth

Microstegium Nees

Species native in tropics and subtropics of the Old World.

1. EULALIA VIMINEA (Trin.) Kuntze
Microstegium vimineum (Trin.) A. Camus
Culms slender, branching, rooting at nodes; habit somewhat like that of a species of *Commelina;* blades tapering to slightly petioled base; racemes solitary or in cluster of 2–6; internodes of rachis thick; lemma awnless or awned. Introduced from Asia, occurring locally from Ohio southward; recorded from Adams County.

70. ANDROPOGON L. BEARD GRASS

About 200 species mostly of tropical and subtropical regions. Culms solid; spikelets in racemes, racemes solitary or two or more digitate on terminal peduncles or on peduncles arising from axils of spathe-like leaves, the whole inflorescence then a panicle or a corymb of racemes; glumes of fertile spikelet firm, fertile lemma awned.

a. Very tall to tall; racemes 2–6 on each peduncle, at least the terminal peduncle well exserted; pediceled spikelet staminate, awnless ....1. *A. gerardi*
aa. Tall to moderately tall; racemes 1–few on each peduncle; axillary peduncles several to many, arising from sheaths, often included; rachis of raceme slender, often fragile; pediceled spikelet usually vestigial.
    b. Racemes solitary on each peduncle; rachis segments and pedicels pilose; fertile spikelet 6–8 mm long, pediceled spikelet smaller, short-awned ....2. *A. scoparius*
    bb. Racemes 2, or rarely 3–4, on each peduncle; rachis segments and pedicels bearing long hairs; pedicel beside fertile spikelet naked or bearing a small rudiment.
        c. Sheaths subtending peduncles much inflated and overlapping, conspicuously reddish-brown; sessile spikelet 4–5 mm long, lemma with loosely twisted awn ....3. *A. elliottii*
        cc. Sheaths subtending peduncles little or not inflated and overlapping, whole inflorescence elongate; sessile spikelet 3–4 mm long, lemma with straight awn
                                              4. *A. virginicus*

1. **Andropogon gerardi** Vitman BIG BLUESTEM. TURKEY-FOOT
A. furcatus Muhl.
Recognized by its distinctive inflorescence of a few clustered racemes and by the tall culm (1–4 m). A dominant grass of the original tall-grass prairies where it grew in dense stands and formed a sod plowed only with great difficulty by pioneers. Prairies, fields, roadsides, throughout the United States except the extreme West.

2. **Andropogon scoparius** Michx. LITTLE BLUESTEM
Culms slender; inflorescence elongate, paniculate, of usually many racemes from axils of upper sheaths. A dominant plant of large areas of original grasslands of the United States, often in drier sites than the preceding. Throughout the United States except the extreme West.

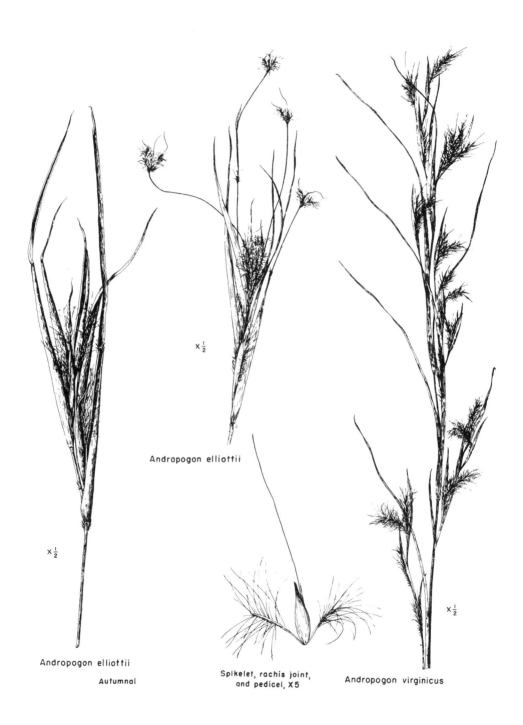

Andropogon elliottii

×½

Andropogon elliottii

Autumnal

Spikelet, rachis joint,
and pedicel, ×5

Andropogon virginicus

×½

3. **Andropogon elliottii** Chapm. ELLIOTT BEARD GRASS

Distinctive because of the crowded fan-like arrangement of the inflated upper sheaths which are glossy burnt-orange within in autumn; racemes usually included in sheaths or the primary exserted. Old fields, open ground; eastern half of the United States from New Jersey to Missouri and southward.

4. **Andropogon virginicus** L. BROOM-SEDGE

Culms tufted, stiffly erect; sheaths somewhat reddish-brown in age, the lower keeled, each enfolding one next above. Roadsides, old fields, pastures; sometimes a weed. Eastern half of the United States.

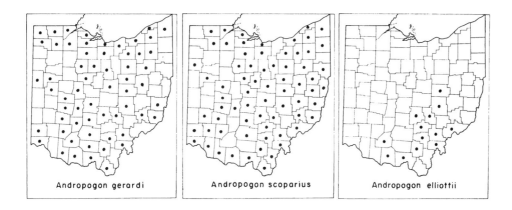

Andropogon gerardi   Andropogon scoparius   Andropogon elliottii

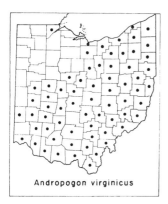

Andropogon virginicus

71. SORGHUM Moench

About 10 species of warm regions of the Old World; tall or moderately tall; glumes of sessile spikelet firm, usually shining; upper node of racemes with 3 spikelets, 2 of them pediceled. Sorghums and Johnson Grass contain a glucoside

Sorghastrum nutans

$\times \frac{1}{2}$

Sorghum halepense

$\times \frac{1}{2}$

Spikelet group, X5

that yields hydrocyanic acid; enough may be present to cause poisoning of live-stock under certain conditions.

Annual ....1. *S. vulgare*
Perennial, with extensive rhizomes ....2. *S. halepense*

1. SORGHUM VULGARE Pers. SORGHUM
Holcus sorghum L.
Cultivated since prehistoric times; consists of many varieties or races planted as source of sweet juice, grain, forage, and materials for making brooms and other industrial products; includes Sorgo, Kafir-corn, Sudan Grass, Broom-corn, and others. Sometimes spontaneous but probably not persistent.

2. SORGHUM HALEPENSE (L.) Pers. JOHNSON GRASS
Holcus halepensis L.
Culms usually more slender than in the preceding; blades mostly less than 2 cm wide, tapering to base, white midrib prominent. Native of Mediterranean region, sometimes cultivated, occurring as an escape in much of the United States; beautiful, but a troublesome weed in Ohio because of the rapidly spreading rhizomes.

72. SORGHASTRUM Nash

A few species of America and Africa. Tall perennials with narrow panicles of few-flowered racemes.

1. **Sorghastrum nutans** (L.) Nash INDIAN GRASS
One of the most beautiful of the prairie grasses with its graceful slender tawny to brown panicle; tall; blades long, tapering to base; spikelets and pedicels covered with long gray hairs. Prairies, open woods; eastern United States as far west as North Dakota and Arizona.

Sorghastrum nutans

Tribe XII.   Tripsaceae (Maydeae)

Spikelets monosporangiate (unisexual), the two kinds in different parts of the same inflorescence or in different inflorescences on same plant; spikelet of 2 florets, 1 sometimes vestigial.

> The two kinds of spikelets in different parts of same inflorescence, the carpellate (pistillate) enclosed in a bead-like involucre, the staminate in a spike or raceme projecting from apex of this involucre ....73.   *Coix*
> The two kinds of spikelets in separate inflorescences; terminal inflorescence (tassel) a panicle of staminate spikelets; one or more axillary branches consisting of axis (cob) and rows of sessile carpellate (pistillate) spikelets, axis and spikelets forming an ear, the whole covered with large bracts (husks) ....74.   *Zea*

73.   COIX L.

About 5 species native in Asia.

1.   COIX LACHRYMA-JOBI L.   JOB'S-TEARS
Much branched tall annual with blades to 4 cm wide; inflorescences peduncled, clustered in leaf axils. Cultivated as an ornamental; the white to blue-gray smooth hard involucres sometimes used as beads.

74.   ZEA L.

A genus of one species the origin of which was somewhere in the Americas.

1.   ZEA MAYS L. Maize.   INDIAN CORN. CORN
Cultivated by the inhabitants of North and South America for centuries before 1492, introduced into the eastern hemisphere, and now one of the most valuable grain crops of the world. A variable species, including several races or varieties, some of which are dent, the common field corn; sweet, with relatively high sugar content, used mostly in the immature state as a vegetable and extensively canned; pop, used in preparing the familiar confection; and pod, mostly a curiosity, each grain enclosed in large glumes. Extensive corn improvement programs have resulted in the appearance of many forms within each of the cultivated races or varieties. A relatively recent result of such programs is the widespread use by farmers of hybrid seed corn.

CYPERACEAE. Sedge Family

Annual or perennial grass-like or rush-like herbs with terete or 3-angled (rarely flattened or 4-angled) usually solid stems (culms); leaves sometimes reduced to bladeless sheaths, more often grass-like, 3-ranked, sheaths closed; flowers perfect or imperfect, each in axil of a scale-like bract, aggregated into spikelets (small spikes), these either solitary and terminal or variously arranged in head-like or umbel-like inflorescences; ovary 1-celled, with 1 ovule, stamens usually 3, perianth none or consisting of 1–6 (–8) hypogynous bristles, and (rarely) of 1–3 petaloid scales; fruit an achene, usually lens-shaped, plano-convex, or 3-angled, with or without apical tubercle.

A very large and cosmopolitan family of some 4,000 species classified into about 75 genera, of which *Carex* is much the largest. A very few species have become troublesome weeds (*Cyperus esculentus* may be so considered when it invades lawns or gardens). Very few are of economic importance except as soil binders; *Cyperus esculentus* var. *sativus* Boeckl., Chufa of Africa, is culti-vated in the South for its edible tubers; many play an important role in the successional development of vegetation, particularly of swamps and bogs; one, *Cyperus alternifolius* L., umbrella plant, of Africa is grown as an ornamental and has become naturalized in Florida. Papyrus or Egyptian paper plant (*Cyperus papyrus* L., *Papyrus antiquorum* Link) of the Nile (and widespread in tropical Africa) is sometimes grown in aquatic gardens; its culms are 3–5 m tall, with a large terminal compound umbel; leaves are radical.

Sedges are most readily distinguished from grasses by the closed leaf-sheath (most grasses have open sheath) and by the 3-ranked arrangement of leaves; from rushes (Juncaceae), by the absence of the liliaceous 6-parted perianth, and by the achene (instead of capsule). For satisfactory identification, mature fruits are necessary; root-system is important in some genera.

a. Scales of spikelets 2-ranked.
  b. Inflorescence terminal, dense and head-like or open and umbelliform; stems leafy at base, otherwise naked except for involucral leaves at summit ....1. *Cyperus*
  bb. Inflorescence axillary, the peduncles scarcely exserted from leaf-sheath; stems leafy, the upper sheaths often overlapping, the lower bladeless ....2. *Dulichium*
aa. Scales of spikelets (or perigynia) spirally arranged, often imbricate.
  b. Achene naked, i.e. not enclosed although often partially wrapped in subtending scale.
    c. Inflorescence a single terminal spikelet without obvious subtending bracts; stems rush-like, terete, flattened or angled, without leaves or leafy bracts, but with essentially bladeless sheaths near base; achene tipped with tubercle ....
        3. *Eleocharis*
    cc. Inflorescence various, with leaf-like or stem-like involucre; if (rarely) a single terminal spikelet subtended by modified sterile scales, then achene without tubercle.
      d. Flowers perfect, all alike; lowest scales of spikelet often sterile.
        e. Spikelets more than 2-flowered, usually many-flowered; empty scales at base of spikelet 1 (more in *Psilocarya* and *Eriophorum*).
          f. Base of style dilated or bulbous, persistent or deciduous; perianth-bristles absent; slender tufted annuals (our species) with narrow or filiform leaves and umbelliform inflorescence.

        g.   Base of style persistent and forming a minute tubercle on apex of achene; leaves and culm capillary ....4. *Bulbostylis*

        gg. Enlarged base of style deciduous, not forming a tubercle on achene; culm flattened; leaves 1–2 mm wide ....
                                                     5. *Fimbristylis*

  ff. Base of style not dilated, style slender, terete; perianth-bristles usually present (absent in *Hemicarpha* and some species of *Scirpus*).

        g.   Perianth-bristles appearing as if numerous (6, 4–6 cleft), much elongated in fruit and silky, the whole inflorescence then appearing like a large tuft (or tufts) of silky hairs ....7. *Eriophorum*

        gg. Bristles and inflorescence not as above.

             h.   Small tufted annual with very slender stems and leaves, rare in Ohio; spikelets 1–3, 2–4 mm long; bristles absent, a small translucent scale between axis of spikelet and flower; achene cylindric ....
                                         8. *Hemicarpha*

             hh. Annual or perennial grass-like or rush-like plants; spikelets solitary and terminal or few to many in heads, inflorescence spike-like or umbelliform; perianth-bristles 1–6 (–8), rarely absent; achene lens-shape, plano-convex, or 3-angled ....6. *Scirpus*

  ee. Spikelets 1–2 flowered; empty scales at base of spikelet 2-many.

      f.   Achene crowned by a tubercle, the enlarged and persistent base or lower part of style; style 2-cleft ....9. *Rhynchospora*

      ff. Achene without tubercle, usually acorn-shape, the tip corky; style 2–3 cleft ....10. *Cladium*

 dd. Flowers imperfect; inflorescence terminal and axillary, of spikes composed of 1-flowered pistillate spikelets among few-flowered staminate spikelets; achenes large, hard and bony, usually white, and exposed ....
                                                11. *Scleria*

bb. Achene enclosed in a sac (perigynium) with small opening at top through which style protrudes; flowers unisexual, in spikes, some of which may be pistillate, others staminate, or spike may be partly staminate, partly pistillate; grass-like sedges with usually 3-angled stems and 3-ranked leaves ....12. *Carex*

1. CYPERUS L. Umbrella Sedge

Best distinguished from other sedges by the 2-ranked arrangement of scales of the spikelets, and terminal leafy-bracted inflorescence. Flowers consist of a single pistil, with 2- or 3-cleft styles, 1–3 stamens, no perianth; each is in the axil of a scale or bract, and arranged in 2 ranks on axis of spikelet (rachilla) which is often winged by decurrent margins of scales; achene lenticular in species with 2-cleft style, 3-angled in species with 3-cleft style; spikelets variously arranged in loose or dense heads, or in more or less umbellate clusters, the rays simple or branched, the whole subtended by leafy bracts. A large genus of about 600 species, 17 or 18 of which occur in Ohio. Name from the Greek *Cypeiros* or *Kupeiros*, hence the accent on the second syllable. For satisfactory identification of species, mature spikelets are necessary, and basal part of plant desirable, to show corm-like thickening of base of stem, fibrous roots, rhizomes, or tuberous enlargements. Two keys are offered: the first is more technical; the second, based to a considerable extent on variable gross characters, may serve to identify average specimens.

a. Achene lenticular, not 3-angled; style 2-cleft; annuals.
  b. Spikelets 1.5–2 mm long, 1-flowered, with 3–4 scales, aggregated into 1–3 small compact sessile heads subtended by 2–4 widely spreading bracts; plants low, 5–15 (–20) cm tall .... 1. *C. tenuifolius*
  bb. Spikelets 8–15 mm long, with several to many flowers, some sessile, or occasionally terminating short rays; involucre of 2-several leafy bracts; plants often taller, 5–30 (–40) cm tall; (3 species, similar in appearance).
    c. Scales of spikelet straw-color, closely imbricated, hiding mature achene; achenes black, broadly obovoid, about as broad as long, minutely transversely wrinkled .... 2. *C. flavescens* var. *poaeformis*
    cc. Scales red-brown or striped or tinged with red-brown; achenes gray or brownish gray, oblong to narrowly obovate, longer than broad, not transversely wrinkled.
      d. Scales usually dull, mature scales red at tip and down along margin, loosely imbricated, achenes visible in dried specimens; styles branched nearly from base, persistent, conspicuously exserted (2–4 mm) in mature spikelets .... 3. *C. diandrus*
      dd. Scales lustrous, red-brown or suffused with red, closely imbricated and hiding achenes, tip hooded; styles branched from about middle, usually deciduous, if persistent then exserted 1–1.5 mm .... 4. *C. rivularis*
aa. Achene 3-angled; style 3-cleft.
  b. Scales of spikelet strongly to slightly outcurved at apex (as seen in profile); spikes more or less spherical or subglobose; cespitose annuals.
    c. Scales long-acuminate, sharp-pointed, the upper ¼–⅓ of scale widely spreading or recurved; plant fragrant when dry .... 5. *C. inflexus*
    cc. Scales only slightly outcurved at apex, acute; spikelets ovate; plants not fragrant .... 6. *C. acuminatus*
  bb. Scales of spikelet straight or slightly incurved at tip (as seen in profile).
    c. Axis of spike short, less than 1 cm long; spikelets more or less crowded into globose heads or short clusters; spikes sessile at top of culm or on elongate rays; culms angled.
      d. Inflorescence of several dense globose (or nearly globose) heads; spikelets radiating in all directions, 2–3 flowered, 4–5 mm long, often only 1 achene maturing; culms smooth, leaves and bracts scabrous-margined .... 7. *C. ovularis*
      dd. Inflorescence less dense; spikelets radiating horizontally and upward, few deflexed, 3–12 (–15)-flowered, mostly 5–15 (–20) mm long.
        e. Culms scabrous, sharply angled; spikelets spreading to ascending, 5–12-flowered, 10–25 mm long; midrib of scales excurrent as an awn .... 8. *C. schweinitzii*
        ee. Culms smooth.
          f. Leaves and bracts scabrous-margined; heads loose or compact, rays (if present) capillary; scales of spikelet oblong-elliptic, blunt, not strongly overlapping; achene narrowly oblong ....
            10. *C. filiculmis*
          ff. Leaves and bracts smooth; scales of spikelet roundish, mucronate; achene short-ellipsoid .... 9. *C. houghtonii*
    cc. Axis of spike elongate, usually 1–3 cm long, spikelets in obovoid, short-cylindric, or longer dense to open spikes.
      d. Spikelets ascending; scales not appressed, large (4 mm long and in half-view about half as wide), upper awned .... 8. *C. schweinitzii*
      dd. Spikelets spreading and ascending or reflexed; scales appressed, smaller.
        e. Mature spikelets reflexed (at least those of lower part of spike), if not reflexed then widely spaced; rachilla with white hyaline wings on internodes.
          f. Spikes open; spikelets widely spaced, only the lower strongly reflexed, 2–6-flowered; scaletip reaching to base of one next above on same side; culms and rays smooth .... 11. *C. refractus*
          ff. Spikes dense; some or all of the spikelets strongly reflexed.
            g. Spike long-obconic; spikelets 1–3-flowered, firm with subulate tips; culms rough, rays smooth .... 12. *C. dipsaciformis*

       gg. Spike cylindric, upper spikelets ascending, lower strongly reflexed, middle horizontal; spikelets more than 2-flowered, soft; culms smooth ....13. *C. lancastriensis*

  ee. Mature spikelets spreading.

    f. Flowers remote, tip of scale not reaching base of scale next above on same side of spikelet ....14. *C. engelmanni*

    ff. Flowers approximate, tip of scale overlapping base of scale next above on same side of spikelet.

      g. Annuals or short-lived perennials with fibrous roots and soft bases.

        h. Rachilla breaking into sections at maturity; scales of spikelet 2–2.5 mm long, overlapping less than half their length; achenes 1–1.5 mm long, obovoid-oblong, yellowish to reddish brown ....15. *C. ferruginescens*

       hh. Rachilla not breaking into sections at maturity.

         i. Scales of spikelet about 1.5 mm long or less, red-brown with green midrib, closely imbricate; achenes less than 1 mm long, ovoid, whitish or pearly ....16. *C. erythrorhizos*

         ii. Scales of spikelet about 4 mm long, yellowish with green midrib, loosely imbricate; achenes 1.5–2 mm long, linear ....18. *C. strigosus*

    gg. Perennials with short woody rhizomes, tubers, or slender scaly stolons; spikes more or less open.

      h. Spikelets acuminate to subulate, not flattened, widely spaced, 2–6-flowered, the lower strongly reflexed; rhizome short and thick ....11. *C. refractus*

      hh. Spikelets flattened or compressed, not subulate, but may be acuminate, usually more than 6-flowered; scales nerved on sides.

        i. Producing slender scaly stolons, not hard-based; scales of spikelet 2.5–3 mm long, deciduous ....

                          17. *C. esculentus*

        i. Without stolons, base hard, corm-like; scales 3–4.5 mm long, green midrib usually prominent; spikelets deciduous ....18. *C. strigosus*

Supplementary Key

Key based on gross characters; should be used in conjunction with illustrations and with more technical key.

a. Spikelets clustered into tight heads or short loose subcapitate or ovoid heads, sessile and/or elevated on short or long slender rays; inflorescence appearing crowded.

  b. Spikelets closely crowded, slender, few (1–3)-flowered, 1.5–7 mm long; scales closely imbricated.

    c. Inflorescence of 1 or 2–3 confluent sessile heads 4–8 mm in diam.; small annual, fragrant when fresh ....1. *C. tenuifolius*

    cc. Inflorescence of 1 or 2 sessile heads and usually several pedunculate heads to 1.5 cm in diam.; perennial, rhizome with tuberous enlargements ....

                           7. *C. ovularis*

  bb. Spikelets clustered into compact umbels or subglobose, ovoid, or hemispheric heads, many-flowered, up to 2 cm long; scales not so closely imbricated.

    c. Scales of spikelets more or less strongly outcurved at tip; tufted annuals.

      d. Scales with long slender recurved tips, spikelets thus appearing fringed; fragrant when dry ....5. *C. inflexus*

      dd. Scales with acute recurved tips, the ovate spikelets thus appearing serrate; spikelets (at least some) with broad faces parallel; not fragrant when dry ....6. *C. acuminatus*

cc. Scales of spikelets not outcurved at tips, regularly imbricate.
   d. Tufted annuals; roots fibrous.
      e. Scales straw-color, thin; mature achenes black, showing through the thin scales as circular dark patches ....2. *C. flavescens*
      ee. Scales with brown-red tip and margin, green midrib, and colorless band between; achenes not completely covered by scales ....
                    3. *C. diandrus*
      eee. Scales red-brown or suffused with red, lustrous; midrib slightly incurved at apex, the tip of scale hooded ....4. *C. rivularis*
   dd. Perennial; rhizome with hard corm-like thickenings or branches.
      e. Spikelets large (1–2.5 cm long), ascending.
         f. Scales acuminate, awned; culms scabrous ....8. *C. schweinitzii*
         ff. Scales obtuse, or upper mucronulate, culms smooth ....
                    9. *C. houghtonii*
      ee. Spikelets smaller, radiating; scales blunt to acute ....10. *C. filiculmis*
aa. Spikelets loosely disposed in clusters on axis (rachis) of spike, thus exposing short internodes of rachis; some or all of the clusters of spikelets on simple or branched rays; inflorescence appearing more or less open.
   b. Spikelets all (or almost all) strongly reflexed, spikes obconic ....12. *C. dipsaciformis*
   bb. Spikelets horizontally spreading, ascending, or some reflexed.
      c. Annuals, with fibrous roots and soft bases.
         d. Spikelets very slender, remotely flowered, tip of a scale reaching only to base of scale next above on same side ....14. *C. engelmannii*
         dd. Spikelets with scales definitely overlapping.
            e. Spikelets easily broken into short sections; scales rusty or yellowish brown ....15. *C. ferruginescens*
            ee. Spikelets not easily broken into sections; scales red-brown ....
                    16. *C. erythrorhizos*
   cc. Perennials, the base hard and corm-like, or if soft, plant with slender scaly stolons.
      d. Spikes with lower, or lower and middle spikelets reflexed (see also no. 18).
         e. Spikes loose, spikelets widely spaced, leaving remote scars on rachis after falling ....11. *C. refractus*
         ee. Spikes dense, spikelets crowded, leaving closely placed scars on rachis after falling ....13. *C. lancastriensis*
      dd. Spikes with horizontally spreading and ascending spikelets (some reflexed on crowded spikes of no. 18).
         e. Plants with slender scaly stolons; spikelets not deciduous; scales of spikelet deciduous, less than 3 mm long ....17. *C. esculentus*
         ee. Plants with hard corm-like bases; spikelets deciduous; scales of spikelet narrower, more than 3 mm long ....18. *C. strigosus*

1. **Cyperus tenuifolius** (Steud.) Dandy
   C. densicaespitosus Mattf. & Kukenth., *Kyllinga pumila* Michx.
   Small annual, often in dense tufts, on muddy creek and pond shores, or sometimes in lawns; fresh plants with strong rather pleasant odor. Culms 2–15 cm tall; leaves soft, 1–2 mm wide, the lower bladeless; inflorescence terminal, leafy-bracted, with 1–3 subglobose or ovoid sessile, closely crowded heads 4–8 mm long and wide; spikelets 1-flowered, flattened.

2. **Cyperus flavescens** L., var. **poaeformis** (Pursh) Fern.
   Slender, often cespitose annual of wet soil; culms mostly 10–15 (5–30) cm tall; inflorescence usually with 3 widely spreading leafy bracts, umbellate, rays usually short, the spikelets crowded, about 1 cm long (0.5–1.5). This and the two following species are similar in appearance, and distinguished by spikelet and achene characters. Scales of spikelet with hard green keel, otherwise straw-

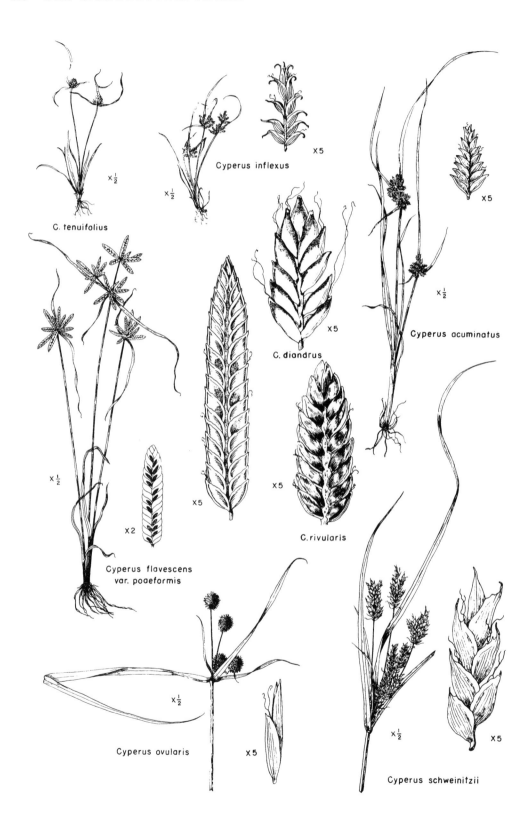

C. tenuifolius

x½

x½

Cyperus inflexus

X5

Cyperus acuminatus

x½

C. diandrus

X5

X5

x½

Cyperus flavescens
var. poaeformis

X2

X5

C. rivularis

X5

X5

Cyperus ovularis

x½

X5

x½

Cyperus schweinitzii

X5

color, thin and often semitransparent, the black subglobose achenes then showing faintly through the scales on mature spikelets.

North American plants are, by some authors, considered to be distinct from the typical widespread tropical plants; thus interpreted, all our plants belong to var. *poaeformis*. More southern in range than the 2 following species.

3. **Cyperus diandrus** Torr.
Similar to the preceding in habit and habitat; apparently rare and local in Ohio. Most readily distinguished by color and shape of scales of mature spikelets: apex pointed, brown-red, the color extending as a marginal band tapering toward base of scale and separated from the green midrib by a colorless band; entire spikelets sometimes fringed with the long-exserted deeply 2-cleft styles; scales of spikelet thin, loosely imbricated, the mature achenes not covered by their bases.

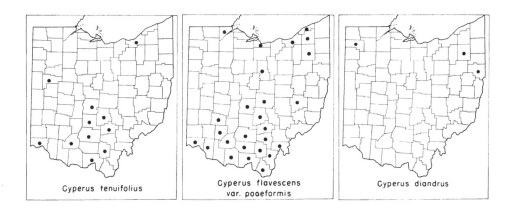

Cyperus tenuifolius

Cyperus flavescens
var. poaeformis

Cyperus diandrus

4. **Cyperus rivularis** Kunth
Similar to the preceding, but much more frequent throughout Ohio. Distinguished from the 2 preceding species by the hard and lustrous scales of the spikelet, which at maturity are red-brown or suffused with pigment throughout; midrib of scale curved inward at tip, producing a hooded or cucullate apex; scales in half-view relatively wider than those of preceding species.

5. **Cyperus inflexus** Muhl.
C. aristatus Rottb.
A small tufted (cespitose) annual with culms mostly about 5 (1–12) cm tall and compact leafy-bracted inflorescence of a few sessile or some peduncled heads. Most readily distinguished by the strongly recurved slender tips of the prominently ribbed scales of spikelet. A wide-ranging but not common sedge of wet soil.

6. **Cyperus acuminatus** Torr. & Hook.
Cespitose annual, culms 5–20 (–25) cm tall; leaves as long or longer than culm; leafy bracts long. Resembling the preceding species; distinguished by the

short-tipped faintly 3-nerved scales curving outwardly at apex only, and arrangement of spikelets in head—many with their flat sides facing one another. A southern species extending northward into Ohio.

7.   Cyperus ovularis (Michx.) Torr.

Perennial, with smooth culms, ours ranging from 2–7 dm tall. Distinguished by the dense globose or globose-ellipsoid heads of the inflorescence (7–15 mm in diam.), with spikelets radiating in all directions. More slender plants with smaller heads are sometimes segregated as var. *sphaericus* Boeckl. of more southern range. However, as stature and size of heads vary greatly even in a single collection, no varieties are distinguished here; our smallest plants are from the northwest corner (Wood County), and from the south-central part (Ross County) of Ohio. A wide-ranging species of wet meadows, fields, and dry slopes.

8.   Cyperus schweinitzii Torr.

Perennial, with short hard knotty rhizomes; culms sharply 3-angled, scabrous, 1–6 dm tall; inflorescence made up of sessile and long peduncled spikes; spikelets loosely flowered. A plant of sandy lake shores and dunes (rarely on railroad embankments), distinguished by the broad strongly veined scales of spikelet

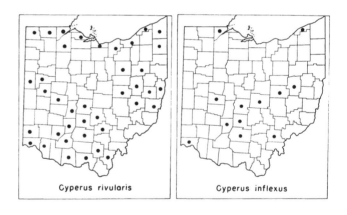

Cyperus rivularis

Cyperus inflexus

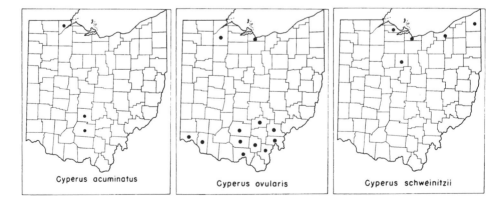

Cyperus acuminatus

Cyperus ovularis

Cyperus schweinitzii

with midrib extended as a prominent mucro or awn, and roughened angles of culm and of leaf-margins.

9. Cyperus houghtonii Torr.

Similar to the preceding but smaller in every way. Leaves, bracts, and culms smooth; scales of spikelet smaller (2–2.8 mm long), but proportionately broader, pointed but not awned (uppermost with acuminate point). Represented in Ohio by a single specimen in herbarium of Bowling Green State University collected by E. L. Moseley in 1895 at Cedar Point, Erie County.

10. Cyperus filiculmis Vahl

Perennial, with hard corm-like branches of the rhizome; culms smooth, slender, mostly 1–5 dm tall; leaves and bracts scabrous margined; spikelets few to many (2–16)–flowered; scales toward base of spikelet oblong (in half-view), obtuse, those near apex narrower and with awn-like tip. A variable species sometimes segregated into 3 varieties on a basis of density and size of heads, number of flowers in spikelet, length and texture of scales. However, specimens from the same county and even from the same collection show a wide range of variability of these characters. Although the extremes are different, the gradation from one to the other and the comparable geographic ranges make varietal designations undesirable. Common and widespread in a variety of situations, such as sandy soil, fields, dry rocky slopes and ridges, and creek-banks.

11. Cyperus refractus Engelm.

Perennial; rhizome with 1-few hard corm-like enlargements; culms and rays smooth, to 8 dm tall; leaves 4–10 mm wide, slightly roughened on margins; inflorescence open with 4–12 rays from 2–12 cm long; spikelets in loose cylindric heads, upper spikelets ascending, lower strongly deflexed; spikelets slender, acuminate, few-flowered; scales with tip of one reaching about to base of scale next above on same side; internodes of rachilla winged, the white hyaline wings partly enclosing linear or linear-oblong achene. Southern in range.

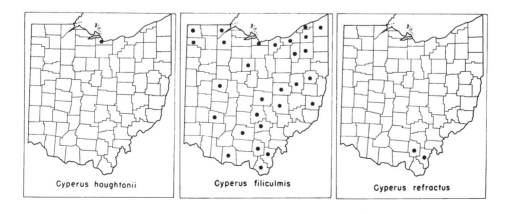

Cyperus houghtonii   Cyperus filiculmis   Cyperus refractus

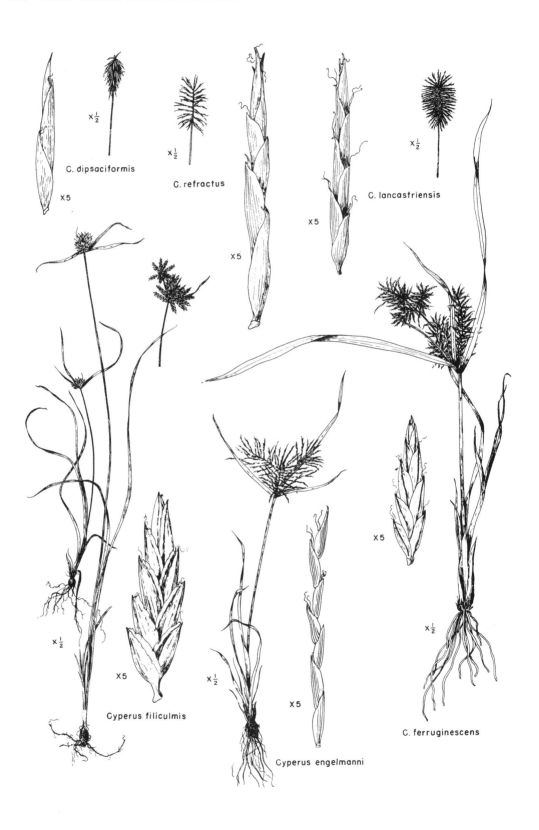

C. dipsaciformis

x ½

x5

C. refractus

x ½

C. lancastriensis

x ½

x5

x5

Cyperus filiculmis

x ½

x5

Cyperus engelmanni

x ½

x5

C. ferruginescens

x ½

x5

12. **Cyperus dipsaciformis** Fern.

C. retrofractus (L.) Torr., var. *dipsaciformis* (Fern.) Kukenth.

Perennial; rhizome with hard corm-like enlargements; culms rough; rays smooth, often 1.5 dm long, terminated by a long obconic spike; spikelets 6–11 mm long, slender, rigid, and sharp pointed; achenes linear, 3 mm long. *C. dipsaciformis* is similar to *C. plukenetii* Fern. (*C. retrofractus* of ed. 7, not L.), which has scabrous rays, smaller (3–8 mm) spikelets, and relatively broader and shorter spikes. Both species are southern in range.

13. **Cyperus lancastriensis** Porter

Perennial; rhizome with hard corm-like enlargements; culms and rays smooth; leaves and bracts 4–10 mm wide; rays mostly elongate, to 12 cm long; spikes dense, short-cylindric; spikelets radiating, pointed, soft; scales with tip reaching beyond base of scale next above; internodes of rachilla winged. Wet or dry meadows and open slopes; southern in range.

14. **Cyperus engelmanni** Steud.

Annual; distinguished from all other species with elongated axis of spikes (i.e., with internodes of rachis of spike distinctly visible) by the very slender or linear spikelets whose scales do not reach bases of those above on same side of rachilla; spikelets breaking off at base, and (at maturity of achenes) separating into one-seeded sections with the wings of the rachilla partially surrounding the achene. Wet, sandy, or mucky lake borders.

15. **Cyperus ferruginescens** Boeckl.

C. speciosus Vahl, in part; *C. speciosus* var. *ferruginescens* (Boeckl.) Britton; *C. ferax* of auth., not Richard; inc. in *C. odoratus* L. by Gleason (1952).

Annual; resembling *C. erythrorhizos*, but distinguished from it by spikelet and achene characters, and, less definitely, by color: spikelets readily breaking into short segments; achenes oblong, 1–1.5 mm long, brownish to golden; inflorescence usually duller and less reddish. Included in *C. odoratus* by Gleason, which species (fide Fernald) inhabits saline and brackish shores along the Atlantic Coast from tropical America to Massachusetts, and along the Pacific Coast to California, while *C. ferruginescens* is more interior in range.

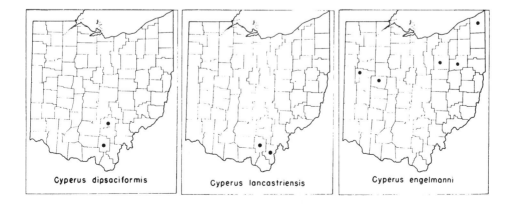

Cyperus dipsaciformis   Cyperus lancastriensis   Cyperus engelmanni

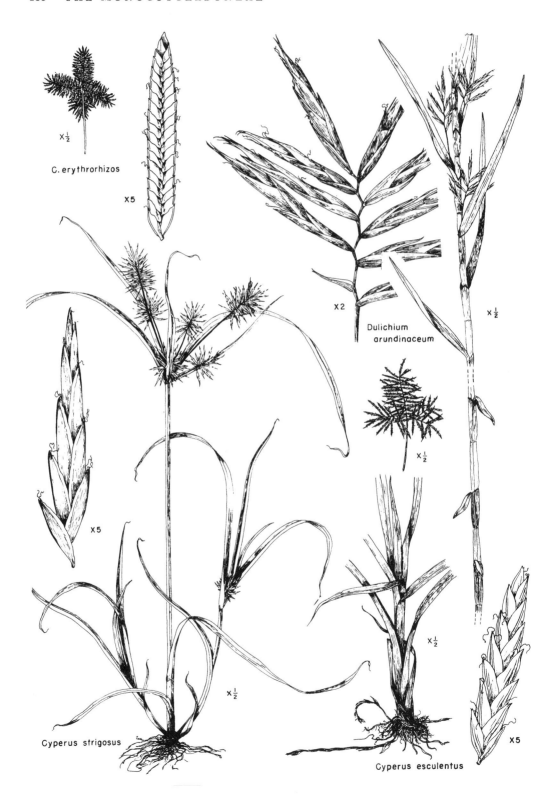

C. erythrorhizos

X½

X5

X2

Dulichium
arundinaceum

X½

X½

X5

X5

X½

X½

Cyperus strigosus

Cyperus esculentus

16. Cyperus erythrorhizos Muhl.

Annual, with red fibrous roots (hence the specific name); culms smooth, from a few cm to almost one m in height; basal leaves and involucral leaves long, paler on lower side, and rough-margined; leaf-sheaths purple at base; inflorescence of several sessile and several peduncled crowded groups of dense spikes; spikelets many-flowered, all or most flowers maturing achenes; scales beautifully imbricate, red-brown with green midrib; rachilla with pale chaff-like deciduous wings; achene small, less than 1 mm long and half as thick, pearly white. A wide-ranging species of alluvial soil, muddy stream and lake shores.

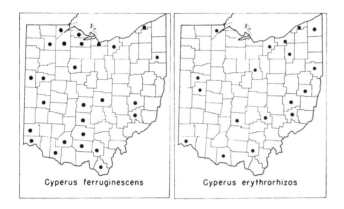

Cyperus ferruginescens   Cyperus erythrorhizos

17. Cyperus esculentus L.   YELLOW NUT-GRASS

Perennial, with numerous slender scaly stolons which ultimately terminate in small tubers; culms and leaves (except for scabrous margins toward apex) smooth; rays mostly short, scarcely exceeding the sessile spikes; spikelets yellowish to golden brown, numerous, widely spreading; scales strongly nerved, obtuse to acute, gradually deciduous from base toward apex; achenes oblong to narrowly obovoid, but often not developing. A common and widespread, almost cosmopolitan, species of damp soil; often a troublesome weed in cultivated fields and in lawns.

18. Cyperus strigosus L.

Our most common species, probably in every county, and ranging almost throughout the eastern two-thirds of the country and on the Pacific Coast in wet soil; perennial, with hard corm-like rhizome and smooth culm. Extremely variable, robust or dwarf; inflorescence with some sessile, some long-peduncled spikes; spikelets (deciduous in age) loosely or densely arranged on rachis, when crowded some spikelets are reflexed and plants then superficially resemble *C. lancastriensis*. Best distinguished by corm-like base, deciduous spikelets, golden to straw-color scales more than 3 mm long, with prominent green midrib (and nerves on sides as well).

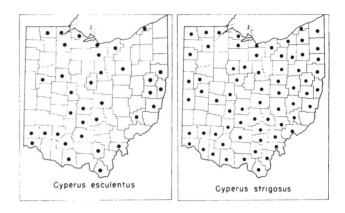

Cyperus esculentus   Cyperus strigosus

2. DULICHIUM Pers.

A monotypic genus.

1. **Dulichium arundinaceum** (L.) Britt.   THREE-WAY SEDGE

Distinguished in flower or fruit by the axillary inflorescences, sometimes little exserted from leaf-sheath, and 2-ranked spikelets; in vegetative condition by the very leafy terete and hollow stems, 3–10 dm tall, the short leaves 3-ranked and widely diverging, the lower with bladeless sheaths, the upper sheaths often overlapping. Spikelets linear, scales hyaline-margined, the lowest wrapping base of spikelet and wide at place of attachment to axis of inflorescence; perianth of 6–9 downwardly barbed bristles; achenes linear-oblong, tipped with persistent style-base. Swamps and margins of ponds, throughout much of temperate North America.

Dulichium arundinaceum

3. ELEOCHARIS R. Br.   SPIKE-RUSH

Perennial or annual plants of wet ground, swamps, or shallow water, some small and delicate with capillary culms but a few cm tall, others coarse, to a meter tall. Inflorescence a single terminal spikelet (spike), hence the common name; scales of spikelet imbricated, all except a few of the lowest fertile; leaves reduced to bladeless sheaths confined to lower part of stem; stems simple, terete, flattened, or rarely angled, rising from matted or creeping rhizomes, or if annual, without rhizomes. Flowers perfect, perianth consisting of 6 bristles (sometimes fewer or more, rarely lacking), stamens 1–3, pistil with 2- or 3-cleft style, its base enlarged and persistent (but jointed with achene), forming a definite tubercle; achene lenticular or 3-angled.

A genus of about 150 species, almost cosmopolitan in distribution, best represented in warmer regions. Stature of plants and thickness of stems, in fact, general aspect, vary greatly within a species, and make casual recognition of most species impossible. Well developed spikes and mature achenes are necessary for identification. Achenes of most of our species are small (0.7–1.5 mm long); those of three species (*quadrangulata, pauciflora, rostellata*) are larger (2–2.5 mm long). Our figures show these enlarged 15–20 times, and, with few exceptions, are adapted from illustrations in the papers cited below. Serious students of the genus should refer to monographic papers by Fernald and Brackett (1929) and by Svenson (1929, 1932, 1934, 1939, 1945) with illustrations and distribution maps. Most of our specimens were determined by H. K. Svenson.

a. Culms coarse, quadrangular, up to 1 m tall and 5 mm in diameter; spikelets little if any thicker than culm ....1. *E. quadrangulata*
aa. Culms slender (rarely stout in no. 5), terete, flattened, or rarely angled; spikelets thicker than culm just below them.
   b. Culms flattened or 2-edged.
      c. Culms strongly flattened, slightly spirally twisted; scales of spikelet (at least upper) acuminate, bifid; tubercle depressed-conic, clearly differentiated from achene ....2. *E. compressa*
      cc. Culms more or less flattened, the elongate outer ones often arched and rooting; scales of spikelet obtuse; tubercle pyramidal, nearly confluent with achene ....3. *E. rostellata*
      ccc. Culms 2-edged, often inrolled and concavo-convex, slightly spirally twisted; lower empty scales of spikelet much enlarged; tubercle obovoid, longitudinally 9-ribbed ....7. *E. wolfii*
   bb. Culms terete, grooved, or somewhat angled, not definitely flattened.
      c. Tubercle not clearly differentiated from achene, appearing to be a narrowed apex of different texture.
         d. Culms very slender, 1–3 (–4) dm tall, in small tufts; spikelets few-flowered (2–7 flowers), ovoid ....4. *E. pauciflora*
         dd. Culms wiry, somewhat flattened, 1–2 mm wide, 2–10 (–20) dm tall, the outer often arched and rooting ....3. *E. rostellata*
      cc. Tubercle distinctly differentiated from achene.
         d. Sheaths blackened at apex, with prominently darkened V-shaped sinus; culms wiry; spikelets ellipsoid-ovoid, acute; scales stiff, acuminate, their tips spreading ....5. *E. smallii*
         dd. Not as above.
            e. Achenes with distinct longitudinal ridges and minute cross ridges.
               f. Culms filiform ....6. *E. acicularis*
               ff. Culms 2-edged, inrolled, concavo-convex ....7. *E. wolfii*
            ee. Achenes without longitudinal ridges, their surfaces smooth or variously roughened.
               f. Cespitose annuals with fibrous roots.
                  g. Tubercle low-conic to cap-like, nearly the width of achene.
                     h. Tubercle broadly deltoid; bristles longer than achene and tubercle; spikelets broad-ovoid, obtuse; common and widespread ....8. *E. obtusa*

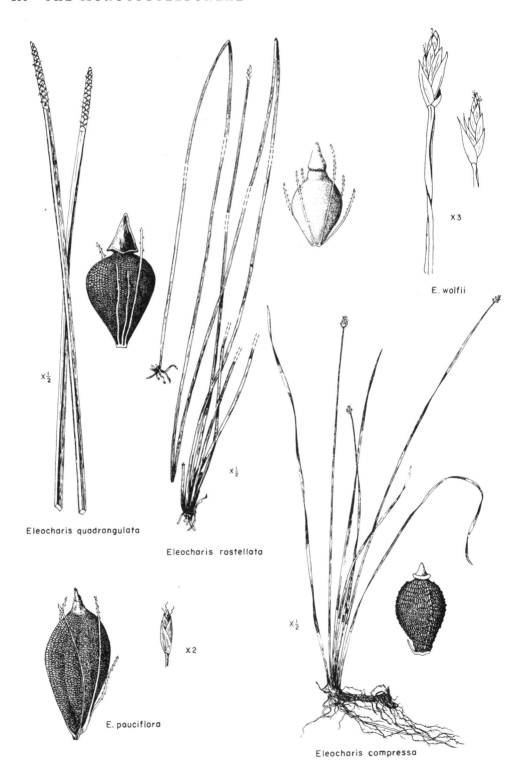

E. wolfii

×3

Eleocharis quadrangulata

×½

Eleocharis rostellata

×½

E. pauciflora

×2

Eleocharis compressa

×½

hh. Tubercle much flattened; bristles short or wanting:
   spikelets slender-cylindric, subacute ....9. *E. engelmanni*
gg. Tubercle conic-subulate, its base about ¼ width of achene;
   bristles unequal; spikelets narrow-ovoid, acute ....
                                                    10. *E. intermedia*
ff. Perennials, with slender or cord-like rhizomes or stolons;
   culms slender to almost filiform.
   g. Spikelets closely many-flowered, slender, cylindric-ovoid,
      8–15 mm long; tubercle conical; bristles delicate, 1–4 or
      wanting; sheaths red, tight ....11. *E. calva*
   gg. Spikelets loosely-flowered; 3–10 mm long.
       h. Achenes biconvex obovoid, lustrous; tubercle with
          flattened base and conic-subulate center; sheaths loose,
          their summits dilated, whitish scarious; culms com-
          pressed filiform ....12. *E. olivacea*
       hh. Achenes trigonous obovoid, coarsely to finely reticu-
           late or rugose; tubercle broad-based, pyramidal to
           depressed-deltoid; sheaths tight, reddish at base,
           truncate at apex, mucronate; culms angled.
           i. Culms 4–5-angled or corrugated; achenes drab
              or olivaceous; tubercle pyramidal or depressed
                                                    13. *E. tenuis*
           ii. Culms 6–8-angled; achenes bright yellow; tubercle
               much depressed, with central point ....
                                                    14. *E. elliptica*

1. **Eleocharis quadrangulata** (Michx.) R. & S.   FOUR-ANGLED SPIKE-RUSH
    E. mutata of Schaffner 1932, not *E. mutata* (L.) R. & S.
    A tall coarse species with sharply 4-angled culms to 1 m in height, growing
in shallow water of lake and pond margins. Basal sheaths brown to red, membranous; spikelet 2–5 (–6) cm long, slender, little if any wider than culm. Widely
distributed in eastern United States west to Wisconsin and Texas. Fernald (1950)
distinguished two varieties, the typical, Coastal Plain in distribution, and var.
crassior Fern., of more interior range which he distinguishes chiefly by its
somewhat larger size.

Members of the series to which this species belongs (Mutatae) occur chiefly
in warm regions; five are found in Florida. Most are coarse, stoloniferous
perennials; two species (from the Orient and from Australia) produce tubers
which are used for food.

2. **Eleocharis compressa** Sulliv.   FLAT-STEMMED SPIKE-RUSH
    E. acuminata of auth., not *Scirpus acuminatus* Muhl.
    One of our most easily recognized species, growing in wet calcareous soil;
the type specimen, on which the species is based, was collected about a century
ago by W. S. Sullivant, an Ohio botanist, in wet places in the Darby Plains,
15 miles west of Columbus. Achene golden-yellow to brown, granular roughened. The strongly flattened culms, which on living plants are slightly spiral,
and the acuminate, broadly white-scarious margined, often bifid, scales of the
spikelet are good recognition characters. Widely distributed in wet calcareous
clay, sand, or gravel of seepage spots and shores, from western Quebec to
Saskatchewan, and southward almost to the Gulf slope.

3.   **Eleocharis rostellata** Torr.

Culms slender, wiry and flattened, up to 1 or sometimes 2 m tall, the more elongate often arching and rooting upon contact with soil; sheaths rigid, straw-color to brown with dark upper margin; spikelets ovoid to fusiform, acute, many-flowered; scales firm, obtuse to acute toward tip of spikelet; tubercle confluent with achene, but differing in color and texture. A widespread species of southern latitudes, in saline marshes of the Atlantic Coastal Plain and locally, calcareous or marly swamps inland, north to Wisconsin, Indiana, and the Mohawk Valley of New York.

4.   **Eleocharis pauciflora** (Lightf.) Link, var. **fernaldii** Svenson

Small slender plants with soft culms, known in Ohio from a single collection from an open marl bog in Cedar Swamp, Champaign County. Spikelets small (4–6 mm long), few-flowered; scales brown with pale margins, the two lowest enlarged. A wide-ranging boreal species of alkaline soils; several varieties are recognized: the European var. *pauciflora*, our eastern American var. *fernaldii*, and others in western America and Asia.

Eleocharis quadrangulata
var. crassior

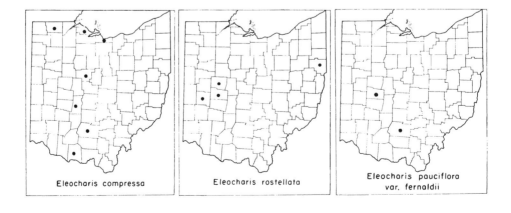

Eleocharis compressa   Eleocharis rostellata   Eleocharis pauciflora
                                                              var. fernaldii

5.   **Eleocharis smallii** Britt.

Tufted to loosely stoloniferous; culms wiry and firm, varying greatly in height and diameter of culms, the coarser plants resembling *E. palustris* (L.) R. & S. (not known in Ohio), the more slender ones resembling *E. calva*. Spikelet ellipsoid-ovoid, acute, scales acuminate, their tips spreading (in well developed spikes). "*E. smallii* is fairly easily recognized by its rigid texture, obvious in the stiff acuminate scales and the hardened character of the sheaths, which are usually black at the apex, with a prominently darkened V-shaped sinus" (Svenson, 1939). This species and *palustris* have 2 or 3 sterile scales at base of spikelet, but differ in shape of tubercle. Gleason (1952) considers *smallii* to be one of three segregates of *palustris*, but not specifically distinct.

6.   **Eleocharis acicularis** (L.) R. & S.   Needle Spike-rush

Our most slender species, culms filiform, 2–20 (–25) cm long, densely tufted and forming mats or carpets in mud in depressions or at margins of ponds; sheaths reddish toward base; spikelets ovate to lanceolate, 2–7 mm long, mostly 5–8-flowered; scales green with reddish brown sides and scarious margins; achenes about 1 mm long, with distinct longitudinal ridges and many minute transverse ridges. Wide-ranging in North America and Eurasia.

7.   **Eleocharis wolfii** Gray

A tufted perennial with slender rhizomes; culms loosely spiral, 3 (1.5–6) dm tall, 2-edged and in places inrolled so as to appear concavo-convex, the edges with minute tooth-like elevations (seen under high magnification), the surfaces slightly striate; sheaths scarious toward apex; spikelets 5–9 mm long, the lower empty scales elongate and acuminate, the fertile shorter, acute to obtuse, with scattered purple striae and scarious margins; achene obovoid, distinctly longitudinally ribbed (ribs 9) and minutely cross-barred; tubercle cap-like with apiculate center; bristles none. A rare species with less than a score of recorded stations widely scattered in the eastern two-thirds of the continent; known in Ohio from a single Ross County collection.

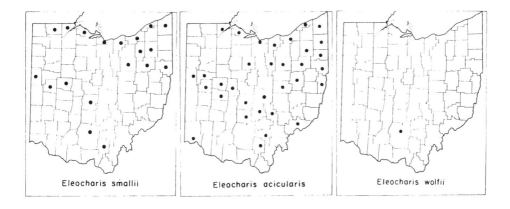

Eleocharis smallii   Eleocharis acicularis   Eleocharis wolfii

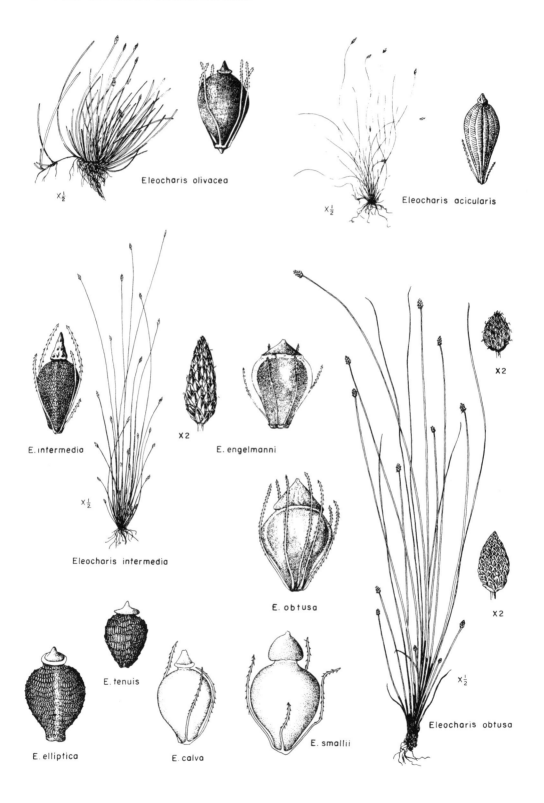

Eleocharis olivacea

$X\frac{1}{2}$

Eleocharis acicularis

$X\frac{1}{2}$

E. intermedia

$X\frac{1}{2}$

X2

E. engelmanni

Eleocharis intermedia

E. obtusa

X2

E. tenuis

E. elliptica

E. calva

E. smallii

X2

$X\frac{1}{2}$

Eleocharis obtusa

8. **Eleocharis obtusa** (Willd.) Schultes. BLUNT SPIKE-RUSH

Our commonest species, growing in wet mud of pond margins, ditches, or seepage areas. Annual, culms slender (to 1.5 mm thick), usually erect, densely tufted, most often 2–3 dm long, sometimes longer (to 5–7 dm) or shorter (only 3–4 cm, then spreading); sheaths with firm subtruncate to oblique apex, short-pointed; spikelets globose-ovoid to ovoid-cylindric, obtuse, many-flowered; achene lustrous, pale to deep brown, the broadly deltoid tubercle distinctive. Common and widely distributed in suitable habitats throughout the Dedicuous Forest region and in the Pacific Northwest and the Hawaiian Islands.

9. **Eleocharis engelmanni** Steud.

Doubtfully present in Ohio: a single collection from Montgomery County (1901) is determined by Svenson as "approaching" this species. Similar to *E. obtusa,* and distinguished from it chiefly by its more strongly depressed tubercle, and shorter (or rudimentary or wanting) bristles.

10. **Eleocharis intermedia** (Muhl.) Schultes. MATTED SPIKE-RUSH
E. reclinata Kunth

Culms tufted, capillary, weak and arching or reclining, very unequal in length, from 2 cm to 4 dm long; sheaths pale, apex spreading; spikelets ovoid to cylindric-ovoid, acute; tubercle slender-conic. Ranging in mid-latitudes west to Iowa and Minnesota, usually in calcareous areas.

11. **Eleocharis calva** Torr. CREEPING SPIKE-RUSH
E. palustris (L.) R. & S., var. *glaucescens* of Amer. auth. not Willd.; *E. erythropoda* Steud.

Perennial, forming mats or tufts, rhizomes slender, reddish; culms very slender, 1–5 (–7) dm tall; lower sheath and often part of longer upper sheath red; spikelets slender, cylindric-ovoid, basal scale rounded, encircling base of spikelet; conical tubercle spongy; bristles delicate, 1–4, or wanting. The red sheaths, rounded basal scale of spikelet and conical spongy tubercle distinguish this species. Wide-ranging through much of temperate eastern North America except the southern tier of states. In Ohio, most frequent in the calcareous section of the state.

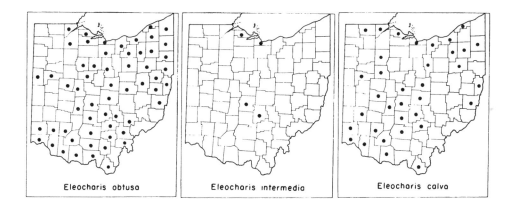

Eleocharis obtusa   Eleocharis intermedia   Eleocharis calva

12.   **Eleocharis olivacea** (Poir.) Urban.   OLIVACEOUS SPIKE-RUSH

E. flavescens (Poir.) Urban, var. *olivacea* (Torr.) Gl.

A low species with culms 2–15 cm long (occasionally to 3 dm), chiefly of the north Atlantic Coastal Plain and Great Lakes–Mohawk area. Cited by Svenson (1929) from Cleveland, Cuyahoga County; one specimen in Ohio herbaria, apparently from the same collection. Tufted, and with slender rootstocks; culms grooved; upper part of sheaths pale, scarious and dilated; spikelets loosely flowered, scales obtuse.

13.   **Eleocharis tenuis** (Willd.) Schultes.   SLENDER SPIKE-RUSH

Including *E. capitata* (L.) Br. var. *typica*, var. *pseudoptera* Weatherby, and var. *verrucosa* Svenson.

Culms very slender, 0.5–6 dm tall, 4–5-angled, arising in small or large tufts from reddish cord-like rhizomes; sheaths obliquely truncate, mucronate; spikelets ellipsoid to ovoid, blunt or acute; scales ovate, rounded to acute, reddish brown with green midrib area and scarious margins. A variable species made up of three intergrading varieties, of which two, the typical and var. *verrucosa*, occur in Ohio; the latter more western in range (chiefly central Mississippi Valley), the former more eastern (mid-Appalachian and coastal); var. *verrucosa* has flatter tubercle and rougher achene.

14.   **Eleocharis elliptica** Kunth

E. tenuis var. *borealis* (Svenson) Gl.; *E. capitata* (L.) Br. var. *borealis* Svenson

Similar to *E. tenuis*, and sometimes not considered a distinct species; culms slender, 6–8-angled (instead of 4–5-angled); achene with prominent transverse bands; tubercle much depressed, apiculate. More northern in range than *E. tenuis*, entirely to the north of the Ohio and Potomac rivers.

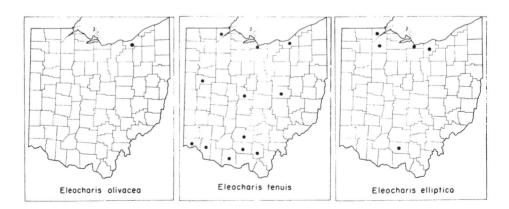

Eleocharis olivacea   Eleocharis tenuis   Eleocharis elliptica

4.   BULBOSTYLIS (Kunth) C. B. Clarke

Stenophyllus Raf.

Delicate tufted annuals, 3–30 cm tall; leaves linear-filiform, sheaths ciliate

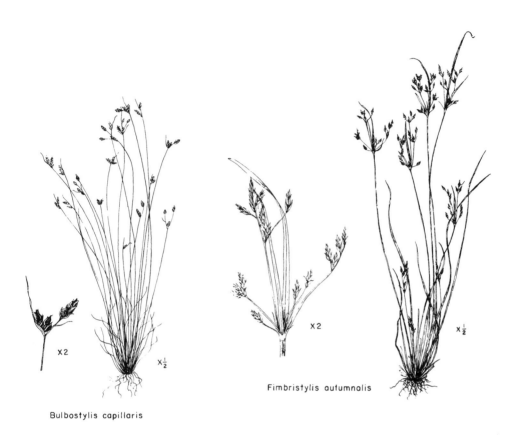

Bulbostylis capillaris

Fimbristylis autumnalis

or long-pilose at upper end; inflorescence of a few dark spikelets, terminal on capillary culms and often also on very short stems among the basal leaves, the filiform bracts dilated at base; flowers perfect, stamens 2 or 3, style 2- or 3-cleft; achene 3-angled obovoid, capped with minute tubercle. A genus of about 70 species, chiefly tropical; one species in Ohio.

1. **Bulbostylis capillaris** (L.) C. B. Clarke
   Stenophyllus capillaris (L.) Britt.

Recognized by its dark purplish brown spikelets with comparatively large obtuse scales with pale midrib, its filiform leaves, much shorter than culm, bracts minutely roughened or serrulate on the margins, and the pilose sheath-orifice. Achenes straw-colored, 3-angled, 1 mm or less long, with minute dark tubercle. A delicate annual, usually in sandy soil or rock crevices. Three varieties, two of which (var. *capillaris* and var. *crebra* Fern.) may occur in our range, are sometimes segregated on a basis of presence or absence of basal spikelets, and comparative length of spikelet pedicels. Most of our specimens can be referred to var. *crebra.* Somewhat similar in appearance to *Fimbristylis,* and distinguished in the field by its filiform culms and leaves, and obtuse spikelet scales.

5.   FIMBRISTYLIS Vahl

A large genus of warm regions, with few northern representatives; ours annual. Similar to *Bulbostylis,* but achenes without tubercles, i.e., without persistent style bases; base of style often dilated, but deciduous.

1.   **Fimbristylis autumnalis** (L.) R. & S.

Tufted annual with slender flattened culms, erect or spreading, leafy toward base; leaves flat, 1–2 mm wide; inflorescence simple or compound, the spikelets mostly sessile or on peduncles which may be again branched; spikelets 3–10 mm long, ovoid to narrow-cylindric; scales not closely appressed, tips free, the lower more slenderly pointed than the upper; achene 3-angled, obovoid. Plants with mostly compound inflorescence and slender spikelets are sometimes segregated as var. *mucronulata* (Michx.) Fern. (*F. autumnalis* of ed. 7 of Gray's Manual) and those with simpler inflorescence and ovoid spikelets referred to the typical var. *autumnalis* (*F. frankii* Steud., *F. geminata* Kunth); however, these varieties intergrade, and are not distinguished on our map.

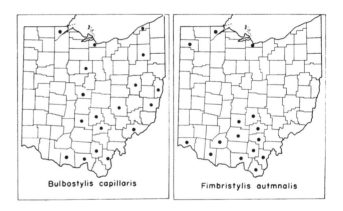

Bulbostylis capillaris

Fimbristylis autmnalis

6.   SCIRPUS L.   Bulrush

A worldwide genus of about 200 species, with no unifying aspect characters, although such characters are very useful in the identification of species or groups of species. Culms hard or soft, cylindric or 3-angled, leafy or with all or most of the leaves reduced to bladeless sheaths; inflorescence terminal, in some species only a solitary spikelet, in some, a few crowded sessile spikelets, in others, peduncled spikelets or peduncles again branched, the whole from a few mm to 2–3 dm long; inflorescence in most species subtended by a leafy bract or bracts (the involucre), when only one, appearing as a continuation of the culm, the inflorescence then appearing lateral. Spikelet, flower, and achene characters distinguish the genus. Spikelets few- to many-flowered, scales imbricate (except 1 or 2 of the lowest), each flower in axil of a scale; flowers perfect, perianth of 1–6 bristles (rarely none, or more), stamens 2–3, style 2–3-cleft, deciduous, or deciduous above base and leaving a point or tip (not tubercle) on achene; achene lenticular, plano-convex, or 3-angled.

With few exceptions, the Ohio species of *Scirpus* inhabit wet soil or shallow water.

a. Inflorescence a solitary terminal spikelet about 5 mm long, without true involucre, the basal scale of spikelet enlarged, long-awned, equaling or exceeding spikelet ....
                                            1. *S. verecundus*
aa. Inflorescence with one or more foliaceous or stem-like bracts.
    b. Inflorescence appearing lateral (unless bract is short); involucre a single bract (rarely a small secondary bract) similar in appearance to culm; culms leafless, or leafy only at base.
        c. Spikelets sessile or nearly so, compactly grouped.
            d. Cespitose annuals with slender terete or somewhat angled culms usually less than 5 dm tall.
                e. Achene transversely wrinkled, strongly 3-angled; style 3-cleft ....
                                  2. *S. saximontanus*
                ee. Achene smooth or more or less pitted, black, not 3-angled.
                    f. Achene smooth and lustrous, plano-convex; style 2-cleft; bristles occasionally fully developed, usually shorter than achene; culms terete ....3. *S. smithii*
                    ff. Achene more or less pitted, unequally biconvex; style 2–3-cleft; bristles usually fully developed and longer than achene; culm obtusely triangular ....4. *S. purshianus*
            dd. Rhizomatous perennials with sharply 3-angled culms usually more than 5 dm tall.
                e. Achene plano-convex, short-pointed, style 2-cleft.
                    f. Involucral bract 3–15 cm long, acute; scales of spikelet notched or 2-cleft at apex, the midvein distinctly excurrent ....
                                    5. *S. americanus*
                    ff. Involucral bract 1–4 cm long, blunt; scales of spikelet rounded at apex, minutely mucronulate ....6. *S. olneyi*
                ee. Achene unequally 3-angled, long-pointed; style 3-cleft; involucral bract 4–15 cm long, tip rounded, oblique, blunt; scales of spikelet ovate, acute, or mucronulate ....7. *S. torreyi*
        cc. Spikelets both sessile and peduncled; inflorescence usually open; rhizomatous perennials with terete culms up to 2–3 m tall; sheaths often without blades; achene thick plano-convex, style 2-cleft.
            d. Culms soft, thick, easily compressed; spikelets ovoid to linear cylindric, 5–9 (–15) mm long; scales round-ovate with prominent green midrib, glabrous except at tip, as long as or slightly shorter than achenes ....
                                    8. *S. validus*
            dd. Culms harder, stiff and firm; spikelets acute, slender-ovoid to cylindric, 10–20 mm long; scales oblong-ovate, dotted and streaked with red, short viscid-pubescent, longer than achenes ....9. *S. acutus*
bb. Inflorescence obviously terminal; involucre of 2 or more flat leaf-like bracts.
    c. Spikelets large, 1.5–4 cm long; scales long-awned; inflorescence of sessile and peduncled clusters of spikelets; culms stout, sharply 3-angled, leafy ....
                                    10. *S. fluviatilis*
    cc. Spikelets smaller, usually less than 1 cm long; scales not long-awned, sometimes mucronate; inflorescence large and much branched, to 2–3 dm long or wide.
        d. Involucral bracts shorter than rays of inflorescence; rays slender, nodding at tip; spikelets sessile and pedicelled, cylindric (rarely ovoid), usually 6–12 mm long, 3–4 times as long as wide; culms terete ....
                                  11. *S. lineatus*
        dd. Involucral bracts (or some of them) longer than inflorescence; spikelets less than 3 times as long as wide.

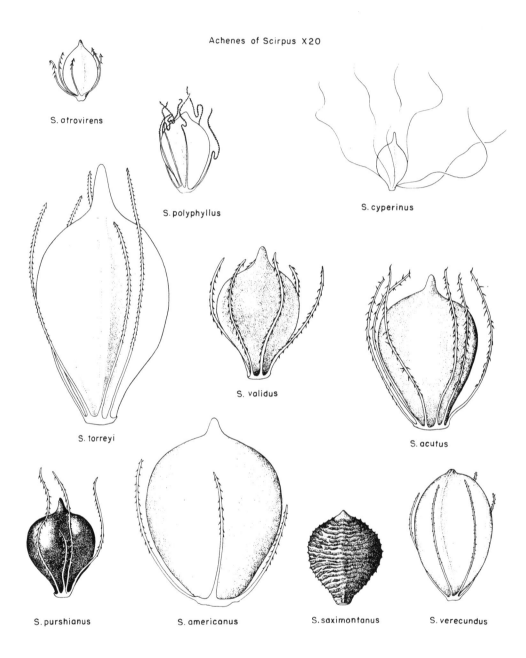

Achenes of Scirpus X20

S. atrovirens

S. polyphyllus

S. cyperinus

S. torreyi

S. validus

S. acutus

S. purshianus

S. americanus

S. saximontanus

S. verecundus

e. Bristles shorter than to somewhat longer than achene, or wanting; spikelets sessile and crowded into small glomerules; culms 3-angled.
    f. Culms stout, about 1 cm thick near base; lower leaf-sheaths reddish, septate-nodulose when dry; achenes brownish ....
                                                    12. *S. expansus*
    ff. Culms more slender; leaf-sheaths green; achenes pale.
        g. Leaves mostly on lower half of culm, inflorescence thus long-exserted; sheaths often septate-nodulose, sheath orifice (ligule) V-shaped, friable; scales dark, ovate to elliptic with green midrib excurrent, longer than achenes; bristles straight or nearly so, sometimes wanting ....13. *S. atrovirens*
        gg. Very leafy, leaves crowded near middle of culm, more or less in 2 vertical ranks; sheath-orifice firm, truncate or upwardly curved; scales rust-color, rotund, little longer than achenes; bristles contorted above middle ....
                                                    14. *S. polyphyllus*
ee. Bristles several times length of achene, very curly, at maturity almost hiding scales of spikelets; spikelets sessile or some pedicellate; culms terete. The Wool-grasses ....
                                        15. *S. cyperinus* and allies (16, 17)

1. **Scirpus verecundus** Fern.
    S. planifolius Muhl.
    A woodland species with the aspect of an *Eleocharis*, but readily distinguished by the abundant flat leaves and by spikelet and achene characters. Distinguished from all other Ohio species of *Scirpus* by the solitary terminal spikelet subtended only by the much enlarged long-awned lowest scale of spikelet. Scales of spikelet shading from whitish at base to golden brown toward tip, the midrib green and excurrent as a short awn or point. Cespitose perennial 1–4 dm tall, the lower leaves reduced to bladeless sheaths, upper equaling or exceeding the 3-angled scabrous culms. Ranging from the middle Atlantic slope westward to Ohio, Kentucky, and Missouri.

2. **Scirpus saximontanus** Fern.
    A rare and local species of pond margins of the Great Plains; known in Ohio only from the borders of a farm pond in Pickaway County, where it may have been introduced from the West (specimens determined by M. L. Fernald in

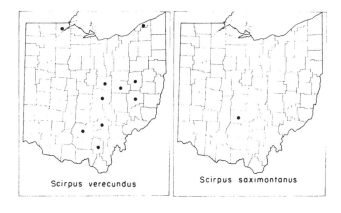

Scirpus verecundus   Scirpus saximontanus

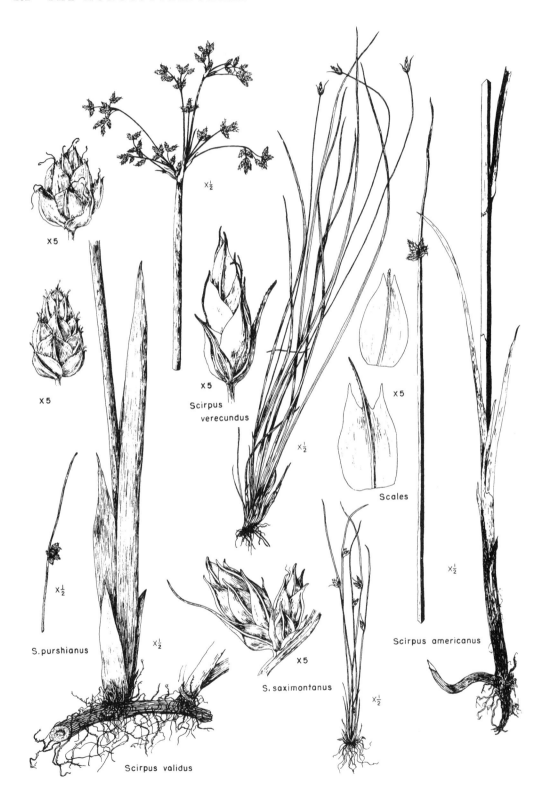

X5

X5

X $\frac{1}{2}$

X5

Scirpus
verecundus

X $\frac{1}{2}$

X $\frac{1}{2}$

Scales

X5

S. purshianus

X $\frac{1}{2}$

X $\frac{1}{2}$

X5

S. saximontanus

X $\frac{1}{2}$

X $\frac{1}{2}$

Scirpus americanus

Scirpus validus

1937). Similar to S. *hallii* Gray, but differing in its strongly 3-angled achene and 3-cleft style.

3. **Scirpus smithii** Gray

A northern species of sandy or muddy shores, reported for northern Ohio by Fernald (1950). Very similar to the next species, to which our Ohio specimens seem to be referable.

4. **Scirpus purshianus** Fern.

S. *debilis* Pursh; S. *smithii* var. *williamsii* (Fern.) Beetle

Similar to S. *smithii* and considered by Gleason to be a variety of that species (see distinguishing characters in key); somewhat more southern in range. Cespitose annual, the 1-several brownish or straw-colored ovoid spikelets closely aggregated and appearing lateral; achenes dark, glossy, obscurely pitted. Muddy or sandy shores and ditches, widely scattered, but infrequent.

5. **Scirpus americanus** Pers. THREE-SQUARE

One of our most easily recognized species; a tall sedge, sometimes more than 1 m high, with sharply triangular culms; a few leaves toward base; the ovoid, acute spikelets 1–2 cm long, bright red-brown, when several appearing to radiate from the *very* short common peduncle which seems to emerge from a slit in culm, the single long involucral bract open for 1–2 cm, then triangular like culm; scales 2-cleft at apex and mucronate; achenes plano-convex, short-tipped. Widely distributed (in various forms) from Newfoundland and Florida across the continent.

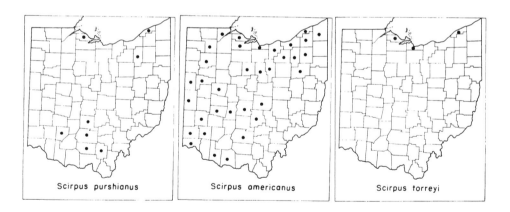

Scirpus purshianus   Scirpus americanus   Scirpus torreyi

6. **Scirpus olneyi** Gray

Somewhat larger and coarser than the preceding, and usually confined to saline and brackish marshes of the Atlantic and Pacific coasts, and saline inland areas. No Ohio specimens seen, but credited to northern Ohio by Fernald (1950).

7. **Scirpus torreyi** Olney

A rare and local swamp species of northern range, resembling S. *americanus* in general appearance; distinguished by the acute or mucronulate scales (not 2-cleft), 3-cleft style, and long-pointed unequally triangular achene.

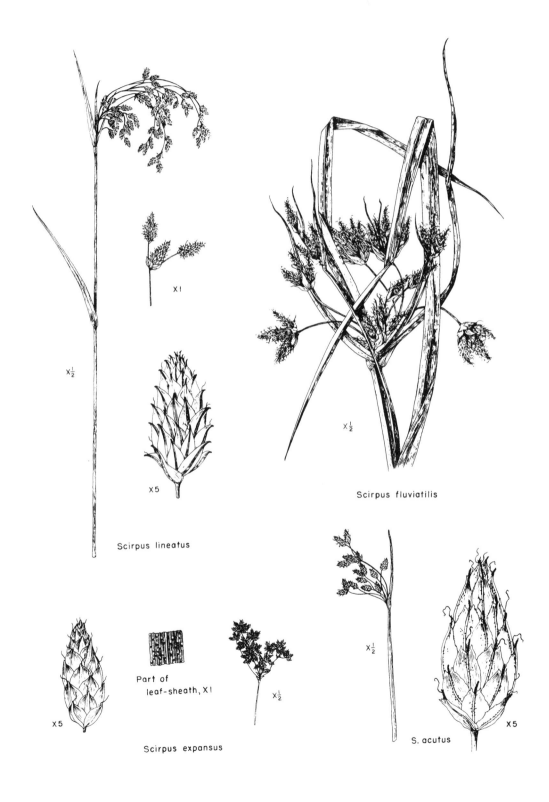

X 1

X½

X 5

Scirpus lineatus

X½

Scirpus fluviatilis

X 5

Part of
leaf-sheath, X 1

X½

X½

X 5

Scirpus expansus

S. acutus

X 5

8. **Scirpus validus** Vahl, var. **creber** Fern. GREAT BULRUSH. SOFT-STEM
BULRUSH

S. lacustris of Amer. author., not L.

A tall bulrush, occasionally 3 m, more often 1–2 m tall, with thick horizontal rhizomes; culms light green, soft (easily compressed), internally white, loosely cellular; inflorescence appearing lateral (unless bract is old and short), of few to several more or less lax branches, the ovoid spikelets (linear-cylindric in forma *megastachyus* Fern.) solitary and in clusters at summit of primary rays and branches; spikelets 5–10 mm long, the broad-ovate golden-brown scales with green midrib, pubescent at apex and on excurrent midrib; mature achenes thick plano-convex, scarcely hidden by scales. A very common and widespread species, essentially transcontinental in range, the typical variety in tropical America. In swamps and shallow water.

9. **Scirpus acutus** Muhl. HARD-STEM BULRUSH

S. occidentalis (S. Wats.) Chase

Similar to the last but with harder culms; distinguished by the usually stiffer inflorescence, darker and more pointed spikelets, lacerate bractlet-tips, red spots, streaks, and viscid globules on bractlets and scales, viscid-pubescent scales (with inconspicuous midrib) much longer than achene. Several Ohio specimens appear to be intermediate between *acutus* and *validus,* suggesting the possibility of hybrids. Somewhat less widespread than the last, and less frequent in Ohio; in similar habitats.

10. **Scirpus fluviatilis** (Torr.) Gray. RIVER BULRUSH

Our largest species; culms 1–2 m tall, stout, sharply 3-angled (one surface to over 1 cm wide), leafy almost to summit; leaves 1–2 cm wide; involucre of 3–5 elongate bracts to 1 cm wide and longer than inflorescence; inflorescence with some sessile spikelets, and others in clusters or solitary at tips of flattened spreading or somewhat drooping peduncles. Spikelets large, 1.5–4 cm long, brown, appearing shaggy because of long curving awns of scales; bristles 6, stiff and backwardly barbed, about equal to the sharply 3-angled whitish to dull brown achene. In wet soil, ditches, ponds, lake shores, and river-bottoms; wide ranging.

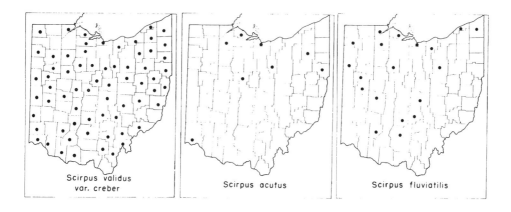

Scirpus validus
var. creber

Scirpus acutus

Scirpus fluviatilis

X½

X5

X2

Scirpus polyphyllus

X½

X5

X2

X½

Scirpus atrovirens

11. **Scirpus lineatus** Michx.

In the field, recognized by the narrowly ellipsoid to cylindric spikelets mostly 7–10 (5–15) mm long, terminal clusters with 1 sessile and 2–3 pedicelled spikelets, slender rays and pedicels, the drooping ultimate branches of the inflorescence (and upper parts of longer rays). Culms slender, often 1 m tall, leaves rather widely spaced, narrow; involucral bracts much shorter than inflorescence; scales light brown to reddish with green midrib and ascending to somewhat spreading mucronate tips; bristles brown, curly, about twice as long as the pale obscurely 3-angled achene, not projecting far beyond scales when mature. In ditches, seepage spots, swamps, and wet ground along small streams; common and widespread.

12. **Scirpus expansus** Fern.

S. sylvaticus of Amer. auth., not L.

One of the several leafy-bracted species with large much-branched inflorescence. Rare and local in northern Ohio. Distinguished in the field by the culm (usually) scabrous in upper 1–5 cm, red tinge of lower sheaths, and wide leaves (1–2 cm) scabrous beneath. Spikelets ovoid, 3–5 mm long, in glomerules; scales greenish (dark with age) with broad green midrib, the keel forming a projecting subulate tip; achene flattened 3-angled. Ranging from Maine to New York and Ohio, south to Georgia.

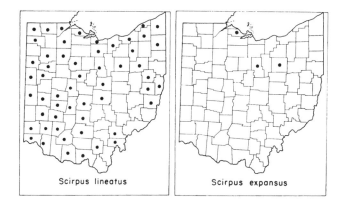

Scirpus lineatus   Scirpus expansus

13. **Scirpus atrovirens** Willd.

Our most common and widespread species, probably occurring in every county in Ohio. Spikelets very numerous, small (2–8 mm long), dark greenish brown, densely crowded into globose heads, some on very short peduncles crowded at base of inflorescence, some elevated on elongate and branched rays; scales with green excurrent midrib; achenes pale; leaves mostly about 1 cm (0.5–2) wide, margins scabrous, more crowded on lower half of culm, more widely spaced above. In autumn, sometimes producing leafy tufts in the inflorescence.

Several varieties are distinguished, of which two, *atrovirens* and *georgianus* (Harper) Fern., with similar range, occur in Ohio. The typical variety is said to have the sheaths and lower part of blades strongly nodulose-septate when dry, leaf-ribs 3–4 per mm, principal glomerules more than 7 mm in diam., and bristles about equalling achene; sheaths not septate in var. *georgianus*, leaf-ribs 5–6 per mm, principal glomerules less than 7 mm in diam. and bristles short or wanting; leaves of var. *atrovirens* average wider, and spikelets larger than those of var. *georgianus*; the latter has longer rays, some as long as 12 cm. Although the extremes are distinct, these varieties intergrade, and many Ohio specimens are intermediate. The occurrence of good examples of var. *georgianus* is shown on a map. Some, with fewer ribs per mm, do not have sheaths septate when dry; others with septate sheaths may have 6 ribs per mm; in a single inflorescence, bristles may be wanting, short, or as long or longer than achene.

14. Scirpus polyphyllus Vahl

Somewhat similar to *S. atrovirens*, but more leafy, with shorter internodes; spikelets red-brown or rust-color, 2–3.5 mm long, scales broad in relation to length, scarcely covering mature achenes, mucronate, with green midrib; bristles about twice length of achene, contorted in upper half. Like the preceding, often proliferating in autumn. A mid-southern species ranging from Massachusetts and New York to Ohio and Illinois, south to Georgia and Tennessee; in wet woods and swamps.

Scirpus atrovirens

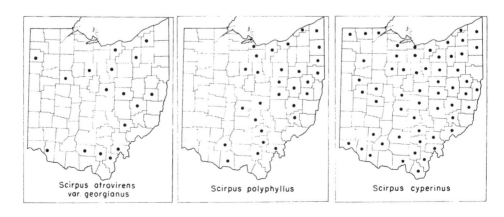

Scirpus atrovirens
var. georgianus

Scirpus polyphyllus

Scirpus cyperinus

15. **Scirpus cyperinus** (L.) Kunth. Wool-grass

Tall perennial with slender, nearly terete culms, and long narrow leaves; inflorescence large, its rays slender, capillary branches upwardly scabrous; bases of leaf-like bracts of involucre (of typical variety) red-brown, lance-attenuate bracts of involucels wholly red-brown; scales of spikelets red-brown, soft. The var. *pelius* Fern. lacks most of the reddish coloration; its involucels are drab to blackish. At maturity, the greatly elongated crisped and flexuous slender bristles give a woolly appearance to the entire inflorescence.

S. *cyperinus*, inclusive, may be considered as representative of the long-bristle section—*Trichophorum*, the wool-grasses—of *Scirpus*. Included in the Ohio flora, are three "species" of this section, distinguished from one another by quantitative (not morphologic) characters: pigmentation of bracts of involucre, involucels, and scales; relative numbers of sessile and pedicellate spikelets; and location of roughened angles of rays (upper or lower part). These species are separated in our manuals as follows:

a. Spikelets mostly sessile, in glomerules of 3–15 .... S. *cyperinus*
   b. Involucres, involucels, and scales reddish brown; bristles rust-color ....var. *cyperinus*
   bb. Involucres, involucels, and scales drab to blackish, without reddish color; bristles drab or smoky .... var. *pelius*
aa. Spikelets mostly pedicellate, usually in small clusters with the central one sessile and the others on pedicels of variable length.
   b. Involucels and scales red-brown; bristles slightly paler than scales; longer rays scabrous except at base; "lateral spikelets on long pedicels" ....16. S. *rubricosus*
   bb. Involucels and scales dull brown or drab; bristles whitish brown; longer rays scabrous only at tip; "lateral spikelets of each group pedicelled" ....
                                              17. S. *pedicellatus*

Examination of approximately 100 Ohio specimens shows that all of these characters (and others) vary. That there is variation in color of *cyperinus* (in restricted sense) is evident in the named var. *pelius* Fern., which has involucels "without red tinge" and scales "with little or no red." The proportion of pedicellate spikelets and length of pedicels vary; the scabrous ray character frequently does not go along with color characters; rays may be scabrous along one angle, smooth on others. And, color may deteriorate with age of specimen; many of the oldest specimens (60 to 75 years) are noticeably duller than recent collections. Inflorescences vary greatly in density. Four specimens in the Oberlin College Herbarium, collected on the same date in the same swamp, show gradation from fairly typical S. *cyperinus* to S. *pedicellatus*. There is little reason to believe that two distinct species are represented here; rather, the collections show variation within a population. Only by intensive study of populations in the field, can the problem of variation or speciation in these "species" be solved. Gleason (1952) regards S. *cyperinus* "as a single polymorphic species." Because of such variation and intergrading, these varieties and so-called species are not separated on the distribution map except that a few Ohio specimens with many distinctly pedicellate spikelets and little reddish color may be referable to S. *pedicellatus*. The occurrence of these is shown on a separate map. Some may prefer to assign a few southern Ohio specimens to S. *rubricosus* (especially one from Jackson County).

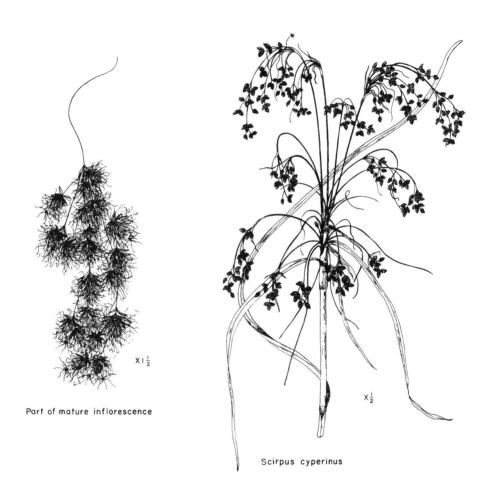

X 1½

Part of mature inflorescence

X ½

Scirpus cyperinus

The ranges of the three "species" overlap; in general, S. *pedicellatus* is more northern, S. *rubricosus* more southern than S. *cyperinus*, which ranges from New England and New York westward across Ohio to Iowa, and southward to North Carolina and Oklahoma.

16. **Scirpus rubricosus** Fern.

S. *eriophorum* Michx.

Doubtfully distinct from S. *cyperinus*; segregated on basis of the variable characters in key above. Ranging from Florida to east Texas and Mexico, northward in the interior to Missouri and Indiana, and in the East to Massachusetts.

17. **Scirpus pedicellatus** Fern.

A northern species of the *cyperinus* complex, distinguished, as is the preceding, on color, roughness of rays, and pedicel-length (see key under S. *cyperinus*). Some specimens appear to have pedicellate spikelets only because lower scales have fallen; others have rays scabrous almost the entire length. Ranging from the Gaspé Peninsula, Quebec, westward to Minnesota and southward into our area.

Scirpus pedicellatus

7. ERIOPHORUM L.   COTTON-GRASS.   BOG-COTTON

A small genus of arctic and north temperate latitudes; Ohio species confined to the northern part of the state and to relic bogs southward, all localities within the area of the last (Late Wisconsin) glacial advance. Spikelets many-flowered, solitary and terminal without foliaceous bracts (the more northern species) or 2-several in a leafy-bracted simple or branched inflorescence (Ohio species); scales membranous, several of the lowest empty; flowers perfect, with perianth of 6 silky bristles, each cleft almost to base into 4–6 parts, thus appearing numerous; bristles greatly elongating in fruit, obscuring all except lowest scales of spikelets, and producing large white to tawny silky-cottony heads which are often a conspicuous feature of swampy margins of subalpine lakes and of open taiga and tundra. Inflorescence at flowering time very different in appearance than at maturity because the bristles have not elongated and the scales are plainly visible. After fall of scales and florets, the deeply pitted rachilla remains.

a. Foliaceous involucral bract 1, erect, short (1–2 cm); leaves 1–2 mm wide, triangular-channeled, blunt, the upper with blade shorter than sheath; inflorescence of 2–5 spikelets; peduncles slender, minutely pubescent ....1. *E. gracile*
aa. Foliaceous involucral bracts 2 or more; leaves 1.5–6 mm wide, flat at least in lower half.
    b. Spikelets 3–30, mostly on slender peduncles of unequal length; midrib of scales prominent, extending to tip, or prolonged in outer scales ....
                     2. *E. viride-carinatum*
    bb. Spikelets several, on short peduncles, thus crowded into a dense head; scales with several prominent ribs, midrib scarcely reaching tip ....3. *E. virginicum*

1.   **Eriophorum gracile** W. D. J. Koch

A circumboreal species growing in bogs and swamps from Newfoundland and Labrador to southern Alaska, southward into our area, where it is represented by a single collection from a bog in Franklin County. Distinguished from

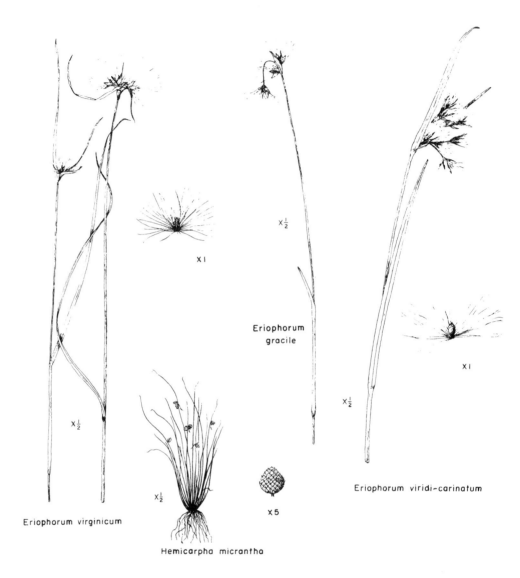

X I

X ½

Eriophorum
gracile

X ½

X ½

X I

Eriophorum viridi-carinatum

X ½

X 5

Eriophorum virginicum

Hemicarpha micrantha

our other species by its weak slender culm, very short culm leaves (much shorter than sheath), and short involucral bracts darkened at base, only one of which has a foliaceous tip.

2. **Eriophorum viride-carinatum** (Engelm.) Fern.

E. polystachion of Amer. auth.

Transcontinental, from Newfoundland and Labrador to southern Alaska and British Columbia, south into the northern tier of states; in bogs and wet meadows. Differs from the preceding species in its stiffer, angled culm, elongate culm leaves, flat except toward tip, and longer foliaceous bracts; from the following species by its more open inflorescence (most of the spikelets on slender peduncles), spikelet-scales with darkened sides and prominent green midrib reaching to apex or beyond, and paler, usually white, bristles.

3. **Eriophorum virginicum** L.   TAWNY COTTON-GRASS

Distinguished by its crowded spikelets forming a dense head 2–6 cm in diam. in fruit, tawny to coppery bristles, greenish to straw-color scales with several strong ribs, bracts divergent to somewhat reflexed in fruit and much longer than inflorescence. An abundant species of bogs and swamps, from Labrador to Ontario and Manitoba, and far southward in the mountains.

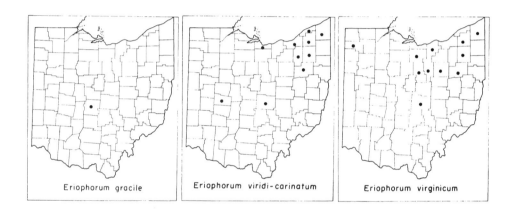

Eriophorum gracile   Eriophorum viridi-carinatum   Eriophorum virginicum

Hemicarpha micrantha

8.   HEMICARPHA Nees

A small genus of the tropics with 2 or 3 species ranging northward in the United States. Flowers and spikelets as in *Scirpus*, from which this genus is distinguished by the presence of a minute translucent scale between axis of spikelet and flower.

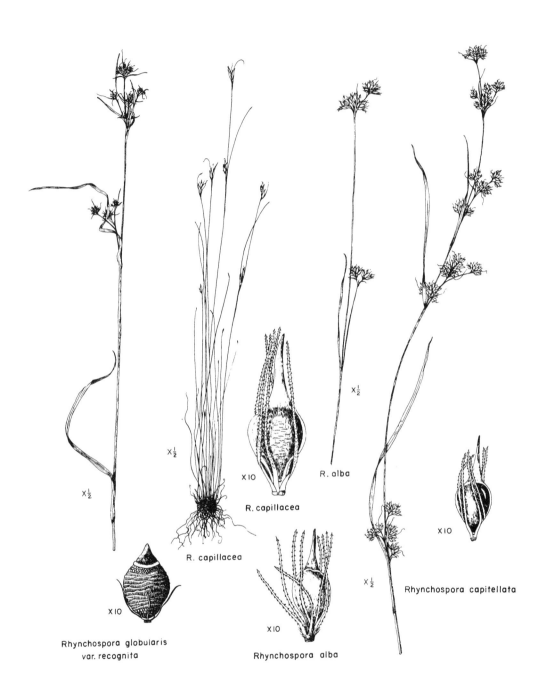

X½

X½

X10

R. capillacea

R. capillacea

X½

R. alba

X10

X½

X10

Rhynchospora capitellata

X10

Rhynchospora globularis
var. recognita

X10

Rhynchospora alba

1. **Hemicarpha micrantha** (Vahl) Pax

Dwarf tufted annual 2–10 (–15) cm tall, with very slender slightly recurved culms, and few basal leaves about half length of culm; involucre of 1 bract which appears like a continuation of the culm; spikelets 1–3, ovoid, 2–4 mm long, appearing lateral; spikelet in shape and aspect like a diminutive pine cone. Wet sandy borders of ponds and ditches, ranging through much of the United States, but local.

9. RHYNCHOSPORA Vahl. BEAK-RUSH

Perennials (our species) with slender, usually 3-angled culms, narrow elongate basal leaves, and 2-several stem-leaves subtending in some species, short-peduncled fascicles of spikelets; spikelets ovoid, subglobose, or fusiform, in terminal and axillary clusters; scales in most species chestnut to dark brown, spirally imbricate, the lower empty, the uppermost often sterile, the fertile larger; achenes flattened, lenticular to globular, with a conspicuous tubercle, and surrounded by the hypogynous bristles; bristles usually 6, sometimes more by division. The most important characters in the identification of species are those of achenes and bristles; stature and shape of flower-clusters are useful field characters, but are not determinative. *Rhynchospora* is a large genus of about 200 species, most of which inhabit warmer regions of both hemispheres; about twice as many are found in the Southeastern States as in the "Manual" range; many are Coastal Plain plants. The generic name, from the Greek, *rhyncos*, a snout, and *spora*, a seed, refers to the characteristic achene.

The key below will not serve to distinguish Ohio species from others that may be found. For more complete treatment see Gale (1944), Fernald (1950), and Gleason (1952). The figures of achenes are after Gale.

a. Scales of spikelet white or whitish when fresh, later tawny; achenes smoothish, but finely granular, somewhat pear-shaped to obovoid, tubercle flattened triangular-lanceolate; bristles 10–12 ....1. *R. alba*
aa. Scales of spikelet chestnut-brown.
    b. Leaves filiform, involute, less than 0.5 mm wide; spikelets lanceolate to spindle-shaped, in narrow erect terminal fascicles (3–8 mm wide) and often a remote short-peduncled lateral fascicle; achenes narrow oblong-elliptic, narrowed to base; tubercle attenuate-subulate; bristles 6 ....2. *R. capillacea*
    bb. Leaves wider, flat; spikelets narrow-ovoid to sub-rotund; lateral fascicles usually more than 1.
        c. Achenes smooth, uniformly brown, pear-shaped; bristles 6, strap-like, retrorsely barbed, about equalling achene and tubercle; tubercle compressed-subulate ....3. *R. capitellata*
        cc. Achenes transversely ridged, broadly obovoid to subglobose; bristles 5–6, shorter than achene, tubercle broad-conic ....4. *R. globularis* var. *recognita*

1. **Rhynchospora alba** (L.) Vahl. WHITE BEAK-RUSH

One of the leafy-stemmed species, with 1–3 small exserted lateral (axillary) fascicles of spikelets; fascicles, when fresh, milk-white, later with brownish red or salmon-color tinge. Leaves slenderly linear, 0.5–2.5 mm wide, flat, becoming slightly keeled; fascicles turbinate, 0.7–1.6 cm wide, the terminal larger than the lateral. In sphagnum bogs and sedge swamps, Newfoundland to Alaska and southward.

2.   **Rhynchospora capillacea** Torr.

As the name suggests, this is a species with very slender (capillary) culms and leaves, commonly growing in more or less dense tufts. Culms 1–4 (commonly 2–3) dm tall; leaves filiform-setaceous, involute; fascicles small, fusiform to narrow-ovoid, usually one lateral, short-peduncled. Locally abundant on wet calcareous ledges and in marly bogs, Newfoundland to Saskatchewan, south to Virginia, Tennessee, and Missouri.

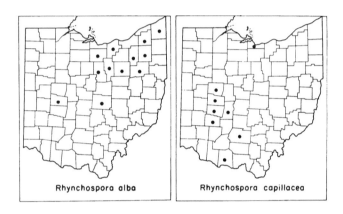

3.   **Rhynchospora capitellata** (Michx.) Vahl

R. glomerata of Amer. auth., not (L.) Vahl

A flat-leaved, rather tall species (2–10 dm) with several (1–5) lateral fascicles. Leaves 1.5–3.5 mm wide; fascicles of terminal inflorescence 1-several, turbinate to subglobose; lateral (axillary) fascicles short-peduncled; scales of spikelet early deciduous, revealing achene; achene brown, lustrous, pear-shaped, with pale compressed-subulate tubercle. A common species (local in Ohio) of swamps, borders of lakes, and stream-banks in much of eastern United States and southeastern Canada. All early Ohio reports of *R. glomerata*, which is a more southern species, should be referred to *R. capitellata*.

4.   **Rhynchospora globularis** (Chapm.) Small, var. **recognita** Gale

R. cymosa of Torr. and later auth.

A fairly wide-leaved (2–4 mm) tufted beak-rush usually about 4–6 dm tall. Most readily distinguished from our other species by its short bristles and very different achene—transversely ridged and with low conic tubercle. A southern species of swamps and bogs, extending north on the Atlantic slope to New Jersey and southeastern Pennsylvania, and in the interior to Tennessee, Missouri, and, very locally, to southern Ohio; also near the Great Lakes in Ohio, Indiana, and Illinois.

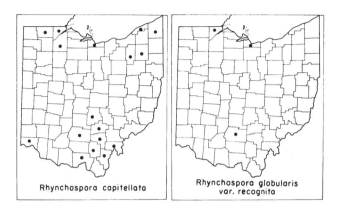

Rhynchospora capitellata

Rhynchospora globularis
var. recognita

10. CLADIUM P.Br.   TWIG-RUSH

Tall perennials, with aspect of *Rhynchospora*, but differing in absence of tubercle on achene. Inflorescence branched and widening upward, terminal and from upper one or more axils, the bracts foliaceous; stems almost terete. A large genus, mostly confined to tropical and warm-temperate regions of both western and eastern hemispheres, only two species entering the United States, one southern, the other ranging far northward.

1. **Cladium mariscoides** (Muhl.) Torr.   TWIG-RUSH
   Mariscus mariscoides (Muhl.) Ktze

   Perennial, up to 1 m tall; stems almost terete; leaves channeled toward base, becoming terete and very slender toward apex. Spikelets bearing 1 perfect flower, rich brown, tapering to slender apex, in small tight clusters at apex of branches of the cymose inflorescence; achene pointed at tip but not bearing a tubercle, truncate at base, more or less acorn-shaped, or contracted at base. Plants of wet poorly drained prairies, marl bogs, or open peat bogs. Wide-ranging but local.

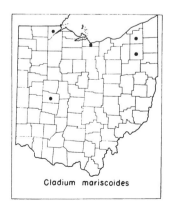

Cladium mariscoides

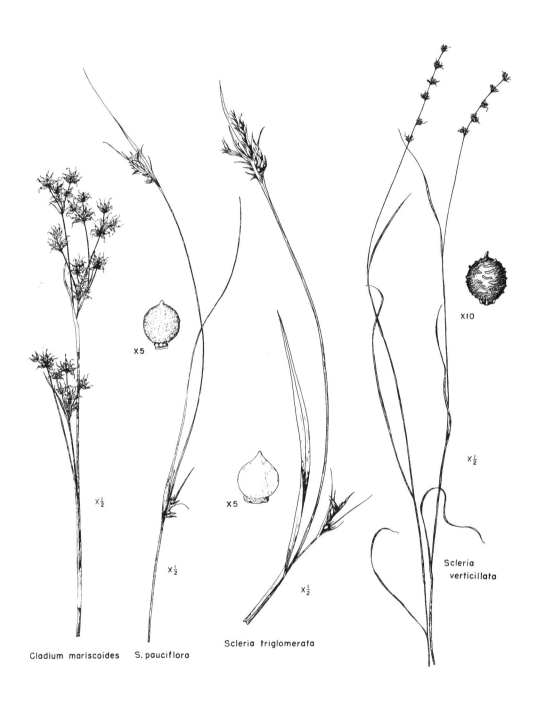

X5

X5

X10

X½

X½

X½

X½

X½

Cladium mariscoides   S. pauciflora

Scleria triglomerata

Scleria
verticillata

11. SCLERIA Bergius. Nut-rush

A large genus, about 200 species for the most part inhabiting tropical and warm-temperate regions, very few ranging into cold-temperate latitudes. Monoecious sedges, the one-flowered pistillate spikelets intermixed with few-flowered staminate spikelets; lower scales of pistillate spikelets empty; style 3-cleft; stamens 1–3; achenes at maturity usually exposed, white and bony, elevated on a disk (hypogynium) which may be plain or ornamented. Culms 3-angled, inflorescences small, terminal and axillary. Name from the Greek, *skleros*, meaning hard, and referring to the bone-like achenes.

Drawings of achenes after Gleason (1952) by permission of the New York Botanical Garden.

a. Achenes smooth, like white enamel, subglobose, nearly 3 mm high, on a white crustaceous or minutely pebbled disk; leaves 5–9 mm wide, abruptly short-pointed; coarse perennial .... 1. *S. triglomerata*
aa. Achenes roughened, wrinkled, or ridged, subglobose; leaves 3 mm wide or less; culms slender.
    b. Perennial, with hard forking rhizome; leaves 1–3 mm wide; inflorescence of 1–3 small terminal fascicles, and occasionally a slender-stalked lateral; achene white, rough; disk with a circle of 6 round tubercles ....2. *S. pauciflora*
    bb. Annual, with filiform culms; leaves about 1 mm wide; spikelets in small head-like sessile fascicles arranged in an interrupted spike; achene white or whitish, transversely irregularly ridged; disk short, stipe-like ....3. *S. verticillata*

1. **Scleria triglomerata** Michx.
Perennial, with sharply 3-angled culms, 0.5–1 m tall, hard knotty rhizomes, harsh scabrous leaves abruptly narrowed to attenuate or acuminate tip, and small terminal and often lateral inflorescences. The hard white and lustrous subglobose achenes (2.5–3 mm long and broad) often protrude conspicuously at maturity. Dry grassy slopes and ridges and woodland openings, local through much of the eastern half of the United States.

2. **Scleria pauciflora** Muhl.
Perennial, with very slender 3-angled culms 2–6 dm tall, hard knotty rhizomes, narrowly linear leaves, and very small terminal inflorescence (sometimes also a very small filiform-peduncled lateral) subtended by an erect leaf-like bract, the inflorescence appearing lateral; achenes white, papillose, 1.5–2 mm long and broad, on a disk with 6 rounded tubercles. Three intergrading varieties are distinguished, the typical with leaf-blades glabrous or thinly pubescent, var. *caroliniana* (Willd.) Wood, with leaves and culms strongly pilose, and a western variety. Our specimens vary from almost glabrous (Adams, Jackson counties) to more or less densely pubescent (Erie, Jackson, Pike counties). In dry or moist soil, meadows and open fields.

3. **Scleria verticillata** Muhl.
Tufted annual with filiform 3-angled culms 1–5 dm tall, fibrous roots, soft linear leaves (the margins soon revolute), and inflorescence a series of widely separated fascicles of spikelets; achenes white or whitish, prominently cross-

ridged, 1–1.5 mm long and broad. Marl bogs and wet sandy soil, local in much of the eastern half of the United States.

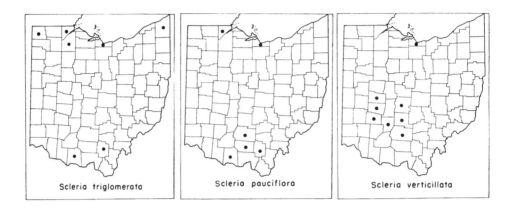

Scleria triglomerata   Scleria pauciflora   Scleria verticillata

12. CAREX L.   SEDGE

Perennial grass-like herbaceous plants; culms (stems) usually 3-angled; leaves 3-ranked, consisting of blade and sheath, the sheath closed (grown together) along its inner band (part opposite leaf-blade), and with a ligule (a usually inconspicuous more or less membranous projection from summit of the sheath). Flowers of two kinds: staminate, consisting of 3 stamens subtended by a small bract, the staminate scale; and pistillate, consisting of a single pistil with a 2-cleft or 3-cleft style and enclosed in a sac (the perigynium) with a terminal opening through which the style projects. Fruit a hard lenticular or 3-angled achene with or without attached style, and enclosed in the mature perigynium, this subtended by a small bract, the pistillate scale. Staminate and pistillate flowers borne in spikes, either in different parts of the same spike or in different spikes on the same culm, or, rarely, on different plants. Spikes with staminate flowers terminal are androgynous, those with staminate flowers basal are gynecandrous. The spikes may be sessile, and separate and distinct, or crowded into a compact head in which the separate spikes are obscured; or spikes may be peduncled and distinct, either short or long.

Carex is a very large genus, with over 1500 species, about 250 of which occur in the area covered by Gray's Manual, and 140 in Ohio, by far the largest genus in the state. World-wide in distribution, the genus is, however, best represented in cooler climates. It will be noted from Ohio distribution maps that there are many more species in northern than in southern Ohio, although even in the southern part of the state, northern species have been collected in bogs and swamps harboring relics of the last glacial migrations.

Few species of the genus are of economic importance; a few act as sand-binders in dune areas; ecologically, their role is great, for many are pioneer species, or species of ponds and swamps, where they take part in the gradual natural filling of depressions, or, in arctic and alpine regions, where they often comprise the dominant meadow vegetation.

The genus is divided into two subgenera, very distinct in appearance, but both with perigynia—the most distinctive feature of the genus. Each subgenus is further subdivided into sections, which are briefly characterized in the text.

Identification of species of *Carex* is difficult; mature perigynia are essential; growth-habit, readily noticed in the field, will help with some species; characters of the leaf-sheath, its orifice and ligule are sometimes diagnostic. Although the species within a section often resemble one another, some of the sections (as *Ovales* in VIGNEA, and *Lupulinae* in EUCAREX), well represented in the Ohio flora, can soon be recognized by the beginner. The 3-ranked leaf-arrangement (a character which distinguishes the vegetative stage of sedges from grasses) is well shown by the summer rosettes of some species (see *C. platyphylla*, p. 258). The illustrations on pp. 242 to 244 will help to familiarize the user with characters of VIGNEA, and to distinguish this from EUCAREX, illustrated on pp. 248 to 291, and to explain characters that must be used in the keys to sections and to species. Reference to the more complete descriptions in the current manuals, in Mackenzie (1931), and to the full-page illustrations in NORTH AMERICAN CARICEAE by K. K. Mackenzie (1940) will be helpful.

The illustrations of perigynia are redrawn from Gleason (1952) and Mackenzie (1940) by permission of the New York Botanical Garden. Many of our Ohio specimens were determined by F. J. Hermann.

Spikes all alike (or nearly so), sessile, with staminate flowers either at apex or base, or sometimes all pistillate or all staminate (but essentially the same in appearance); stigmas 2 and achenes lenticular ....Subg. I. VIGNEA
Spikes of two kinds, staminate and pistillate, or staminate flowers in a spike which is partly pistillate; stigmas 3 and achenes 3-angled, or, if stigmas 2 and achenes lenticular, then lateral spikes peduncled ....Subg. II. EUCAREX

Subg. I. *Vignea*

a. Spikes (at least the terminal) with staminate flowers (or staminate scales) at tip, i.e., androgynous*; some species occasionally dioecious.
  b. Perigynia with thin or wing-margin; rootstocks long-creeping.
    c. Perigynia with thin margin, not winged; spikes androgynous, staminate, or inflorescence sometimes all staminate, dense; inner band of sheath green-striate ....1. INTERMEDIAE, p. 225
    cc. Perigynia wing-margined; spikes various; inner band of sheath not green-striate ....2. ARENARIAE, p. 225
  bb. Perigynia with blunt to sharp, but scarcely winged margins; rootstocks not long-creeping, plants cespitose.
    c. Culms not soft, not flattened in drying.
      d. Inflorescence not branched, spikes sessile along main axis of inflorescence, few (2–12), crowded to well separated; perigynia spongy at base in some small species with narrow (1–3 mm) leaves and moniliform inflorescence ....3. BRACTEOSAE, p. 225

* Terminal staminate flowers are often difficult to find; if no basal staminate flowers can be seen, it may be assumed that spike is androgynous, i.e., that staminate flowers are terminal, and hidden by perigynia.

      dd. Inflorescence more or less paniculate (but usually dense), spikes numerous, on lateral (usually short) branches as well as main axis; leaf-sheaths close.

          e. Leaf-sheaths with inner band usually transversely wrinkled; perigynia thin, low-plano-convex; scales awned .... 4. MULTIFLORAE, p. 231

          ee. Leaf-sheaths with inner band not transversely wrinkled; perigynia thick, plano- to biconvex; scales blunt, acute, or mucronate, not awned .... 5. PANICULATAE, p. 232

    cc. Culms soft and loosely cellular, flattened in drying; sheaths loose; perigynia corky or spongy at base, tapering to beak (or in no. 22, more abruptly contracted to beak); plants coarse .... 6. VULPINAE, p. 233

aa. Spikes (at least the terminal) with staminate flowers (or their scales) at base, i.e., gynecandrous.

    b. Perigynia without winged margins, although often thin-edged, spongy or corky at base.

      c. Perigynia 1.5–4 mm long; spikes little longer than broad; inflorescence usually moniliform.

          d. Perigynia minutely whitish-puncticulate, not thin-edged, elliptic .... 7. HELEONASTES, p. 235

          dd. Perigynia not puncticulate, thin-edged, usually broadest below middle 8. STELLULATAE, p. 237

      cc. Perigynia 4–5.5 mm long; spikes oblong to linear-oblong, overlapping to approximate or the lower remote .... 9. DEWEYANAE, p. 238

    bb. Perigynia with winged margins, not spongy-thickened at base .... 10. OVALES, p. 239

Subg. II. *Eucarex*

a. Spike solitary (additional spikes may be born on capillary almost basal peduncles), small, mostly less than 5 mm thick, staminate above.

    b. Spike spire-like, without leafy bracts; perigynia beakless .... 11. POLYTRICHOIDEAE, p. 247

    bb. Spike with oblong or broad-ovoid pistillate part and slender staminate part; lowest pistillate scales enlarged into foliaceous bracts; perigynia long-beaked .... 12. PHYLLOSTACHYAE, p. 247

aa. Spikes more than one, or if solitary, then 1–2 cm thick and staminate at base.

    b. Inflorescence corymbiform, 1–2 cm wide and high, pale, bracts spathe-like, whitish; perigynia at first pale, later brown or black, lustrous; mat or sod-forming plants .... 16. ALBAE, p. 253

    bb. Without this combination of characters.

      c. Style jointed at or near its base and soon breaking away, leaving achene beakless or apiculate.

          d. Perigynia tightly enclosing achenes (see also sections 18, 19).

              e. Perigynia obtusely or obscurely 3-angled, with convex sides, finely pubescent; bracts sheathless, scale-like, or the lowest needle-like; low plants growing in dense tussocks or mats .... 13. MONTANAE, p. 249

              ee. Perigynia sharply 3-angled, the sides flat or concave (sometimes convex above).

                  f. Leaves and culms glabrous; sheaths red-brown; bracts bladeless; perigynia abruptly contracted at apex to minute beak, glabrous or thinly pubescent; plants with leaves crowded near base .... 14. DIGITATAE, p. 250

                  ff. Leaves and culms pubescent; sheaths not purple-tinged; bracts with well developed blades; perigynia rounded to slender beak, very hairy .... 15. TRIQUETRAE, p. 253

          dd. Perigynia not tightly enclosing achenes, except toward base.

              e. Achenes lenticular or biconvex; stigmas 2.

   f.   Lateral spikes short (5–15 mm) and thick; pistillate scales
        and perigynia markedly different in color; perigynia plump,
        white-puberulent or orange-yellow; low plants with narrow
        leaves ....17. BICOLORES, p. 253
  ff.   Lateral spikes longer (usually more than 2 cm), cylindric;
        scales and perigynia not strongly contrasting in color, or scales
        reddish-brown with green midrib; perigynia flattened; coarse
        taller plants with leaves to 1 cm wide.
        g.   Pistillate scales obtuse to acuminate; achenes lens-shape,
             not contorted; pistillate spikes mostly sessile, erect, or lowest
             peduncled and drooping ....33. ACUTAE, p. 277
       gg.   Pistillate scales long-awned; achenes constricted on one
             side near middle; pistillate spikes peduncled, arching to
             drooping ....34. CRYPTOCARPAE, p. 279
ee. Achenes triangular in cross-section; stigmas 3.
   f.   Lowest bract with long closed or tubular sheath (only about
        5 mm in Extensae).
        g.   Perigynia beakless or short-beaked, beak if present not
             2-toothed, at most only emarginate.
             h.   Terminal spike staminate at base, pistillate above ....
                              22. GRACILLIMAE, p. 267
            hh.   Terminal spike staminate throughout, or if partly
                  pistillate, then perigynia toward base of spike.
                  i.   Perigynia with raised nerves.
                       j.   Perigynia tapering or narrowed at base,
                            triangular in cross-section, more or less closely
                            enclosing achenes.
                            k.   Plants stoloniferous; bract-sheaths slender
                                 tubular; pistillate scales often dark
                                 brown-purple .....18. PANICEAE, p. 254
                           kk.   Plants not stoloniferous; bract-sheaths
                                 often loose; pistillate scales usually green
                                 with hyaline borders ....
                                 19. LAXIFLORAE, p. 255
                       jj.   Perigynia rounded at base, almost circular
                             in cross-section, loosely enclosing achenes ....
                                 20. GRANULARES, p. 263
                  ii.   Perigynia with fine impressed nerves, rounded or
                        tapering at base; pistillate scales awned ....
                                 21. OLIGOCARPAE, p. 264
       gg.   Perigynia with well developed beak, this (in mature peri-
             gynia) distinctly 2-toothed.
             h.   Perigynia densely pubescent; beak deeply 2-toothed....
                              27. HIRTAE, p. 274
            hh.   Perigynia glabrous (occasionally minutely puberulent).
                  i.   Pistillate spikes cylindric, not compact, spread-
                       ing to drooping on slender peduncles; perigynia
                       not reflexed.
                       j.   Plants red-tinged at base; pistillate spikes
                            3–4 mm thick, linear-cylindric; perigynia
                            tapering to beak.....23. SYLVATICAE, p. 268
                       jj.   Plants not red-tinged at base; pistillate spikes
                             8–10 mm thick, oblong-cylindric; perigynia
                             abruptly contracted to long slender beak
                             usually bent at base ....
                                 24. LONGIROSTRES, p. 269
                  ii.   Pistillate spikes compact, short-cylindric to glo-
                        bose, sessile, or lower short-peduncled; at least
                        the lower perigynia reflexed, somewhat inflated,
                        strongly ribbed, beaked ....
                                 25. EXTENSAE, p. 270

ff. Lowest bract sheathless or nearly so (occasionally with sheath
in *C. lasiocarpa* in Sec. 27).
  g. Perigynia distinctly beaked.
    h. Perigynia densely pubescent, abruptly short-beaked,
beak deeply 2-toothed; staminate spikes usually 2,
long-peduncled .... 27. HIRTAE, p. 274
    hh. Perigynia scabrous, with curved beak about as long
as body, its tip hyaline; staminate spike solitary, the
upper pistillate spikes extending beyond its base ....
      28. ANOMALAE, p. 274
  gg. Perigynia beakless or very short beaked.
    h. Perigynia or foliage (or both) pubescent; terminal
spike staminate or staminate only at base ....
      26. VIRESCENTES, p. 271
    hh. Perigynia and foliage glabrous.
      i. Terminal spike staminate; pistillate spikes slender-
peduncled ....30. LIMOSAE, p. 275
      ii. Terminal spike pistillate above, staminate below.
        j. Perigynia brown, broadly obovoid; pistillate
scales ovate, acute or mucronate, usually
shorter than perigynia, brown with green
midrib ....29. SHORTIANAE, p. 275
        jj. Perigynia glaucous green, pale, obovoid;
pistillate scales lanceolate, long-acuminate or
aristate, usually longer than perigynia, dark,
with light midrib .... 31. ATRATAE, p. 277
cc. Style not jointed, continuous with achene, persistent, firm.
  d. Perigynia less than 1 cm long (about 1 cm in no. 136 in Sec. 38, with
broad-ovoid, few-ribbed perigynia).
    e. Perigynia obconic, truncate or depressed-truncate, inflated, abruptly
contracted to long slender 2-toothed beak; spikes 1-several, terminal
staminate or abortive (or with a few perigynia), or staminate only
at base ....37. SQUARROSAE, p. 285
    ee. Perigynia not obconic, tapering or curving to beak, inflated or
subcoriaceous and firm.
      f. Perigynia firm, thick in texture, glabrous or pubescent, ovoid
to slender-ovoid; beak short; basal sheaths often purple-tinged ....
        36. PALUDOSAE, p. 282
      ff. Perigynia thin or papery, the ribs strong, glabrous (except in
hispidulous var. of no. 140); beak long.
        g. Pistillate scales with long rough awns, awn equal to or
longer than blade; perigynia with 12–17 nerves; lower
spikes often nodding .....35. PSEUDO-CYPEREAE, p. 281
        gg. Pistillate scales acute to acuminate, or (in no. 133) the
lower awned; perigynia with 10 or fewer nerves; lower
spikes not nodding ....38. VESICARIAE, p. 286
  dd. Perigynia more than 1 cm long, inflated; beak long.
    e. Perigynia lance-ovoid, tapering from near base to beak, finely many-
nerved ....34. FOLLICULATAE, p. 281
    ee. Perigynia ovoid, tapering or curving into beak, coarsely nerved
(nerves 15–20) ....39. LUPULINAE, p. 290

Subgenus I. VIGNEA

This subgenus is represented in Ohio by over 50 species classified into
ten sections.

1. *Intermediae*

A small group of swamp sedges with one American representative; included in the Arenariae by Fernald (1950).

1. **Carex sartwellii** Dewey
Culms arising singly or in small groups from tough, black, fibrillose root-stocks, dark toward base; spikes crowded into an elongate-ovoid or oblong head, the middle and upper often staminate; scales conspicuous, silky, bright brown with green midrib and hyaline borders; perigynia not wing-margined (but thin), ovate-orbicular.

2. *Arenariae*

Characters similar to those of the Intermediae, but with perigynia narrowly wing-margined, oblong-lanceolate. A small group of dry-soil species; one Ohio representative.

2. **Carex foenea** Willd.
C. siccata Dewey
Transcontinental in the North. Spikes vary in composition, the lowest small and pistillate, the middle staminate, the upper pistillate at tip; scales pale brown with white margin; bracts similar to scales.

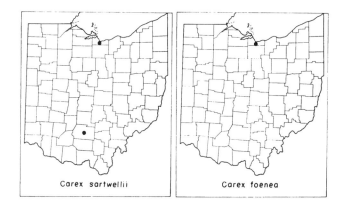

Carex sartwellii   Carex foenea

3. *Bracteosae*

A large section of mostly woodland, cespitose species maturing in spring or early summer. Inflorescence of 3–10 (–15) spikes, closely crowded to widely separated, the lowest rarely compound; bracts often little developed; perigynia plano-convex or biconvex, with conspicuous beak.

a. Sheaths fitting closely around culm, usually not septate-nodulose or green-mottled.
   b. Basal third or half of perigynia spongy- or corky-thickened, at maturity widely spreading to reflexed.
      c. Margins of beak and perigynia smooth; scales about equal in length to body of perigynium; spikes (except lowest) approximate.

     d. Perigynia broadly ovoid, biconvex, distinctly nerved or striate toward base ....3. *C. retroflexa*

     dd. Perigynia narrower, plano-convex, nerveless or nearly so ....
                            4. *C. texensis*

  cc. Beak finely serrulate; scales shorter; spikes (except upper) well separated.

     d. Stigmas slender, little twisted, pale reddish; perigynia tapering to beak, yellow-green; leaves 1–2 mm wide ....5. *C. rosea*

     dd. Stigmas stout, more strongly twisted or contorted, deep red; perigynia contracted to beak, the neck distinctly concave, deep green.

        e. Culms erect or nearly so; leaves about 2.5 (1.5–3) mm wide ....
                        6. *C. convoluta*

        ee. Culms reclining or widely spreading, thread-like; leaves about 1 mm wide ....7. *C. radiata*

bb. Perigynia not spongy- or corky-thickened at base (or slightly so in no. 10 with perigynia 4–6 mm long).

  c. Inflorescence ovoid to oblong-ovoid, less than 2 cm long; spikes densely crowded, scarcely distinguishable.

     d. Perigynia nerveless or faintly nerved; stigmas short.

        e. Perigynia broadest near or below middle, round-tapering at base, with raised margins on inner face; beak finely but distinctly serrulate; leaves 2–4.5 mm wide.

           f. Perigynia 2.5 (–3) mm long; culms 3–6 dm tall ....
                       8. *C. cephalophora*

           ff. Perigynia 3–3.5 mm long; culms to 1 m tall ....
                       9. *C. mesochorea*

        ee. Perigynia truncate-cordate at base, flat on inner face, 2–3.5 mm long; beak smooth; leaves 1–3 mm wide; culms 2–5 dm tall ....
                      10. *C. leavenworthii*

     dd. Perigynia strongly ribbed, at least on back, broadly ovate, to suborbicular, plano-convex, margins slightly raised; beak rough-margined; leaves 2–4 mm wide ....11. *C. muhlenbergii*

  cc. Inflorescence linear- to ellipsoid-cylindric or slender-ovoid, 1.5–4 cm long; spikes not densely crowded, at least the lowest distinct.

     d. Perigynia 3–3.5 mm long, strongly ribbed, at least on back, broadly ovate to suborbicular; beak rough-margined ....11. *C. muhlenbergii*

     dd. Perigynia 4–6 mm long, nerved on back, nerveless on inner face, ovate, lustrous; beak serrulate; scales red-tinged ....12. *C. spicata*

aa. Sheaths loose, usually septate-nodulose or reticulate on back, often green and white mottled; beak of perigynium serrulate.

  b. Scales about as long as perigynia, and ¾ as wide, acuminate to awn-tipped; perigynia ovate, pale, suffused with brown; beak strongly 2-toothed, sutures prominent; leaves 3.5–5 (–8) mm wide, sheaths truncate or concave at orifice; inflorescence dense to slightly open ....13. *C. gravida*

  bb. Scales shorter than perigynia, and mostly hidden by them, acute to obtuse; perigynia deep green.

     c. Spikes distinctly separated, often widely so; inflorescence 3–15 cm long; perigynia concave on inner face, with raised borders, wing-margined to base; stigmas short; leaves 5–10 mm wide (rarely narrow) ....
                       14. *C. sparganioides*

     cc. Spikes closely spaced, inflorescence scarcely interrupted, usually less than 5 cm long; perigynia flat on inner face.

        d. Stigmas short, slender; leaves 5–8 mm wide; sheaths truncate at orifice, not thickened, pale and fragile ....15. *C. cephaloidea*

        dd. Stigmas long, slender; leaves 3–7 mm wide; sheaths concave at orifice, thickened, reddish brown ....16. *C. aggregata*

3. Carex retroflexa Muhl.

Densely tufted sedge, with stiff slender culms 2–5 dm tall and narrow leaves (1–3 mm wide) about equaling culms. Rare in Ohio, usually in dry or rocky woods.

4. **Carex texensis** (Torr.) Bailey

   C. retroflexa var. *texensis* (Torr.) Fern.

   More slender than the last; occasional in lawns and open grassy places; southern in range and perhaps adventive in Ohio.

5. **Carex rosea** Schkuhr

   The name refers to the arrangement of perigynia—radiating in all directions, each spike a small rosette. Densely tufted, culms 2–5 dm tall, usually exceeding the narrow (1–2 mm) leaves. A very common, wide-ranging woodland species maturing in May or June. Closely related to the next two species.

6. **Carex convoluta** Mackenz.

   C. rosea of Gray, ed. 7.

   Not always considered specifically distinct from *C. rosea*, under which species this and the next are treated by Gleason (1952) and Gleason & Cronquist (1963). Usually larger than the last, with deeper green wider leaves (about 2.5 mm), deeper green perigynia, and deep red, short, contorted stigmas. When with *C. rosea*, the color contrast is distinct. A common and widespread woodland species.

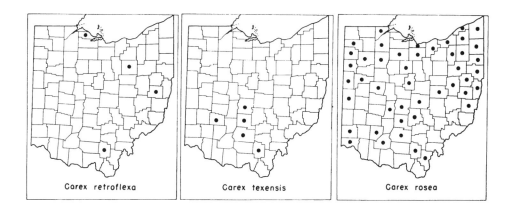

Carex retroflexa   Carex texensis   Carex rosea

Carex convoluta

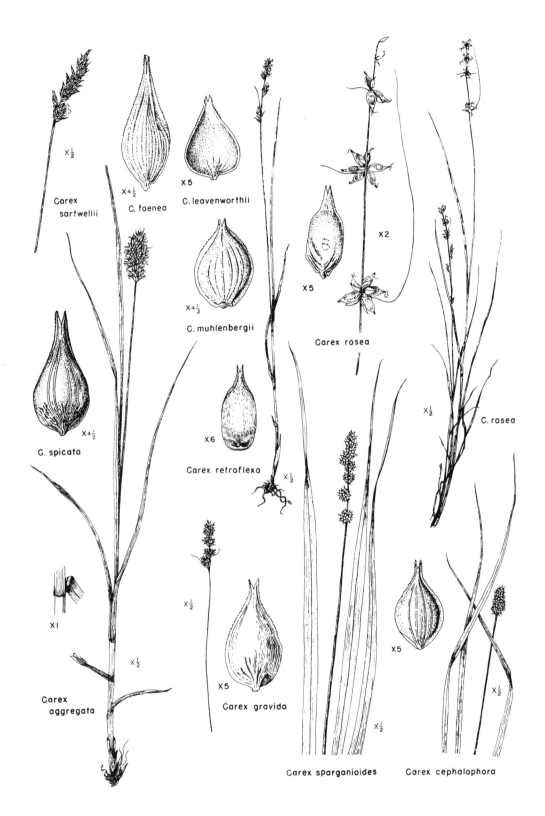

$\times\frac{1}{2}$

Carex
sartwellii

$\times 4\frac{1}{2}$

C. foenea

$\times 5$

C. leavenworthii

$\times 4\frac{1}{2}$

C. muhlenbergii

$\times 5$

$\times 2$

Carex rosea

$\times 4\frac{1}{2}$

C. spicata

$\times 6$

Carex retroflexa

$\times\frac{1}{2}$

$\times\frac{1}{2}$

C. rosea

$\times 1$

$\times\frac{1}{2}$

Carex
aggregata

$\times\frac{1}{2}$

$\times 5$

Carex gravida

$\times 5$

$\times\frac{1}{2}$

$\times\frac{1}{2}$

Carex sparganioides

$\times 5$

$\times\frac{1}{2}$

Carex cephalophora

7. **Carex radiata** (Wahlenb.) Dewey

   C. rosea var. *radiata* Dewey

   More slender than *C. rosea*, with capillary, widely spreading or reclining culms, and very narrow (1 mm) leaves. Rare in Ohio; a northeastern and Appalachian species.

8. **Carex cephalophora** Muhl.

   A densely tufted, rather wide-leaved species with culms 2–5 dm tall, about equaling or taller than leaves; head of spikes compact, about 1–2 cm long, distinctly broadened near base. Common and widespread.

9. **Carex mesochorea** Mackenz.

   C. cephalophora var. *mesochorea* (Mackenz.) Gl.

   Similar to the last, but larger, and culms taller (to 1 m); distinguished chiefly by its larger perigynia and distinctly 3-nerved pistillate scales.

10. **Carex leavenworthii** Dewey

    Similar to *C. cephalophora*; distinguished by its cordate-deltoid, smooth-beaked perigynia, and narrower leaves. Widespread, but apparently infrequent in Ohio, in woods and woodland borders, or occasionally on roadsides.

Carex radiata

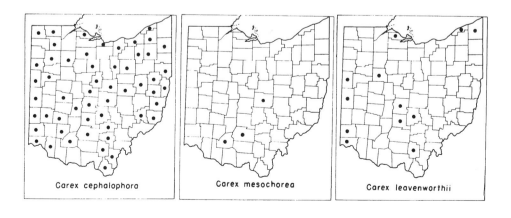

Carex cephalophora   Carex mesochorea   Carex leavenworthii

11. Carex muhlenbergii Schkuhr

Inflorescence densely capitate to slightly interrupted, at least the lowest spike easily distinguishable; culms densely tufted from short rhizomes, interspersed with brownish dried-up leaves of the previous year. Two intergrading varieties (not distinguished on map) may be recognized: the typical variety, with perigynia strongly many-ribbed on back, and many-ribbed on inner face; var. *enervis* Boott, with perigynia nerveless or short-nerved at base only on inner face. Wide-ranging, but apparently infrequent in Ohio.

12. CAREX SPICATA Huds.

C. muricata of auth., not L.

Inflorescence interrupted, the distinct spikes bristling with the long-beaked large lustrous perigynia; pistillate scales tinged with bright red. Naturalized from Eurasia, in dry fields and on roadsides.

13. Carex gravida Bailey

Local in woods in the western part of Ohio, mostly in or near the Prairie Peninsula area. A variety, *lunelliana* (Mackenz.) Hermann, more western in range is distinguished by its more abruptly tapering and more strongly ribbed perigynia. Both this and the typical variety range westward in the Prairie Region.

14. Carex sparganioides Muhl.

Distinct in appearance from all other Ohio representatives of the section, because of its long wide leaves with loose over-lapping sheaths, and its long open inflorescence with spikes (except uppermost) widely separated. A common species in woods throughout much of eastern United States except in the South.

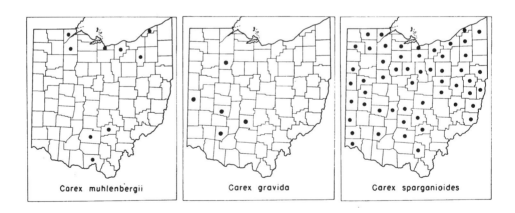

Carex muhlenbergii   Carex gravida   Carex sparganioides

15. Carex cephaloidea Dewey

Similar in appearance to the last, but inflorescence more compact, and leaves usually narrower. Northern in range—New Brunswick to Minnesota and southward into our area; apparently confined to northeastern Ohio.

16. Carex aggregata Mackenz.

Distinguished from other species of the section with more or less compact inflorescence by the leafy lower part of culms, loose sheaths that are pale dorsally and with concave, thickened, brownish orifice, and other characters given in key. Apparently infrequent in Ohio.

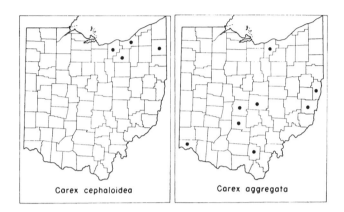

Carex cephaloidea   Carex aggregata

4. *Multiflorae*

A small section of North and South America, and eastern Asia; ours in low woods, swamps or wet fields, and roadside ditches. Inflorescence of numerous closely crowded, or somewhat separated spikes, usually with well developed but bristle-like bracts; perigynia plano-convex, beaked, usually serrulate on margins toward apex; leaf-sheaths close, usually with transversely wrinkled inner band.

a. Leaves usually equaling or exceeding culm, 2–6 mm wide; perigynia ovate to narrow-ovate in outline, less than 2 mm wide; beak nearly as long as body ....
17. *C. vulpinoidea*
aa. Leaves usually shorter than culm, 2–4 mm wide; perigynia broad-ovate; beak shorter ....
18. *C. annectens*
   Perigynia 2 mm or more wide, broadest at base; beak plainly 2-toothed ....
var. *annectens*
   Perigynia less than 2 mm wide, broadest below middle; beak obscurely 2-toothed ....
var. *xanthocarpa*

17. Carex vulpinoidea Michx.

A very common and widespread sedge, probably in every county in Ohio. Variable in size of inflorescence and prominence of lateral spikes; very leafy, leaves taller than culms. Forms transitional between this and the next occur.

18. Carex annectens Bickn.

Much less widespread than the preceding; usually represented by the typical variety; var. *xanthocarpa* (Bickn.) Wieg. (*C. brachyglossa* Mackenz.) is rare (Carroll, Hamilton, and Pickaway counties, where the typical variety also occurs). Inflorescence usually more compact than in the preceding species, and lateral spikes shorter; culms taller than leaves.

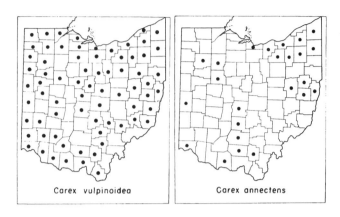

Carex vulpinoidea   Carex annectens

5. *Paniculatae*

A small section, its American representatives in swamps and bogs of cool temperate latitudes, often in calcareous soil. Inflorescence more or less open, paniculate; bracts usually shorter than spikes; perigynia biconvex to thick plano-convex, plump; leaf-sheaths close, red-dotted on the thin ventral face.

a. Perigynia broad-obovoid, tapering at base, very abruptly contracted to short beak; inflorescence decompound, the panicle open, dark in color; leaves 3–8 mm wide ....
                                                               19. *C. decomposita*
aa. Perigynia ovate to lanceolate, rounded at base, tapering to beak; inflorescence less obviously paniculate; leaves 1–3 mm wide.
  b. Sheath-orifice pale; perigynia widely spreading, not concealed by scales ....
                                                               20. *C. diandra*
  bb. Sheath-orifice rusty or copper-color; perigynia more or less appressed, nearly concealed by scales ....21. *C. prairea*

19. Carex decomposita Muhl.

Known in Ohio only from Cranberry Island in Buckeye Lake; a southern species—Florida to Louisiana and northward on the Atlantic seaboard to Maryland, and (west of the mountains) to Missouri, Michigan, and western New York.

20. Carex diandra Schrank

C. teretiuscula Gooden.

A circumboreal species, ranging south locally in our area; in peaty or marly swamps and bogs.

21. Carex prairea Dewey

C. diandra var. *ramosa* (Boott) Fern.; *C. teretiuscula* var. *prairea* Britton

Similar to the last, and in similar habitats; Quebec to Saskatchewan, southward to mid-temperate latitudes. The darkened, rusty or coppery sheath-orifices often conspicuous.

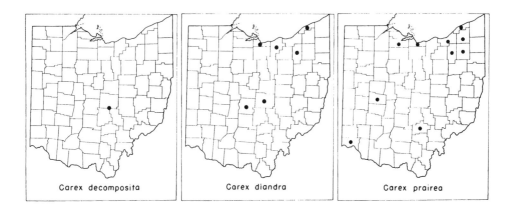

Carex decomposita   Carex diandra   Carex prairea

6. Vulpinae

Culms strongly triangular or concave-triangular in cross-section, often soft and flattened when pressed; leaf-sheaths loose; perigynia spongy or corky at base (sometimes enlarged at base), tapering or incurving to bidentate beak. Characters of leaf-sheaths are important in determination of species. The section is represented on all continents except South America; best developed in North America.

a.   Perigynia ovate, 3.5–4 mm long, not enlarged at base, incurving or contracted to beak half to nearly as long as body; leaves 5–10 mm wide; thin inner band of sheaths finely cross-puckered, fragile, convex at orifice ....22.   *C. conjuncta*
aa. Perigynia narrower, lanceolate, 4–9 mm long, tapering to beak longer than body.
   b. Perigynia with enlarged, corky, disⁱ-like base, and very slender above, beak 2–3 times length of body; leaves 5–12 mm wide; inner band of sheaths thin, not cross-puckered, not thickened at orifice ....23.   *C. crus-corvi*
   bb. Perigynia without enlarged disk-like base, but plainly corky or spongy at base, tapering to long beak.
      c. Leaf-sheaths with thin, easily torn, cross-puckered inner band not thickened at orifice; leaves 4–8 (–15) mm wide ....24.   *C. stipata*
      cc. Leaf-sheaths with inner band firmer, not cross-puckered, cartilaginous-thickened at orifice; leaves 3–6 mm wide ....25.   *C. laevivaginata*

22. Carex conjuncta Boott

Culms and leaves about equal in height (to 1 m); inflorescence interrupted, cylindric, to 7 cm long. Wet soil, open woods, creek valleys, and ditches, ranging from the Atlantic westward to Missouri and Kansas in middle latitudes.

23. Carex crus-corvi Shuttlw.

Occasional, in swampy thickets and sloughs, in much of eastern United States. A tall, wide-leaved species, recognized by its peculiar perigynia with basal disk and body subulate above rounded base.

24. Carex stipata Muhl.

Common and widespread; varies greatly in size—height, leaf-width, stem-

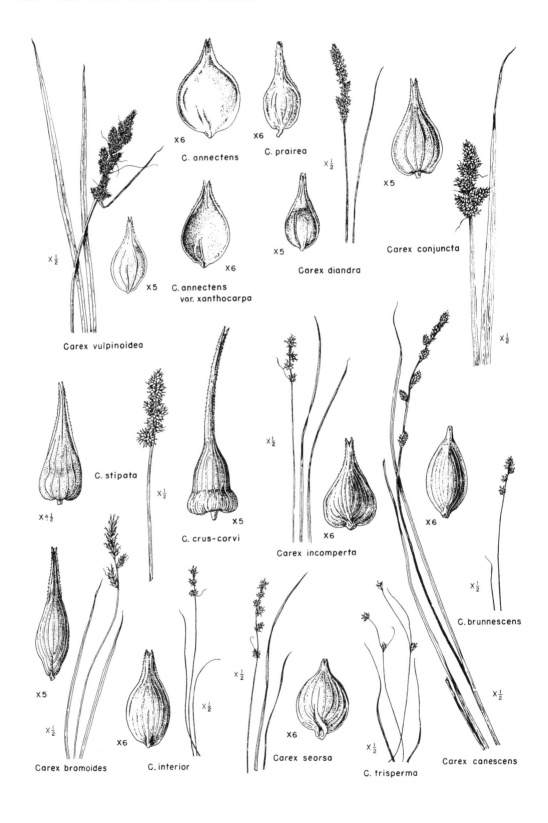

C. annectens
×6

C. prairea
×6

×½

×5

Carex conjuncta
×5

Carex diandra
×5

C. annectens
var. xanthocarpa
×6

Carex vulpinoidea
×½

Carex conjuncta
×½

C. stipata
×4½

×½

C. crus-corvi
×5

Carex incomperta
×6

C. brunnescens
×½

Carex bromoides
×5
×½

C. interior
×6
×½

Carex seorsa
×6

C. trisperma
×½

Carex canescens
×½

thickness, and complexity of inflorescence, from slender cylindric to obviously paniculate. Large plants with perigynia 5–6 mm long and leaf-blades 8–15 mm wide may be distinguished as var. *maxima* Chapm. (*C. uberior* (Mohr) Mackenz.), which intergrades with the typical variety (perigynia 4–5 mm long, leaves 4–8 mm wide); recorded from Fairfield, Jackson, and Knox counties.

25. **Carex laevivaginata** (Kükenth.) Mackenz.

    C. stipata var. *laevivaginata* Kükenth.

Similar to *C. stipata*, and best distinguished from it by the cartilaginous thickening of the concave sheath-orifice. Widespread, in the eastern half of the United States.

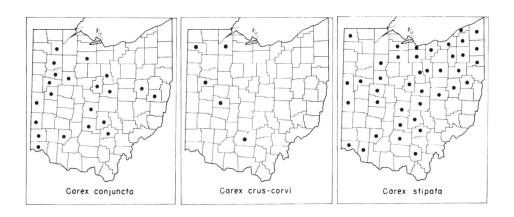

Carex conjuncta   Carex crus-corvi   Carex stipata

Carex laevivaginata

7. *Heleonastes*

A group of cold-climate sedges, growing in wet mossy woods, swamps, and sphagnum-bogs, rarely in drier sites. Cespitose or stoloniferous plants, culms usually exceeding leaves; inflorescence slender, spikes well separated or upper approximate, the terminal (and sometimes lateral) staminate at base (rarely at apex); perigynia finely puncticulate, plano-convex or rarely unequally biconvex, elliptic, angled above.

a. Spikes with staminate flowers basal or scattered; perigynia plano-convex.
    b. Lowest bract long, bristle-like; upper 1 or 2 spikes on slender widely divergent peduncles; spikes 1–5-flowered; culms and leaves very slender ... 26. *C. trisperma*
    bb. Lowest bract shorter, not over 2–3 times length of spike; spikes sessile, more than 5-flowered.
        c. Leaves glaucous, 2–4 mm wide; spikes 4–12 mm long, oval to short-oblong; perigynia 10–30, appressed ....27. *C. canescens*
        cc. Leaves green, 1–2.5 mm wide; spikes 3–7 long, subglobose; perigynia 5–10, loosely spreading ....28. *C. brunnescens*
aa. Spikes with staminate flowers terminal; perigynia unequally biconvex, 1-few in sessile, widely separated spikes, brown and lustrous when ripe; culms very slender ....
                                        29. *C. disperma*

26. Carex trisperma Dewey

Loosely tufted, the filiform culms weak, 2–7 dm long; ranging from Newfoundland and Labrador to Saskatchewan, southward into northern Ohio.

27. Carex canescens L.

A circumboreal species, comprising several intergrading varieties of which two occur in Ohio: var. *canescens,* with all but lowest spikes approximate or only slightly separated, and var. *disjuncta* Fern., with all but uppermost spikes well separated, the lowest 2–3 cm apart. Most Ohio specimens are referable to var. *disjuncta,* only 2 (Lake and Portage counties) represent the typical variety, but both occur in these counties.

28. Carex brunnescens (Pers.) Poir.

A high-latitude circumboreal species extending south in the mountains; a variety, *sphaerostachya* (Tuckerm.) Kükenth., is sometimes segregated: spikes subglobose, well separated. Greener and more slender than *C. canescens.*

29. CAREX DISPERMA Dewey

Erroneously reported from Ohio; a northern species.

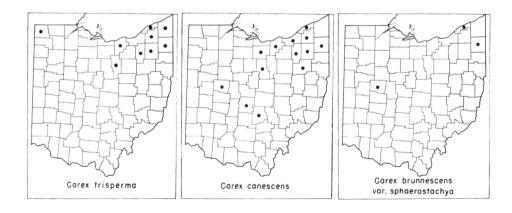

Carex trisperma   Carex canescens   Carex brunnescens var. sphaerostachya

8. *Stellulatae*

Swamp and bog sedges of temperate latitudes. Densely cespitose; culms equaling, exceeding, or shorter than leaves; inflorescence cylindric, of approximate or well separated spikes; staminate flowers usually basal, but variable in position; perigynia thin-edged but not winged, spongy at base, ovoid, beaked.

a.   Perigynia broadest near middle, strongly nerved; beak smooth, short; spikes globular, 3–6 mm thick, all except upper separated; terminal spike usually distinctly staminate at base; scales with keel-like midrib extending to tip ....30.   *C. seorsa*
aa. Perigynia broadest near base; beak serrulate.
    b.   Pistillate scales with midrib ending below apex, not keeled.
        c.   Beak of perigynium notched at apex, scarcely bidentate; scales ovate, obtuse to acute, about ½–⅔ length of perigynium-body.
            d.   Leaves 1–3 mm wide, firm; scales brownish with wide hyaline margins
                                          31.   *C. interior*
            dd. Leaves less than 1 mm wide, flaccid; scales pale, whitish ......32.   *C. howei*
        cc. Beak of perigynium evidently 2-toothed; scales obtuse to cuspidate, longer than perigynium-body, lustrous, chestnut-brown with paler center and hyaline margins ....33.   *C. sterilis*
    bb. Pistillate scales with midrib extending to apex, keeled; leaves 1.5–2.5 mm wide.
        c.   Perigynia very broad near base, suborbicular (body as wide as long), soon recurved, strongly nerved on inner face, all but central nerves bending abruptly outward at base; beak short; scales ⅔ to as long as perigynium-body ....34.   *C. incomperta*
        cc. Perigynia ovate, ascending to spreading, finely nerved on inner face near base; beak about half length of body; scales equaling or longer than perigynium-body ....35.   *C. cephalantha*

30. Carex seorsa Howe
Culms usually exceeding leaves; leaves 2–4 mm wide. A species of the Atlantic slope from Massachusetts to Georgia; local inland, New York to Indiana and Michigan, in swamps and wet woods.

31. Carex interior Bailey
    C. scirpoides Schkuhr, not Michx.
The slender but firm culms and stiffish leaves mostly erect. Ranging across the continent in the North; local in Ohio.

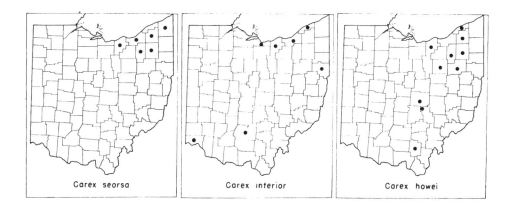

Carex seorsa   Carex interior   Carex howei

32. Carex howei Mackenz.

C. scirpoides var. *capillacea* (Bailey) Fern.

Similar to the last, but very slender, its culms weak and often arching, usually exceeded by the narrow, flaccid leaves. In sphagnum bogs and mossy thickets, Florida to Louisiana and northward.

33. Carex sterilis Willd.

C. muricata L. var. *sterilis* (Carey) Gl.

Inflorescence more compact than that of other Ohio species, usually exceeding leaves; spikes variable, staminate at base or apex, or staminate or pistillate. Northern interior in range, in wet, usually calcareous soil.

34. Carex incomperta Bickn.

C. stellulata var. *excelsior* Fern.

The strongly recurved perigynia with reniform to suborbicular body and sharply 2-toothed beak, and the sharply keeled scales with prominent midrib extending to apex distinguish this species. More widely scattered in Ohio than other species of the section, in swamps and sphagnum bogs.

35. Carex cephalantha (Bailey) Bickn.

C. stellulata var. *excelsior* sensu Fern.; var. *cephalantha* (Bailey) Fern.; *C. laricina* Mackenz.; *C. muricata* L. var. *cephalantha* Bailey, and var. *laricina* (Mackenz.) Gl.

A slender, densely tufted species of acid swamps and bogs; northern in range.

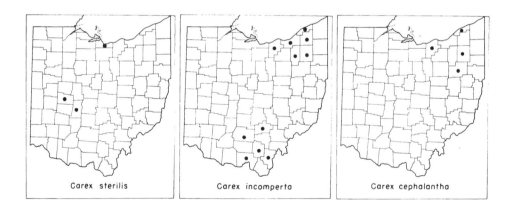

Carex sterilis   Carex incomperta   Carex cephalantha

9. *Deweyanae*

A section of four North American woodland species, 2 eastern, 2 western. Culms clothed near base with dried sheaths, lower sheaths bladeless; spikes pistillate throughout, staminate at base, or rarely staminate; perigynia (in ours) narrowly ovoid or linear-oblong, tapering to serrulate beak half to two-thirds as long as body, almost equaled by scales.

Culms very rough on angles above; leaves 1–2.5 mm wide, green; perigynia narrowly

lanceolate; scales straw-color to brownish with green midrib and hyaline border ....
36. *C. bromoides*
Culms more or less roughened just below inflorescence; leaves 2–5 mm wide, glaucous;
perigynia ovate-lanceolate; scales thin, whitish, with green midrib ....
37. *C. deweyana*

36. Carex bromoides Schkuhr

Culms slender, equaling or exceeding leaves; inflorescence different in appearance from all members of the subgenus *Vignea* previously treated: very slender, lustrous brownish, the spikes lance-cylindric, ascending, approximate or overlapping, or lowest distinct. A common and widespread species ranging through much of temperate eastern North America.

37. Carex deweyana Schwein.

More northern than the last, rare in Ohio. Inflorescence pale green to silvery; culms exceeding leaves.

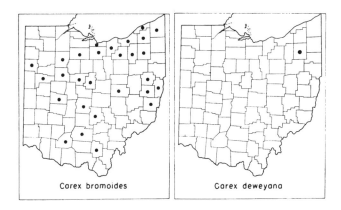

Carex bromoides   Carex deweyana

10. *Ovales*

Tall, usually densely tufted sedges (often 4–10 dm) of swamps, wet woods and meadows, or (a few species) dry soil. Inflorescence of crowded or separated spikes (i.e., capitate or moniliform); terminal spike (and often others) staminate at base; perigynia *with winged margins*, the beak flattened, serrulate; pistillate scales (except in no. 39) shorter than perigynia and narrow above so that upper part of perigynium is exposed. A very large section, best developed in temperate latitudes, but represented in colder regions and in the tropics.

a. Spikes 15–25 mm long, pointed; perigynia flat and scale-like, narrowly lanceolate, 7–10 mm long, the wing much narrowed below middle; pistillate scales pale brown with broad hyaline margins ....38. *C. muskingumensis*
aa. Spikes less than 15 mm long (rarely to 20 mm); perigynia 3–8 mm long (if 7–8 mm, then broad-ovate).
   b. Scales concealing perigynia, or nearly so, acute or acuminate, pale; inflorescence more or less moniliform and flexuous; spikes ovoid, narrowed to staminate base ....
39. *C. argyrantha*
   bb. Scales not concealing perigynia, margins of perigynia exposed, at least above.
      c. Perigynia obovate, widest at or above summit of achene.

   d. Scales lanceolate to ovate-lanceolate with short rough awns; perigynia
      2.5–4 mm long; inflorescence of 3–8 (–11) broad-ovoid crowded spikes
      about 1 cm long ....40. *C. alata*
  dd. Scales ovate, obtuse to acute.
      e. Leaf-sheaths tight, slender; perigynia 1.5–2.5 mm wide, with body
         abruptly narrowed to beak .... 41. *C. albolutescens*
      ee. Leaf-sheaths loose; perigynia 2–3.5 mm wide, with body rounded
         to beak ....42. *C. cumulata*
cc. Perigynia not obovate, widest near middle or base.
   d. Wing of perigynium narrowed and almost lacking from middle to base;
      leaf-sheaths loose.
      e. Perigynia thin and scale-like, only slightly distended over achene,
         lanceolate.
         f. Tips of perigynia appressed or strongly ascending; inner band
            of leaf-sheaths veiny almost to concave cartilaginous-thickened
            orifice; inflorescence dense ....43. *C. tribuloides*
         ff. Tips of perigynia loosely ascending to recurved; inner band
            of sheaths white-hyaline, easily broken, prolonged slightly at
            orifice; inflorescence moniliform ....44. *C. projecta*
      ee. Perigynia thicker, firm, obviously distended over achene, narrowly
         ovate; tips of perigynia widely spreading; leaf-sheaths with inner
         band hyaline near orifice, rusty at mouth and continued beyond
         base of blade ....45. *C. cristatella*
  dd. Wings of perigynia not narrowed below middle, perigynia winged to base.
      e. Perigynia thin and scale-like, lanceolate, 4–7 mm long.
         f. Scales gradually long-pointed, little hyaline at sharp tip; spikes
            lustrous; sheaths close, leaves 1–3 mm wide ..... 46. *C. scoparia*
         ff. Scales usually blunt, with short hyaline tip; sheaths loose,
            leaves 3–8 mm wide ....43. *C. tribuloides*
      ee. Perigynia not thin and scale-like, firm.
         f. Perigynia usually less than 2 mm wide and 4 mm long.
            g. Inflorescence usually moniliform.
               h. Perigynia narrow- to broad-ovate.
                  i. Leaves 1.5–2.5 mm wide; sheaths tight; spikes
                     rounded to staminate base; inflorescence curved or
                     nodding above lowest spike ....47. *C. tenera*
                  ii. Leaves 2.5–6 mm wide; sheaths loose; spikes sub-
                     globose; inflorescence erect ....48. *C. normalis*
               hh. Perigynia with suborbicular body; leaves 1–5 mm
                  wide; sheaths tight; spikes narrowed to base ....
                     49. *C. festucacea*
            gg. Inflorescence usually of crowded spikes; perigynia ovate;
               spikes subglobose to broad-ovoid.
               h. Sheaths loose; leaves 2.5–6 mm wide ....48....*C. normalis*
               hh. Sheaths tight.
                  i. Perigynia about half as wide as long; leaves
                     2–4.5 mm wide ....50. *C. bebbii*
                  ii. Perigynia broader; leaves 2–3 mm wide ....
                     ·   55. *C. molesta*
         ff. Perigynia usually more than 2 mm wide and often more than
            4 mm long.
            g. Scales with prominent midrib excurrent as a short awn
               or cusp, brownish-tinged; perigynia round-ovate to suborbi-
               cular, spreading to ascending; beak half as long as body;
               inflorescence moniliform, flexuous .... 51. *C. straminea*
            gg. Scales not awn-tipped.
               h. Perigynia flat and thin, translucent, 4–7 mm long;
                  spikes globose to ovoid, crowded or separated; leaves
                  2–4 mm wide ....52. *C. bicknellii*
               hh. Perigynia not translucent, firm.

i. Spikes more or less pointed at apex, perigynia appressed; leaf-sheaths green and veiny almost to orifice; spikes crowded or separated ....
53. *C. suberecta*

ii. Spikes rounded at apex; perigynia not appressed.

   j. Perigynia flat, thin, but not almost transparent (except the wings), ovate to suboribcular; scales ovate, obtuse to short-cuspidate ....54. *C. merritt-fernaldii*

   jj. Perigynia plano-convex, thick, coriaceous or submembranaceous.

      k. Spikes rounded at base, more or less crowded; scales blunt, reaching only to base of perigynium-beak; perigynia broadest near base ....55. *C. molesta*

      kk. Spikes usually narrowed at base; inflorescence usually moniliform; scales acuminate, reaching to middle or tip of beak; perigynia broadest near middle
56. *C. brevior*

38. Carex muskingumensis Schwein.

Distinguished from all other species of the *Ovales* by its large spikes and long (7–10 mm) slenderly lanceolate perigynia; culms usually more than half a meter tall. The type specimen of this species was collected near the Muskingum River in Tuscarawas County; it ranges from southern Ontario and Manitoba southward to Kentucky, Missouri, and eastern Kansas; on floodplains and in swampy woods.

39. Carex argyrantha Tuckerm.

C. foenea of Gray, ed. 7, not Willd.

The specific name refers to the pale, silvery pistillate scales which conceal the perigynia. A northern species ranging southward into northern Ohio, in dry woods and on rocky ledges.

40. Carex alata T. &. G.

One of the few species with obovate perigynia; perigynia strongly winged nearly to base, abruptly contracted above to beak; ligule tinged with yellowish-brown, slightly thickened, surrounding culm and extending beyond base of blade. Wide-ranging, but apparently local.

41. Carex albolutescens Schwein.

C. straminea of Mackenz., not Willd.

A species of the Atlantic slope northward to Nova Scotia (inland across New York to northeastern Ohio) and of the Mississippi Valley northward to southern Indiana and to southwestern Michigan.

42. Carex cumulata (Bailey) Mackenz.

No specimens seen; a very local species which Fernald credits to Ohio.

43. Carex tribuloides Wahlenb.

A very common and wide-ranging leafy species; leaves shorter than to overtopping culms, 2.5–8 mm wide; inflorescence dense, or interrupted at base, spikes blunt.

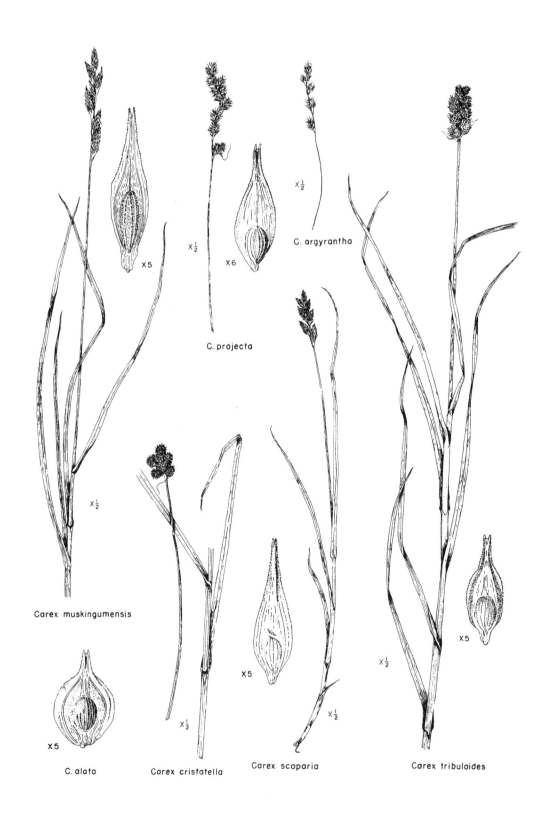

C. argyrantha

C. projecta

Carex muskingumensis

C. alata

Carex cristatella

Carex scoparia

Carex tribuloides

44. **Carex projecta** Mackenz.
    C. tribuloides var. *reducta* Bailey
    A northern species, rare in Ohio.

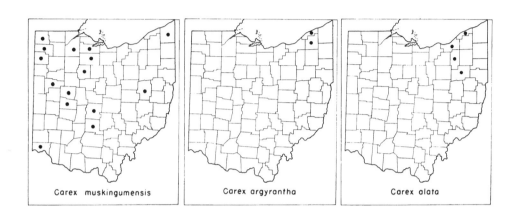

Carex muskingumensis   Carex argyrantha   Carex alata

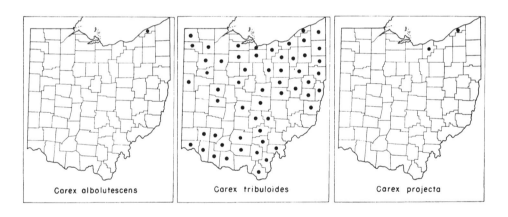

Carex albolutescens   Carex tribuloides   Carex projecta

45. **Carex cristatella** Britt.
    C. cristata Schwein.
    A rather common species of mid-temperate range, in swamps, wet woods, along streams and roadside-ditches. Inflorescence dense, the spikes almost spherical, perigynia widely spreading or radiating in all directions.

46. **Carex scoparia** Schkuhr.
    A highly variable and wide-ranging species in which a number of varieties or forms are sometimes distinguished. Inflorescence compact to moniliform, spikes clearly defined, pointed to rounded at apex, rather lustrous.

47. **Carex tenera** Dewey
    C. straminea of Gray, ed. 7, not Willd.
    A polymorphic species; the moniliform inflorescence with slender axis

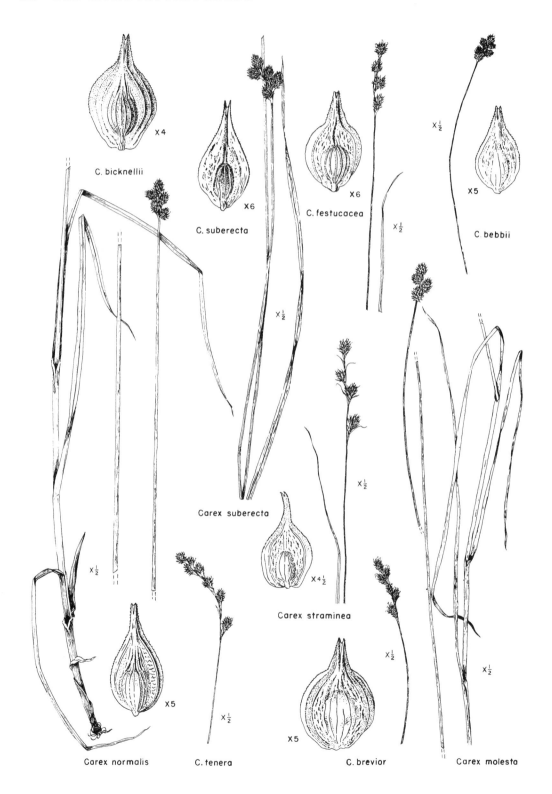

C. bicknellii ×4

C. suberecta ×6

C. festucacea ×6

C. bebbii ×5

×½

×½

Carex suberecta ×½

×½

Carex straminea ×4½

Carex normalis ×5

C. tenera ×½

C. brevior ×5

×½

Carex molesta ×½

flexuous or sometimes nodding above, and the spreading tips of the perigynia give this sedge a somewhat distinctive aspect. Widely scattered in Ohio, but not common.

48. **Carex normalis** Mackenz.

C. mirabilis Dewey, not Host.

A common and widespread leafy woodland species; culms exceeding leaves, leaves mostly 3–6 mm wide; inflorescence more or less crowded, but spikes distinct; perigynia with conspicuous spreading-ascending beaks; scales ovate, nearly as wide as but shorter than body of perigynium.

49. **Carex festucacea** Schkuhr.

C. straminea var. *festucacea* Tuckerm.

Widely scattered in Ohio, but not common; in wet woods, shaded ponds, and ditches; a rather slender species.

50. **Carex bebbii** Olney.

A northern species, rare in Ohio. Somewhat resembling *C. cristatella* but perigynia less widely spreading and scales longer and usually narrowed to tip.

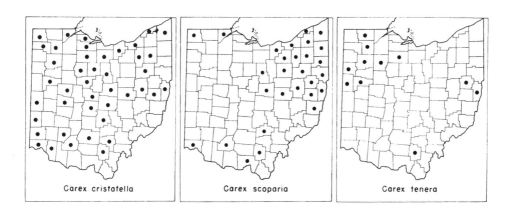

Carex cristatella   Carex scoparia   Carex tenera

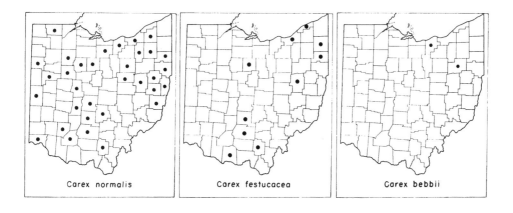

Carex normalis   Carex festucacea   Carex bebbii

51. Carex straminea Willd.

C. richii Mackenz., *C. hormathoides* var. *richii* Fern.

The moniliform, flexous inflorescence, broadly rounded spikes tapering to staminate base, suborbicular perigynia with slender beak, and acuminate scales with excurrent midrib distinguish this local species.

52. Carex bicknellii Britt.

Northern and western in range; the large translucent perigynium is distinctive.

53. Carex suberecta (Olney) Britt.

North-central in range; in Ohio, apparently confined to the glaciated area, as in Indiana. Similar in aspect to other species with crowded or approximate spikes and appressed perigynia, but spikes pointed at apex.

54. Carex merritt-fernaldii Mackenz.

C. festucacea of Gray, ed. 7.

Similar to the next two species, and grouped with *C. brevior* by Gleason (1952). Circumscribed in range, and local; only one Ohio specimen seen (this determined by Mackenzie).

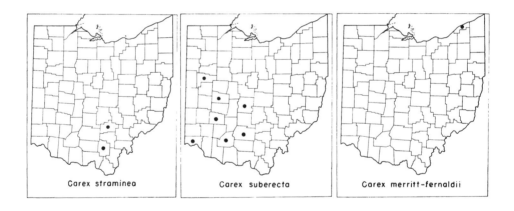

Carex straminea   Carex suberecta   Carex merritt-fernaldii

55. Carex molesta Mackenz.

Similar to the next, from which it is distinguished by characters given in the key; the spikes rounded instead of tapering at staminate base; perigynia faintly few-nerved dorsally. Dry open woods and ditches.

56. Carex brevior (Dewey) Mackenz.

C. festucacea var. *brevior* (Dewey) Fern.

The prominent staminate basal portion of spikes gives the inflorescence of this species a somewhat different aspect than that of the two preceding species.

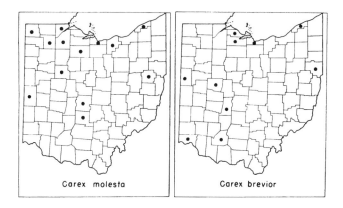

Carex molesta   Carex brevior

Subgenus II. EUCAREX

The larger of the two subgenera of *Carex;* represented in Ohio by almost 90 species classified into 29 sections.

11. *Polytrichoideae*

Densely tufted sedges with capillary culms and very narrow leaves; spikes solitary, bractless, with terminal portion staminate. One species.

57. Carex leptalea Wahlenb.

Transcontinental in range; in swamps and boggy woods, and on wet cliffs. Spikes spire-like, 5–15 mm long; scales of staminate portion wrapping around rachis, their edges overlapping; pistillate scales hyaline, acute, shorter than narrowly ellipsoid perigynia.

12. *Phyllostachyae*

Low, tufted woodland sedges with short culms much overtopped by the leaves; spikes appearing solitary (others basal on elongate peduncles on which spikes appear solitary and terminal), terminal part staminate, its scales tightly clasping rachis, lower part pistillate; lower pistillate scales bract-like, subtending spike; rachis of spike zigzag; perigynia prominently beaked, base stipe-like.

> Perigynia with globose body abruptly contracted to 3-angled beak about half its length; staminate part of spike very slender, less than 0.5 mm thick ....58. *C. jamesii*
> Perigynia with obovoid-oblong body tapering to stout 3-angled beak serrate on its angles and ⅓–½ its length; staminate part of spike more than 0.5 mm thick ....
> 59. *C. willdenowii*

58. Carex jamesii Schwein.

A common sedge of rich woods, mid-western in range; one of the most easily recognized Ohio species, although the rather short-peduncled spikes are hidden among the leaves.

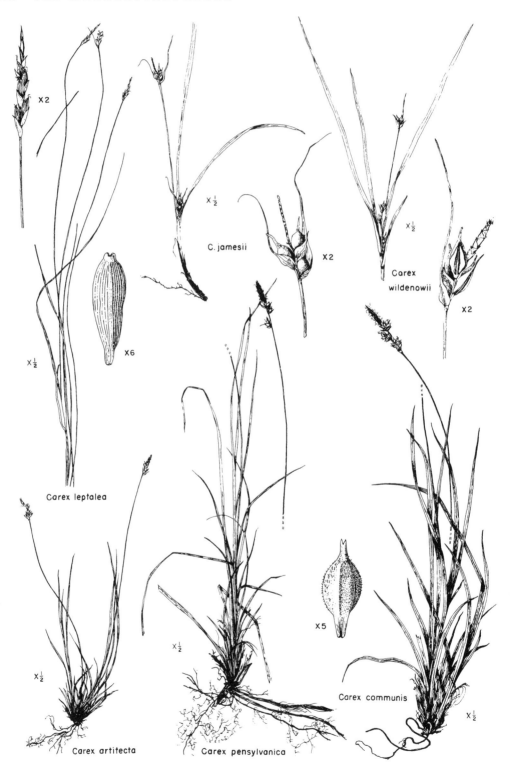

X2

C. jamesii

X½

X2

Carex
wildenowii

X½

X2

X½

X6

Carex leptalea

X½

X½

X½

X5

Carex communis

X½

Carex artitecta

Carex pensylvanica

59. Carex willdenowii Schkuhr

Much wider-ranging than the last, but more local in Ohio; in dry oak woods, clearings, and prairie patches.

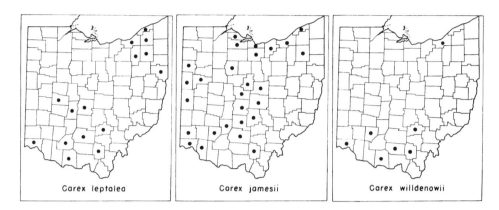

Carex leptalea   Carex jamesii   Carex willdenowii

13. *Montanae*

Woodland sedges of low stature (usually less than 4 dm tall), growing in dense tussocks or mats; flowering in early spring and maturing in May; culms slender; pistillate bracts scale-like, or the lowest prolonged (2–3 cm) and needle-like, without sheathing bases; staminate spike 1, sessile or short-peduncled; pistillate spikes 2 or more, crowded, or the lowest separated, short (usually less than 1 cm); perigynia finely pubescent or puberulent, tightly fitting around achene, obtusely 3-angled, with spongy stipe-like base, and distinct beak ¼–½ length of body; achenes 3-angled. A large group of North America and Eurasia, with one species in the Andes of South America and one in the Azores.

a. Body of perigynium subglobose (above stipe-like base); staminate spikes to 2 cm long, 1.5–3 mm thick, sessile or short-stalked, rising distinctly above pistillate spikes.
    b. Leaves 1–2.5 (–3) mm wide; lowest pistillate bract usually gradually narrowing, acuminate; ligule short; plants with long slender fibrillose stolons ....
                                     60. *C. pensylvanica*
    bb. Leaves 3–5 mm wide; lowest pistillate bract often truncate or bifid, with midrib excurrent as an awn; ligule conspicuous; plants not stoloniferous ....
                                     61. *C. communis*
aa. Body of perigynium longer than wide, ellipsoid to slenderly obovoid.
    b. Perigynia evident in pistillate spikes, not concealed by scales, 2.5–3 mm long; leaves narrow, 0.5–1.5 mm wide.
        c. Inflorescence not crowded, lowest pistillate spikes 3–13 mm apart; midrib of staminate scales usually not extending to tip; culms erect........62. *C. artitecta*
        cc. Inflorescence crowded, pistillate spikes close together (or lowest separate); midrib of staminate scales extending to tip; culms weak, arched ....
                                     63. *C. emmonsii*
    bb. Perigynia inconspicuous in pistillate spikes, mostly concealed by the dark scales, 3–4 mm long; pistillate spikes approximate; leaves 1–4 mm wide, much longer than culms, some of which are almost hidden by leaf-bases ....
                                   64. *C. nigromarginata*

The 3 species, *artitecta*, *pensylvanica*, and *communis*, are all common in Ohio, and are often confused with one another. After one has found all three, the quantitative and qualitative characters given in key and text will usually serve to distinguish them.

60. Carex pensylvanica Lam.

A variable species in which several varieties have been distinguished; Ohio plants belong to the typical variety. Leaves intermediate in width between those of *C. artitecta* and *C. communis*, usually equaling or exceeding culms; stoloniferous. Often forms extensive mats in dry oak woods; wide-ranging.

61. Carex communis Bailey

Usually strongly purple-based; staminate spikes usually purple. The wider leaves and non-stoloniferous habit distinguish this species from the preceding.

62. Carex artitecta Mackenz.

C. varia Muhl., not Lamnitzer; *C. nigromarginata* var. *muhlenbergii* (Gray) Gl.

A common woodland species, with culms usually longer than the very narrow leaves. When compared with the 2 preceding species, the narrow leaves and short staminate spike appear distinctive.

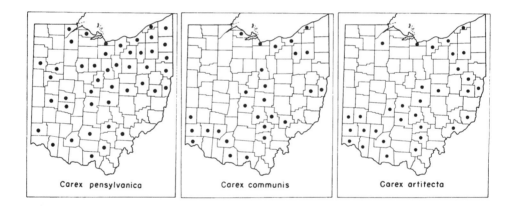

Carex pensylvanica   Carex communis   Carex artitecta

63. Carex emmonsii Dewey

C. nigromarginata var. *minor* (Boott) Gl.; *C. albicans* of Mackenz.; *C. varia* in part, of Gray, ed. 7.

Confined to the Lake area in Ohio.

64. Carex nigromarginata Schwein.

Different in appearance from other members of the section because of its short culms and dark purple-brown spikes; a southern and eastern species.

14. Digitatae

Woodland species of cool-temperate latitudes; leaves basal or nearly so, sheaths red-brown; bracts bladeless; terminal spike staminate or with a few perigynia toward base; pistillate spikes 2–5 on a culm, the upper 1–2 near apex, sometimes with others on slender peduncles almost basal; pistillate scales red-brown; perigynia 3-angled, tapering to stipitate base, abruptly contracted at apex to minute beak, glabrous or thinly pubescent toward tip.

Plants with lower spikes on long, capillary, often basal peduncles; terminal spike pistillate at base; pistillate scales cuspidate or short-awned ....65. *C. pedunculata*
Plants without basal spikes; terminal spike entirely staminate; pistillate scales obtuse, bright brown with broad white hyaline margin ....66. *C. richardsonii*

65. Carex pedunculata Muhl.

A northern species, ranging from Newfoundland to Saskatchewan and southward in the Appalachian Upland; very local in Ohio, flowering in early spring. The often numerous basal or nearly basal red-brown pistillate spikes on capillary peduncles, and leaves much exceeding culms, give this species a distinctive aspect.

66. Carex richardsonii R. Br.

A northern species of calcareous soil, very rare in Ohio; long-sheathing, bladeless purple-brown bracts are conspicuous; peduncles of pistillate spikes included in bract-sheaths.

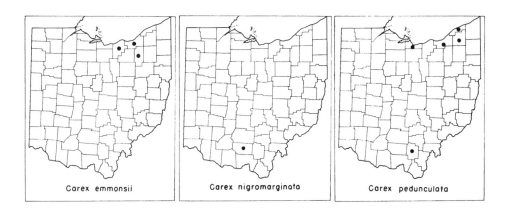

Carex emmonsii   Carex nigromarginata   Carex pedunculata

Carex richardsonii

×2

C. eburnea

X½

C. aurea

X½

C. tetanica

X½

Carex eburnea

X½

C. garberi

X4½

X½

X½

Carex pedunculata

C. hirtifolia

X4½

X½

Carex meadii

15. *Triquetrae*

Woodland or dry-soil species of temperate and warmer latitudes; leafy, more or less tufted sedges with pubescent foliage; terminal spike staminate, pistillate spikes 2–4, short-cylindric; lowest bract with well developed blade; perigynia 3-angled, pubescent, with (in our species) well developed slender beak.

67. **Carex hirtifolia** Mackenz.

A common woodland species maturing in late spring or early summer. The pubescent foliage, culms, and perigynia are distinctive; lowest bract of inflorescence with well developed blade, but sheathless.

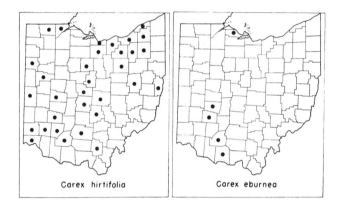

Carex hirtifolia   Carex eburnea

16. *Albae*

A small section with one North American species; stoloniferous, forming mats or sods; culms and leaves very slender; inflorescence corymbiform, pale, the terminal staminate spike inconspicuous, much over-topped by pistillate spikes; pistillate spikes 2–5 mm long; spathe-like bracts and scales whitish; perigynia 3-angled, at first pale, later dark brown or black, lustrous, minutely beaked.

68. **Carex eburnea** Boott

In dry calcareous soil or rock crevices; common (in Adams County) on rocky dolomite slopes under red-cedars, and around red-cedars in prairie patches where it often forms large dense patches. Leaves and culms usually less than 2 dm tall; flowering in late March or April, but the perigynia persistent throughout the summer, when they are dark and lustrous. Ranging from Newfoundland to Alaska, and southward beyond our area.

17. *Bicolores*

Low sedges of wet, chiefly calcareous soil; culms slender, usually less than 3 dm tall; leaves narrow (1–4 mm wide); bracts sheathing, leaf-like; pistillate spikes erect or ascending, loosely to densely flowered, 5–15 mm long; perigynia smooth and golden, or white-puberulent, turgid; achenes lens-shaped, stigmas 2.

Mature perigynia golden-yellow, fleshy (darkening in drying), not crowded; scales pale, short-pointed; terminal spike staminate ....69. *C. aurea*
Mature perigynia white-puberulent, crowded; scales brown or purplish, rounded at tip; terminal spike pistillate above ....70. *C. garberi*

69. Carex aurea Nutt.

A far-northern species—Newfoundland to Alaska, and southward into New England and the northern tier of states; occasional in northern Ohio, but no specimens more recent than 1926. Inconspicuous except at maturity of the golden-yellow or orange perigynia.

70. Carex garberi Fern.

C. bicolor of auth., not All.; *C. hassei* of auth.; not Bailey.

Most frequent near the Great Lakes, in southern Canada and in the states bordering the Lakes on the south. Recorded for Ohio only from the shores of Lake Erie in Ashtabula County.

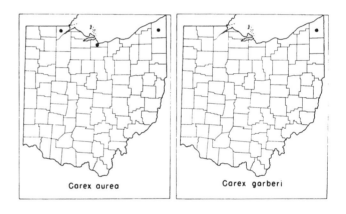

Carex aurea   Carex garberi

18. *Paniceae*

Sedges of woodlands, meadows, and prairies, medium in height; Ohio species somewhat similar in appearance to *Oligocarpae*, but distinguished by the stoloniferous habit, and raised instead of impressed nerves of perigynia; bracts long-sheathing; terminal spike staminate; lower pistillate spikes usually long-peduncled; perigynia obtusely 3-angled, with short outwardly curved tips; achenes 3-angled.

a.　Plants with basal sheaths bladeless, red or "purple;" leaves thin, 1.5–4 mm wide; pistillate spikes loose, alternately flowered, 3–5 mm thick ....71. *C. woodii*
aa.　Basal sheaths blade-bearing, brownish or dull purplish.
　　b.　Leaves thin, 2–4 mm wide; pistillate spikes with perigynia overlapping except toward base of spike, 3–5 mm thick ....72. *C. tetanica*
　　bb.　Leaves stiffish, 3–7 mm wide; pistillate spikes dense, 5–10 mm thick ....
　　　　　　　　　　　　　　　　　　　　　　　　73. *C. meadii*

71. Carex woodii Dewey

C. tetanica var. *woodii* (Dewey) Bailey; *C. colorata* Mackenz.

The purple or red bladeless sheaths, shallow, often purplish stolons, and loosely flowered pistillate spikes distinguish this mid-latitude species of open woods.

72. Carex tetanica Schkuhr

Bases sometimes purplish, usually drab, all sheaths (or all except lowest) blade-bearing. Widely scattered in Ohio, but not common. Typical specimens are rare; this and the next may be considered varieties of *C. tetanica*.

73. Carex meadii Dewey

C. tetanica var. *meadii* (Dewey) Bailey

The compact pistillate spikes and stiff, grayish leaves distinguish this species. Calcareous soil of meadows and prairies, ranging through much of eastern United States and southern Canada, but infrequent except locally.

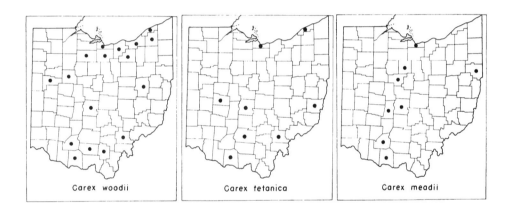

Carex woodii   Carex tetanica   Carex meadii

19. *Laxiflorae*

The name of this section, *Laxiflorae*, meaning loosely flowered, suggests one of its characters: the perigynia are not crowded into dense spikes, and in most species the rachis of the spike, at least toward base, is visible between them. Plants with basal tufts of leaves (wide in some species), the culms often lateral to the tufts; lower bracts (or all) long-sheathing; terminal spike staminate; pistillate spikes linear- to oblong-cylindric, at least the lowest peduncled; perigynia glabrous (or rarely minutely hispidulous), sharply or obtusely 3-angled, short-beaked or beakless; achenes 3-angled. All species of this section grow in woods; perigynia mature in late spring, culms shriveling soon after.

a. Leaves and bracts on fertile culms reduced to bladeless sheaths, these and staminate spikes red-purple; basal leaves 1.5–3 cm wide ....74. *C. plantaginea*
aa. Leaves and bracts on fertile culms with green blades.
   b. Blades of lowest bracts 2–3 times length of sheath, or shorter.
      c. Bases of tufts, lower sheaths, and staminate spikes strongly tinged with red-purple; blades of basal tufts bright green, the basal 2–4 dm long, 7–15 mm wide; perigynia 5–6 mm long ....75. *C. careyana*
      cc. Bases of tufts, sheaths, and staminate spikes not red-purple, mostly straw-color to brownish; blades of basal tufts glaucous, 1–2 dm long, 10–25 mm wide; perigynia 3–4.5 mm long ....76. *C. platyphylla*

bb. Blades of lowest bracts 3 or more times length of sheath.
  c. Perigynia sharply 3-angled; blades of basal leaves 2–12 mm wide.
      d. Basal leaves glaucous, 4–12 mm wide, 1–3 dm long; pistillate spikes 4–5 mm thick, with 1–2 staminate flowers or empty scales at base; perigynia 3–4 mm long ....77. *C. laxiculmis*
      dd. Basal leaves not glaucous, 2.5–5 mm wide, 1–2 dm long; pistillate spikes 3–4 mm thick, without basal staminate flowers; perigynia 2.5–3 mm long, widely spaced ....78. *C. digitalis*
  cc. Perigynia obtusely 3-angled (at least below), tapering to stipe at base.*
      d. Pistillate scales broadly wedge-shaped or fan-shaped, the green midrib not excurrent, or barely projecting; culms wing-angled; basal leaves 1–3.5 (–4) cm wide, culm-leaves and lowest bract (which much exceeds inflorescence) to 2 cm wide; pistillate spikes usually equaling or exceeding staminate spike ....79. *C. albursina*
      dd. Pistillate scales not fan-shaped, the midrib excurrent as a long or short awn or mucro; culms not wing-angled.
         e. Bract-sheaths smooth or nearly so on edges (serrulate in a var. of no. 80); mature perigynia more or less symmetric, with straight or slightly slanted beak, close together or overlapping except toward base of spike.
            f. Perigynia with small straight or curved beak with white or whitish orifice.
               g. Leaf-blades beneath with longitudinal broken lines of very fine whitish dots or streaks (best seen in slanting light); leaves thin, smooth or slightly roughened; staminate spike sessile or short-peduncled; staminate scales pale, with (usually) excurrent midrib ....80. *C. laxiflora*
               gg. Leaf-blades not longitudinally marked with white beneath; leaves thick, rough; staminate spikes with short or long peduncle; staminate scales papery, white, midrib not excurrent ....81. *C. striatula*
            ff. Perigynia with minute, very short beak, the tip therefore conic; culms often purplish at base; staminate scales obtuse to acute, lightly suffused with yellow-brown .... 82. *C. ormostachya*
         ee. Bract-sheaths with erose or serrulate white edges; mature perigynia asymmetric, with beak almost at right angle with longitudinal axis of perigynium.
            f. Bracts much exceeding inflorescence; upper pistillate spikes sessile or nearly so, close together, equaling or overtopping staminate spike; rachis of pistillate spikes wing-angled; perigynia 3–4 mm long; staminate scales greenish white or slighty brown-tinged ....83. *C. blanda*
            ff. Bracts usually shorter than inflorescence; pistillate spikes scattered, often with slender peduncles; rachis of pistillate spikes not winged; perigynia 2.5–3.5 mm long; staminate scales reddish tinged ....84. *C. gracilescens*

74. Carex plantaginea Lam.

One of the sedges with wide basal leaves; readily recognized by its red-purple bladeless sheaths, crowded near base, more widely spaced above, and

---

* Members of this subsection not only vary, but also to some extent intergrade, suggesting that they all may belong to one polymorhpic species—an interpretation used by many authors (Gray, ed. 7, 1908; Gleason, 1952; Gleason & Cronquist, 1963). The species, then is divided into a number of varieties, necessitating the use of trinomials to distinguish the variants. In my treatment, I am following Fernald (1950) and Hermann (1940), and designating these as species; in all cases, the first synonym given is the name used by Gleason, and by Gleason & Cronquist. A large number of specimens are more or less intermediate; occurrence of these is not shown on the distribution maps, unless omission would result in apparent absence from a county; occurrence then is shown by a circle instead of a dot.

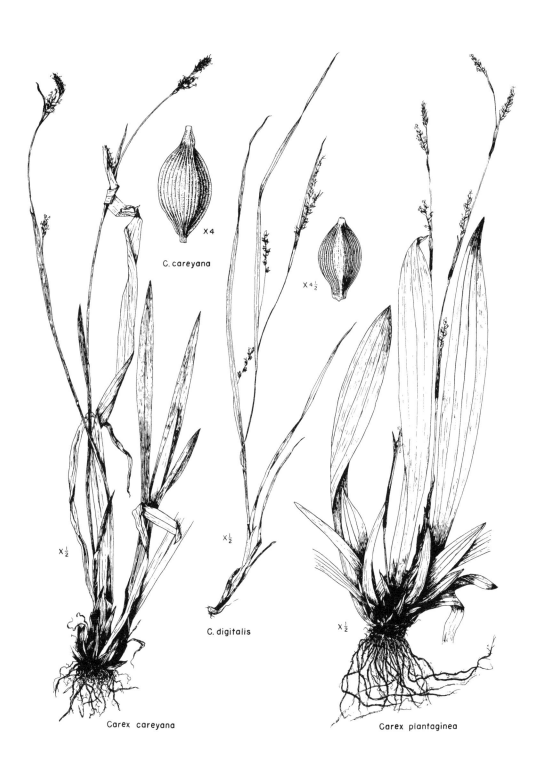

C. careyana

X 4

X 4½

C. digitalis

X½

X½

X½

Carex careyana

Carex plantaginea

C. platyphylla

Corex platyphylla

Carex laxiculmis

red-purple staminate spikes; sheaths produced at apex into concave pointed tip; basal leaves dark green, rather abruptly narrowing to acute apex; perigynia obovoid, 3–5 mm long, with erect or outwardly curving beak. A northern species ranging south in the Appalachian Upland to North Carolina, Georgia, and Tennessee.

75. Carex careyana Torr.

Of the wide-leaved species of the section, this has the largest perigynia (5–6 mm long); beak short, erect; plants red-purple at base, like the last, but distinguished from it by bright green basal leaves tapering to acuminate apex and by the green blades of pistillate bracts; from *C. albursina*, by its narrower basal leaves, and shorter pistillate bracts. Interior in range, usually local.

76. Carex platyphylla Carey

The glaucous basal tufts or rosettes of this species are conspicuous throughout the summer, after the fertile culms have disappeared; viewed from above (see fig.), the 3-ranked arrangement of leaves (common to all Carices) is readily seen; by the following spring, when fertile culms develop, these overwintering leaves are usually ragged at tips; fertile culms usually numerous, often little longer than basal leaves; perigynia obovoid, narrowed to short, outwardly curving beak. Quebec to Michigan, south to North Carolina, Kentucky, and Illinois.

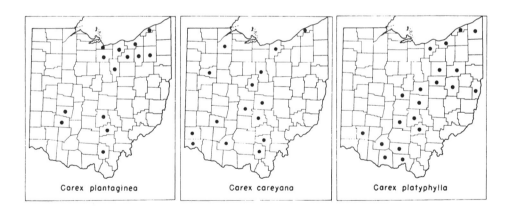

Carex plantaginea   Carex careyana   Carex platyphylla

77. Carex laxiculmis Schwein.

Basal leaves 4–12 mm wide, long, almost equaling culm; the presence of 1 or 2 staminate flowers or empty scales at base of pistillate spikes is a specific character; perigynia distinctly nerved, very short-beaked, much exceeding scales. Maine to Wisconsin, south to North Carolina and Missouri. Apparently hybridizes with *C. digitalis* producing ×*C. copulata* (Bailey) Mackenz. (*C. laxiculmis* var. *copulata* (Bailey) Fern.), scarcely distinguishable from *C. laxiculmis*.

78. Carex digitalis Willd.

A variable species in which 3 intergrading varieties have been distinguished by Fernald. Leaves narrower, culms more slender, pistillate spikes looser or more open, and perigynia smaller than those of other species with acutely 3-angled perigynia; perigynia beakless, almost twice as long as scales. Ranging farther south than any of the preceding species of the section—to Florida and Texas.

79. Carex albursina Sheldon

C. laxiflora var. latifolia Boott

The large basal leaves suggest C. careyana, from which this is distinguished by its bracts (longer in relation to sheath), obtusely 3-angled perigynia, and fan-like pistillate scales. A common woodland species, east-central in range.

Three specimens (1 from Jackson, 2 from Auglaize County) may be hybrids with C. careyana; one of the latter is so annotated by Mackenzie.

80. Carex laxiflora Lam.

C. laxiflora var. laxiflora; C. anceps Muhl.; C. heterosperma Wahl.; C. laxiflora var. patulifolia (Dewey) Carey.

A common and extremely variable "species;" in general appearance, often similar to the last, but distinguished from it by characters given in key, and by its usually smaller bracts; staminate spike occasionally long-peduncled. The white-striolate lower leaf-surface (obscure on old discolored specimens) and white hyaline orifice of perigynium-beak are distinctive.

The var. serrulata F. J. Herm. differs from the typical variety in having the angles of the bract-sheaths serrulate. Originally described from Indiana; Ohio specimens from Auglaize, Jackson, and Scioto counties, determined by F. J. Hermann.

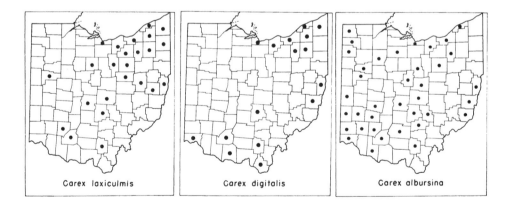

Carex laxiculmis   Carex digitalis   Carex albursina

81. Carex striatula Michx.

C. laxiflora var. angustifolia Dewey, in part; C. laxiflora var. michauxii Bailey.

Intermediates between this and the preceding species are so numerous that recognition as a species is probably unwarranted (included in C. laxiflora by Hermann, 1940). One specimen (Lawrence County) was determined by Mac-

X5

X½

X5

Carex blanda

C. laxiflora

X½

X½

C. gracilescens

X5

Carex albursina

X4½

X½

X5

Carex crawei

Carex granularis

kenzie; another (from Scioto County) may be referred here. Southern in range (where it may be more distinct from *C. laxiflora*), extending north to southern Ohio and southern Indiana.

82. Carex ormostachya Wieg.
  C. laxiflora var. *ormostachya* (Wieg.) Gl.
  A northeastern species, stated by Fernald (1950) to occur in northern Ohio.

83. Carex blanda Dewey
  C. laxiflora var. *blanda* (Dewey) Boott; *C. laxiflora* var. *varians* of auth., not Bailey.
  A common and widespread species—the widest ranging of this subsection —growing in woods in large grass-like tufts. The dense (for the section) pistillate spikes, usually 2 or 3 close together near base of staminate spike, wing-angled rachis of pistillate spikes, awned white-hyaline scales (green midrib excurrent as awn sometimes as long as body of scale), erose whitish margins of bract-sheaths, large perigynia (3–4 mm long) and abruptly bent short perigynium-beak (best seen in profile view) characterize this species, of which Gleason (1952) states "there is evidence that it is a tetraploid."

Carex laxiflora

Carex striatula

Carex blanda

Carex gracilescens

84. **Carex gracilescens** Steud.

C. laxiflora var. *gracillima* (Boott) Robins. & Fern.; *C. laxiflora* of auth., not Lam.

Similar to *C. blanda* and freely integrading with it; staminate spikes usually long-peduncled, elevated well above uppermost (solitary) pistillate spikes; bracts rarely overtopping inflorescence; staminate scales often reddish tinged; lower sheaths of fertile culms often purplish when fresh.

20. *Granulares*

Plants mostly less than 5 dm tall, more or less tufted; bracts with tubular sheaths; terminal spike staminate, pistillate spikes dense, oblong-cylindric, erect or ascending; perigynia beakless or very short-beaked, distinctly nerved; achenes 3-angled, broadest above middle. Differing from the next section chiefly in having elevated instead of impressed nerves. A small section of temperate eastern and central North America.

Staminate spike long-peduncled; pistillate spikes far apart, the lowest often basal ....
85.   *C. crawei*

Staminate spike sessile or nearly so; pistillate spikes short-peduncled (except lower), the upper 1–3 sessile at base of staminate spike ....86.   *C. granularis*

85. **Carex crawei** Dewey

Culms solitary or few together from widely spreading rhizomes, 3–30 cm tall; leaves stiff, glaucous, 1–4 mm wide; pistillate scales much shorter than perigynia, obtuse to short cuspidate, reddish brown with hyaline margins and 3-nerved green center. In calcareous soil, Ohio specimens in prairie patches; wide-ranging, but local.

86. **Carex granularis** Muhl.

Taller than the preceding species, and growing in dense tussocks, the rhizomes short; leaves (except overwintering ones) soft, elongate, to 10 mm wide; pistillate scales from half to as long as perigynia; perigynia puncticulate (best seen in side light), varying in shape and strength of nerves, characters used to distinguish two varieties of similar range; var. *haleana* (Olney) Porter

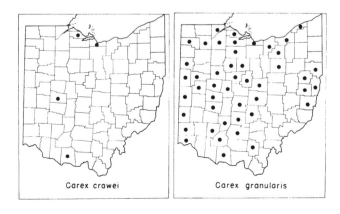

Carex crawei

Carex granularis

(*C. shriveri* Britt.) has more slender perigynia than var. *granularis* (1–1.5 mm thick instead of 1.5–2.5 mm), and, often, weaker nerves (drawing to right); a few specimens (from Clinton, Defiance, Highland, and Lake counties) are referred to var. *haleana*. Wide-ranging, in moist woods, swamps, roadside ditches, and prairie patches, usually in calcareous soil.

21. *Oligocarpae*

Leafy sedges of rich woods and meadows, or less frequently along streams and ditches; less than 1 m tall (2–8 dm); bract-sheaths long (to 6 cm), tubular; terminal spike staminate, pistillate spikes oblong-cylindric; perigynia with impressed nerves, beakless or constricted to beak-like apex with entire or nearly entire orifice; achenes 3-angled. Often divided into two sections, as indicated in key below. All are plants of eastern North America.

a.   Perigynia fitting closely around achenes, narrowed toward base, obtusely 3-angled ....
                                                                                (OLIGOCARPAE)
      b.   Sheaths rough-pubescent; leaves 3–7 mm wide; pistillate scales about equaling
           perigynia, wide, and abruptly contracted to long attenuate, rough, awned tip ....
                                                                    87. *C. hitchcockiana*
      bb. Sheaths glabrous, ventrally prolonged; leaves 2–4 mm wide; pistillate scales
           longer than perigynia, awns as long or longer than body of scale ....
                                                                    88. *C. oligocarpa*
aa. Perigynia not fitting closely around achene and extending well beyond it, rounded at
      base, almost circular in cross-section .... (GRISEAE)
      b.   Staminate spike long-peduncled; bract-sheaths and peduncles rough ....
                                                                    89. *C. conoidea*
      bb. Staminate spikes sessile or nearly so; bract-sheaths and peduncles smooth.
           c.   Leaves green, usually not at all glaucous, thin; pistillate spikes 3–12-flowered.
                d.   Bases of culms and lower sheaths red-purple; leaves 2–4 mm wide;
                     pistillate spikes 3–10-flowered, the lowest often nearly basal; perigynia
                     tapering to base and apex ....90. *C. amphibola*
                dd. Bases of culms and lower sheaths brownish; leaves 4–10 mm wide;
                     pistillate spikes 5–20-flowered, the lowest near middle of culm; perigynia
                     rounded to base and apex ....91. *C. amphibola* var. *turgida*
           cc. Leaves very glaucous, thick and firm, 4–10 mm wide; pistillate spikes 10–40-
                flowered ....92. *C. glaucodea*

87. Carex hitchcockiana Dewey

The rough-pubescent sheaths distinguish this species from others with beakless or short-beaked perigynia and impressed nerves; pistillate spikes 1–2.5 cm long; perigynia 4–5 mm long; achene with bent-apiculate tip. Central in range; widely scattered in Ohio and apparently absent from southeastern counties.

88. Carex oligocarpa Schkuhr

Similar to the last, but with glabrous sheaths, narrower leaves, shorter pistillate spikes (0.5–1.5 cm), smaller perigynia, and pistillate scales mostly longer than perigynia, awns as long as body of scale; achene with straight tip.

89. Carex conoidea Schkuhr

A northern species, rare in Ohio; resembles small specimens of *C. amphibola* var. *turgida*, but distinguished by its long-peduncled staminate spikes and rough peduncles; leaves 2–4 mm wide; perigynia ellipsoid to oblong-conic, 3–4 mm long.

90. **Carex amphibola** Steud.

A southern species, mostly on the coastal plain, and ranging northward in the Mississippi Valley to Missouri and southern Indiana; rare in Ohio. The typical variety differs from the more common *C. amphibola* var. *turgida* (*C. grisea*) in characters given in the key; var. *rigida*, recorded from Lake and Lorain counties, has firmer and stiffer leaves.

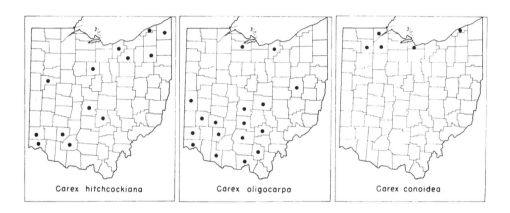

Carex hitchcockiana   Carex oligocarpa   Carex conoidea

Carex amphibola
var. rigida

91. **Carex amphibola** var. **turgida** Fern.

C. grisea Wahlenb.

A common and widespread, very leafy sedge occurring throughout much of the eastern half of the United States and southern Canada; pistillate spikes 1–3 cm long; perigynia 4–5.5 mm long; pistillate scales equaling or longer than perigynia, with awn about as long as body.

92. **Carex glaucodea** Tuckerm.

C. flaccosperma Dewey, var. *glaucodea* (Tuckerm.) Fern.

Conspicuous in the field because of its very glaucous leaves, which, even in dried specimens contrast strongly with the yellowish perigynia; perigynia

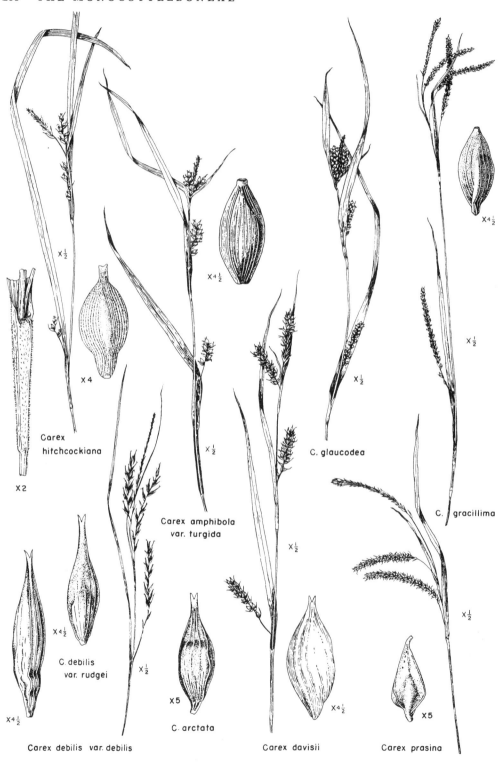

Carex
hitchcockiana

X2

$\times\frac{1}{2}$

$\times 4$

$\times 4\frac{1}{2}$

$\times\frac{1}{2}$

$\times\frac{1}{2}$

C. glaucodea

$\times 4\frac{1}{2}$

$\times\frac{1}{2}$

C. gracillima

Carex amphibola
var. turgida

$\times 4\frac{1}{2}$

C. debilis
var. rudgei

$\times\frac{1}{2}$

$\times 5$

C. arctata

$\times\frac{1}{2}$

$\times\frac{1}{2}$

$\times 4\frac{1}{2}$

$\times\frac{1}{2}$

$\times 5$

$\times 4\frac{1}{2}$

Carex debilis var. debilis

Carex davisii

Carex prasina

3–5 mm long; pistillate scales about half as long, acute, or lower with short awns. South-central and eastern in range; in oak woods, open slopes, and along woodland paths.

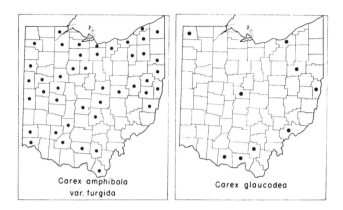

Carex amphibola
var. turgida

Carex glaucodea

22. *Gracillimae*

Woodland species of wet soil; lowest sheaths of fertile culms without blades; bracts at base of inflorescence long-sheathing (0.5–5 cm); spikes 2 or more, terminal spike pistillate except at base (or, in *C. prasina*, pistillate only at tip, or all staminate); pistillate spikes linear- to oblong-cylindric, spreading or ascending, or the lowest drooping; perigynia beakless or short-beaked; achenes 3-angled.

a. Perigynia beakless, bluntly 3-angled, obtuse at apex, 2–3.5 mm long; pistillate scales ovate to obovate, obtuse (rarely cuspidate); bract at base of inflorescence with long (3–4 cm) tubular sheath; leaves glabrous ....93. *C. gracillima*
aa. Perigynia with beak.
    b. Sheaths and leaf-blades glabrous (lower sheaths may be dorsally hispidulous); perigynia sharply angled at lateral ribs, bluntly so on back, tapering to curved flattened beak, 3–4 mm long; pistillate scales acute to short-awned, shorter than perigynia; sheath of bract at base of inflorescence short (to 5 mm) ....
                                   94. *C. prasina*
    bb. Sheaths and often leaf-blades pubescent; perigynia not sharply angled at lateral ribs; sheath of bract at base of inflorescence long.
        c. Pistillate scales long-attenuate or awned, equaling or exceeding perigynia; lateral spikes sessile or short-peduncled ....95. *C. davisii*
        cc. Pistillate scales obtuse to acute or short-cuspidate, shorter than perigynia; lateral spikes on long filiform peduncles ....96. *C. formosa*

93. **Carex gracillima** Schwein.
A common and widespread species of wet woods and shaded swamps; purplish at base, lowest bladeless sheaths often purple throughout; leaf-blades 3–9 mm wide; perigynia 2–3.5 mm long; peduncles slender, shorter than to equaling spikes.

94. **Carex prasina** Wahlenb.
Less widespread than the preceding; in wet woods or along small streams

in woods. Best recognized by the peculiar flattened (later 3-angled) beak tapering from the prominent lateral ribs of the perigynium.

95. Carex davisii Schwein. & Torr.

More or less interior in range; found farther south than the two preceding species. Pistillate spikes thicker (4–6 mm) and perigynia larger (4.5–6 mm long) than those of the two preceding species. A woodland species of neutral or calcareous soil.

96. Carex formosa Dewey

A northern species; the filiform peduncles suggest *C. gracillima*, but perigynia much larger (3.5–5 mm long) and peduncles longer than spikes.

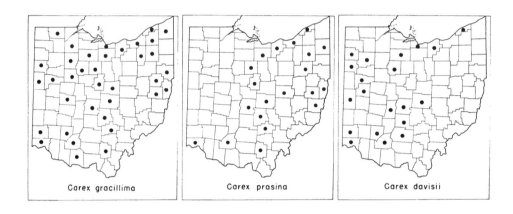

Carex gracillima   Carex prasina   Carex davisii

Carex formosa

23. *Sylvaticae*

Mostly woodland species; bases purple or purplish, lower sheaths without green leaves; spikes 2 or more, terminal spike staminate, pistillate spikes (of Ohio species) linear-cylindric, at least the lower peduncled and spreading to drooping; bract at base of inflorescence (or all) sheathing; perigynia lance-ovoid to ovoid, tapering to beak; achenes 3-angled. A large section of temperate North America, Eurasia, and the highlands of eastern Africa.

a. Pistillate scales cuspidate or awned, more than half as long as perigynia; perigynia 3-angled; achenes sessile to short-stipitate; culm-leaves 3–5 mm wide, those of basal shoots to 10 mm wide .... 97. *C. arctata*
aa. Pistillate scales obtuse to short acuminate, sometimes the lowest short-awned, half to less than half as long as perigynia; perigynia obscurely 3-angled; achenes stipitate; basal and culm-leaves narrow, 2–4 (–7) mm.
    b. Pistillate scales pale; tip of perigynium-beak white-hyaline ....
                                         98. *C. debilis* var. *debilis*
      bb. Pistillate scales straw-color or greenish; tip of perigynium-beak not white-hyaline ....
                                          99. *C. debilis* var. *rudgei*

97. Carex arctata Boott

A northern species, locally south into Ohio. Spikes very slender, loosely flowered, the lower distinctly separated; perigynia narrowly ovoid.

98. Carex debilis Michx. var. debilis

Southern in range, on the coastal plain north to Massachusetts and in the Mississippi Valley to southern Indiana and southern Ohio. Basal and culm leaves narrow, 2–4 (–7) mm. The whitish scales and white-hyaline tip of beak are distinctive characters; perigynium narrowly lanceolate. Represented in Jackson County where some specimens approach var. *rudgei* (fide F. J. Hermann).

99. Carex debilis var. rudgei Bailey

    C. flexuosa Muhl.; *C. tenuis* Rudge

More northern than the typical variety, ranging from Newfoundland to Ontario and Minnesota southward to Missouri, Kentucky, Virginia, and the mountains of Tennessee and North Carolina.

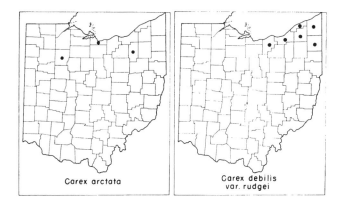

Carex arctata

Carex debilis
var. rudgei

24. Longirostres

Similar in most characters to the preceding; bases not purple, but brownish, lower sheaths with green blades; pistillate spikes oblong-cylindric; perigynia prominently beaked. A small section, 2 species in eastern North America, one in eastern Asia.

100. Carex sprengelii Dewey

Pistillate spikes very different in appearance from other sedges, because of the widely spreading perigynia (rachis visible), the rounded body contracted to a long, usually basally bent, beak 2-cleft at apex; achene with abruptly bent apiculate tip. A northern species ranging from New Brunswick to Alberta, and locally south into our area; on rock ledges in woods.

Carex sprengelii

25. Extensae

Sedges with leafy sheathless bracts and several (usually 3–5) pistillate spikes; subglobose to short cylindric, about 1 cm thick or less, mostly sessile; perigynia strongly nerved, inflated, with 2-toothed beak half to as long as body, divergent to reflexed; achenes 3-angled, loosely enveloped by perigynia. In appearance, suggesting diminutive members of the *Lupulinae*, but style is not persistent on achene. The Ohio species are all widely distributed, northern sedges, local as far south as Ohio. A large section, represented on all continents.

a. Perigynia 3.5–6 mm long, spreading or the lower reflexed; beak as long as body; pistillate spikes subglobose, 7–12 mm thick.
    b. Leaf-blades 3–5 mm wide; pistillate scales reddish brown with green central stripe, plainly visible in mature spikes ....101. *C. flava*
    bb. Leaf-blades 1–3 mm wide; pistillate scales not reddish brown, mostly hidden by perigynia ....102. *C. cryptolepis*
aa. Perigynia 2–3 mm long, not conspicuously reflexed; beak about half as long as body; pistillate spikes oblong, 4–7 mm thick; leaf-blades 1–3 mm wide; pistillate scales tinged with brown ....103. *C. viridula*

101. Carex flava L.

A species of marly bogs, thus far found in Ohio only in Cedar Swamp, Champaign County. The slightly inflated, strongly ribbed, long-beaked perigynia about 5–6 mm long, conspicuous reddish brown pistillate scales, and soft, relatively wide leaves distinguish this species, as here interpreted, i.e., exclusive of var. *fertilis* (see *C. cryptolepis*).

102. Carex cryptolepis Mackenz.

C. flava var. *fertilis* Peck.

Following Gleason, 1950, Gleason & Cronquist, 1963, and Hermann (in

Deam, 1940), I am treating this as a species, distinguished from *C. flava* in the narrower sense (i.e., var. *flava*) by characters given in the key. Also, the perigynia are smaller, 3.5–4.5 mm long. Fernald (1950) states that this is "well marked in extreme development but too freely confluent with true *C. flava.*" Differences between true *C. flava* and this species (or variety) are well marked in Ohio. Except for one southern Ohio occurrence (Adams County, in wet marly clay), *C. cryptolepis* seems to be confined to northeastern Ohio.

103. Carex viridula Michx.

C. irregularis Schwein.; *C. oederi* var. *pumila* of Gray, ed. 7, not Retz., incl. *C. chlorophila* Mackenz.

Terminal spike sometimes partly or wholly pistillate, peduncled or sessile. Because of the short perigynium-beak, the oblong, compact pistillate spikes lack the spiny appearance of the other two species.

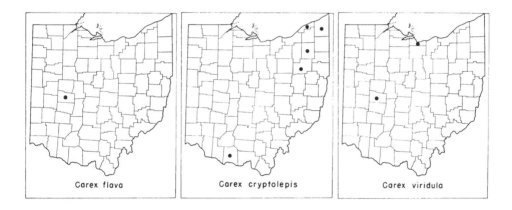

Carex flava   Carex cryptolepis   Carex viridula

26. Virescentes

A group of woodland and meadow sedges of North America, Eurasia, and the mountains of northern South America. Sheaths and often leaf-blades pubescent; leaves shorter than to equaling or exceeding culms; bracts sheathless, at least the lower leaf-like; terminal spike staminate, or staminate at base only; pistillate spikes relatively small, mostly less than 2 cm long and 7 mm thick; perigynia ellipsoid to obovoid, glabrous or pubescent, beakless (in Ohio species); achenes 3-angled.

a.   Terminal spike staminate; pistillate spikes 2–3, thick-cylindric (5–20 mm long, 5–7 thick); perigynia glabrous, faintly nerved; lowest bract with a series of transverse waves near base ....104.   *C. pallescens*
aa.   Terminal spike staminate at base only.
   b.   Perigynia glabrous.
      c.   Leaf-blades and sheaths pubescent; perigynia flattened on inner, weakly-nerved face, convex and strongly nerved on outer face, appressed-ascending ....
                                                 105.   *C. hirsutella*
      cc.   Leaf-blades glabrous, sheaths pubescent; perigynia plump, rounded, equally nerved dorsally and ventrally, spreading-ascending ....106.   *C. caroliniana*

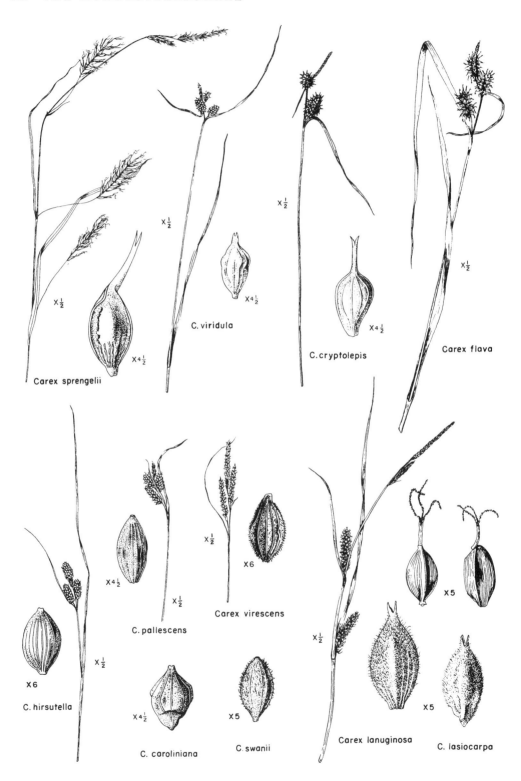

$x\frac{1}{2}$

$x\frac{1}{2}$

$x4\frac{1}{2}$

Carex sprengelii

C. viridula

$x4\frac{1}{2}$

$x\frac{1}{2}$

$x4\frac{1}{2}$

C. cryptolepis

$x\frac{1}{2}$

Carex flava

$x\frac{1}{2}$

$x4\frac{1}{2}$

$x\frac{1}{2}$

$x6$

Carex virescens

$x6$

C. hirsutella

$x\frac{1}{2}$

C. pallescens

$x\frac{1}{2}$

$x4\frac{1}{2}$

C. caroliniana

$x5$

C. swanii

$x\frac{1}{2}$

Carex lanuginosa

$x5$

$x5$

C. lasiocarpa

$x5$

    bb. Perigynia pubescent, obscurely 3-angled; leaf-blades pubescent.
        c. Pistillate spikes oblong to subglobose, compact, 5–18 mm long, 3–5 mm thick, rachis of spike hidden; perigynia broad-obovoid ....107. *C. swanii*
        cc. Pistillate spikes linear-cylindric, loose, 15–40 mm long, 2–4 mm thick; perigynia ellipsoid or narrowly obovoid; rachis of spike visible between perigynia, especially toward base ....108. *C. virescens*

104. Carex pallescens L.

Similar in general appearance to the next three species, but at once distinguished by its all-staminate terminal spike and transversely undulate bract. Fernald (1950) segregates the eastern American from the Old World race as var. *neogaea* Fern.

105. Carex hirsutella Mackenz.

C. complanata Torr. & Hook., var. *hirsuta* (Bailey) Gl.; *C. triceps* var. *hirsuta* (Willd.) Bailey

The most abundant sedge of this section, usually in open woods, clearings, and dry fields; throughout much of the eastern half of the United States.

106. Carex caroliniana Schwein.

C. triceps var. *smithii* Porter

Similar in appearance to the last; distinguished by characters given in key, and, usually, by color of sheaths—reddish brown tinged or spotted, especially on thinner part of sheath (sheaths of *C. hirsutella* are paler); achene with very abruptly bent apiculate tip (achene of *C. hirsutella* with somewhat bent apical tip). Southeastern in range.

107. Carex swanii (Fern.) Mackenz.

C. virescens var. *swanii* Fern.

Similar in appearance to the two preceding; distinguished by its pubescent perigynia. North-central in range.

108. Carex virescens Muhl.

Distinguished from other species of the section by its linear-cylindric, loosely flowered spikes, the lower perigynia scarcely overlapping. East-central in range.

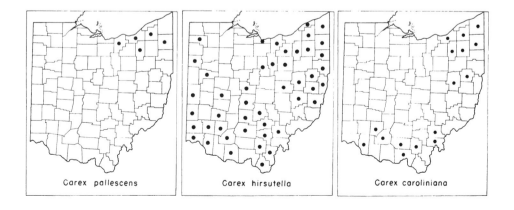

Carex pallescens   Carex hirsutella   Carex caroliniana

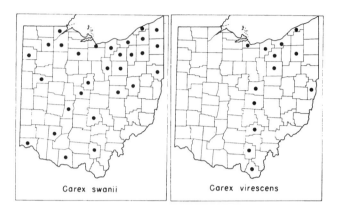

Carex swanii   Carex virescens

27. *Hirtae*

Slender leafy sedges with sessile or nearly sessile cylindric pistillate spikes and pedunculate staminate spikes; bracts leaf-like, equaling or exceeding culm; perigynia plump, many-ribbed, short-beaked, glabrous or (in Ohio species) pubescent, the pubescence dense and partially hiding the ribs; achenes 3-angled.

Leaves 2–5 mm wide, flat, or margins revolute, rough; teeth of perigynium-beak usually more than 0.5 (0.3–0.8) mm long; achene 3-angled, obovoid, with straight short tip; culms sharply 3-angled and serrulate above ....109. *C. lanuginosa*
Leaves 2 mm wide or less, convolute-filiform, smooth except sometimes near tip; teeth of perigynium-beak less than 0.5 (0.2–0.5) mm long; achene 3-angled ellipsoid-obovoid, with short bent tip; culms obtusely angled ....110. *C. lasiocarpa*

109. Carex lanuginosa Michx.

C. lasiocarpa var. *latifolia* (Böck.) Gl.

The densely pubescent, short-beaked, plump-ovoid perigynia distinguish this (and the next species) from other sedges. Transcontinental in the North, ranging far southward; widespread in Ohio but more frequent northward.

110. Carex lasiocarpa Ehrh. var. americana Fern.

C. filiformis of Gray, ed. 7

Distinguished from the last by its quill-like smooth culms, tightly rolled thread-like leaves, and crooked tip on achene. Ranging across the continent in the North, and southward into our area.

28. *Anomalae*

Sedges with leafy sheathless bracts and several (usually 3–5) pistillate spikes; perigynia scabrous, strongly beaked, loosely enveloping the 3-angled achene. A group best represented in eastern Asia; only 2 North American species, one eastern, one western.

111. Carex scabrata Schwein.

Culms 3–8 dm tall; leafy bracts and leaves rough, wide, those of culm 4–8 mm, those of sterile basal shoots up to 18 mm wide; perigynia obovoid, abruptly narrowed to slightly toothed outwardly curved beak about as long as body, minutely pubescent or scabrous. Open woods and swamps, northeastern in range and south in the mountains to South Carolina and Tennessee.

29. *Shortianae*

Inflorescence of several compact narrow-cylindric spikes, the terminal spike pistillate above, staminate below; perigynia 3-angled but flattened (one side larger than others), broad-obovate, about as wide as long, without beak; achenes 3-angled, obovoid, with bent-apiculate tip. A single species.

112. Carex shortiana Dewey

About 4–8 dm tall; leaves rough, 4–8 mm wide. The brown, transversely wrinkled, wide beakless perigynia, brown, broad-ovate pistillate scales with green midrib, and partly staminate terminal spike make this one of the most easily recognized sedges. In wet soil of stream or pond borders, roadside ditches, and open woods; interior in range, common and widespread in Ohio.

30. *Limosae*

Slender bog sedges with yellow felt-covered roots, narrow leaves, and narrow bracts shorter than to about as long as inflorescence; terminal spike staminate, pistillate spikes 1–3, on slender peduncles; perigynia compressed triangular, with very short beak or beakless.

Pistillate scales broad, ovate, about as wide and long as the perigynia ....113. *C. limosa*
Pistillate scales lanceolate, usually longer but narrower than the perigynia ....
114. *C. paupercula*

113. Carex limosa L.

Widespread in the North, and in Eurasia; usually growing in peat bogs in sphagnum. Pistillate spikes ovoid, thick in proportion to length; pistillate scales brown, about as wide as long, with strong, slightly excurrent midrib; perigynia pale, flattened, almost equaled by the dark scales.

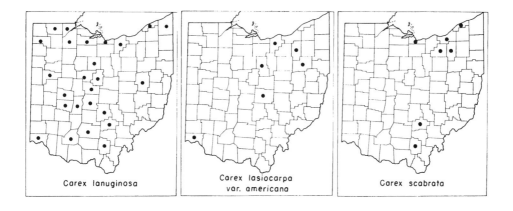

Carex lanuginosa

Carex lasiocarpa
var. americana

Carex scabrata

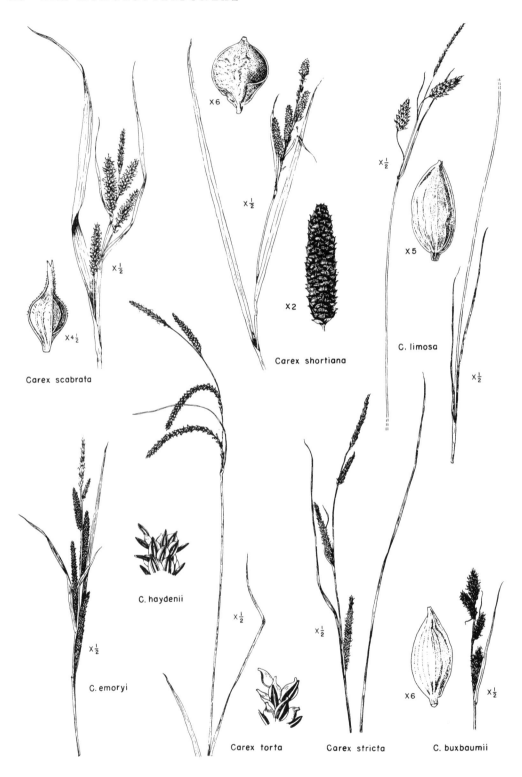

×6

×½

×4½

×½

Carex scabrata

×½

×2

Carex shortiana

×½

×5

C. limosa

×½

C. haydenii

×½

C. emoryi

×½

×½

Carex torta

×½

Carex stricta

×6

×½

C. buxbaumii

114. Carex paupercula Michx.

C. magellanica of auth., not Lam.

A variable boreal species, of which var. *pallens* Fern. is said to grow in northern Ohio (Fernald, 1950); no specimens seen.

31.   *Atratae*

Similar to the last section, but with terminal spike pistillate above, perigynia beakless or with 2 minute teeth, and roots without felt-like cover. A large group of cool temperate and mountainous regions, especially of North America and Asia; one in southern South America.

115. Carex buxbaumii Wahlenb.

C. polygama Schkuhr, not Gmel.

The pale or glaucous, sharply keeled narrow (2–3 mm) leaves, slender culms with few basal (2–4) leaves, and old sheaths disintegrating into fibers are vegetative characters of this species; pistillate spikes 2–4, sessile or nearly so, terminal spike pistillate in upper half, staminate below. The pale or whitish-granular elliptic and flattened beakless (or 2-toothed) perigynia contrast strongly with the dark brown pistillate scales (pale in forma *dilutior* Kükenth. from Fairfield County) with pale midrib prolonged into short point or awn. A widespread northern American and Eurasian species of wet meadows and swamps.

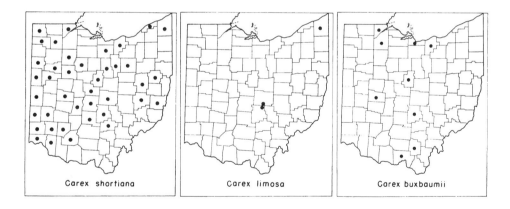

Carex shortiana   Carex limosa   Carex buxbaumii

32.   *Acutae*

Culms mostly less than 1 m tall, leafy below; bracts sheathless or nearly so, leafy blades equaling or exceeding inflorescence, auricled and often darkened at base; terminal one or several spikes staminate; pistillate spikes several, upper erect, lower sometimes drooping; scales obtuse to acuminate, not awned; perigynia compressed, biconvex or plano-convex, beakless or minutely beaked, loosely enveloping the lens-shaped achene; stigmas 2. Plants of shallow water or swamps, borders of creeks, and roadside ditches, of cooler parts of the northern and of the southern hemispheres and mountains of the tropics.

   a. Culms leafy, lowest leaves with well developed blades; bracts usually overtopping inflorescence; pistillate spikes 3–5, dense, erect, separated or the upper overlapping ......
                          116.   *C. aquatilis*
  aa. Culms with lowest leaves reduced to bladeless sheaths.
      b. Perigynia distinctly beaked, the beak bent or twisted; pistillate spikes slender, curving, or the lower drooping; pistillate scales dark red-purple with green midband, obtuse, shorter than to as long as perigynia; culms smooth above ....
                          117.   *C. torta*
    bb. Perigynia very short-beaked or beakless, beak not twisted; pistillate spikes not drooping; culms rough above (at least on angles).
        c. Pistillate scales appressed ascending; achenes longer than broad; pistillate scales obtuse to acute or acuminate.
           d. Lower sheaths (or some of them) becoming filamentous on inner face; ligule longer than wide; pistillate spikes usually 3 (2–6), the lower often remote ....118.   *C. stricta*
          dd. Lower sheaths not becoming filamentous; ligule wider than long; pistillate spikes usually 4 (3–5), erect, overlapping ....119.   *C. emoryi*
      cc. Pistillate scales widely spreading at maturity, acuminate, longer than perigynia; achenes suborbicular; upper bracts much reduced, with hyaline auricles at base ....120.   *C. haydenii*

116. Carex aquatilis Wahlenb.

A variable circumpolar species ranging from Arctic America southward into our area, where, according to some authors, it is represented by var. *altior* (Rydb.) Fern. (var. *substricta* Kükenth). Bases of fertile culms surrounded by old dried leaves of previous year; perigynia flattened, often purple-dotted, minutely beaked.

117. Carex torta Boott

Distinguished from other Ohio species of the section by the bent or twisted beak of perigynium, and by its more slender drooping pistillate spikes. Central in range; widely scattered in Ohio, but apparently not common.

118. Carex stricta Lam.

C. stricta var. *angustata* (Boott) Bailey

As interpreted here, this includes var. *strictior* (Dewey) Carey, which is said to differ in having the leaf-sheaths hispidulous. Inflorescence much less crowded than that of *C. emoryi*, and spikes generally more slender, and leaves narrower (1.5–3.5 mm); staminate spike peduncled, slender, often with 1 or 2

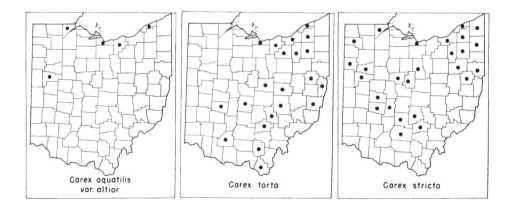

Carex aquatilis
var. altior

Carex torta

Carex stricta

small sessile spikes at its base. A wide-ranging species, and the most frequent species in Ohio.

119. Carex emoryi Dewey

C. stricta var. *elongata* (Böck.) Gl.

The generally more crowded spikes, often overlapping half or more their length, and wider leaves (3–5 mm) give this species an aspect different from that of the similar *C. stricta*. Although this is generally considered a more southern species (Florida, Texas, and New Mexico northward), all Ohio records are from the glaciated area, as is also the case in Indiana (Hermann, 1943).

120. Carex haydenii Dewey

C. stricta var. *decora* Bailey

Rare in northern Ohio; distinguished by its less flattened perigynia and widely spreading (divaricate) pistillate scales.

33. *Cryptocarpae*

Tall sedges (to 1.5 m) of wet woods and swamps, with long, wide (to 12 mm) leaves, the lowest reduced to bladeless (at length filamentous) sheaths; bracts long; staminate spikes slender, 1–3; pistillate spikes peduncled, arching to pendulous; pistillate scales (in ours) long-awned, much exceeding perigynia; perigynia broad-ellipsoid to obovoid, the beak minute; achenes (in ours) contorted and deeply constricted near middle on one side.

121. Carex crinita Lam.

A striking and beautiful species, with long (to 10 cm) arching or pendulous pistillate spikes shaggy with the long-awned scales; awns flat, to 10 mm long, 2–4 times as long as perigynia, rough-margined. Two varieties occur in Ohio: var. *crinita*, with smooth sheaths and body of lower pistillate scales truncate or retuse; and var. *gynandra* (Schwein.) Schwein. & Torr., with sheaths of lower leaves rough, and body of pistillate scales acute; the latter is recorded from Harrison and Medina counties, as well as from several (Carroll, Coshocton, Jefferson) from which the typical variety is also known.

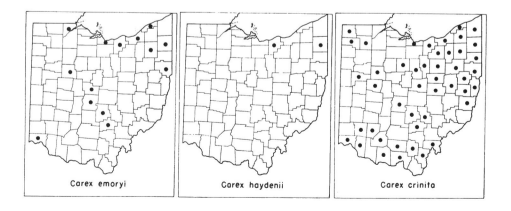

Carex emoryi   Carex haydenii   Carex crinita

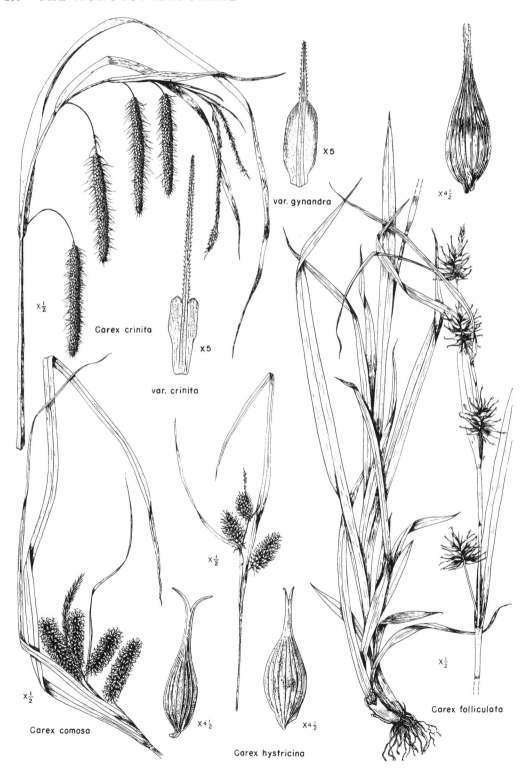

X5

var. gynandra

X4½

X½

Carex crinita

X5

var. crinita

X½

X½

X½

X½

Carex comosa

X4½

X4½

Carex hystricina

X½

Carex folliculata

34. *Folliculatae*

Densely tufted sedges with basal leaves shorter than culms; perigynia inflated, lanceolate, tapering from near base into beak, many-nerved; teeth long (1 mm), erect; achenes 3-angled, continuous with straight or bent style. One species in Ohio.

122. Carex folliculata L.

This species includes two varieties, of which only the typical occurs in Ohio. In growth habit, unlike other of the sedges with large perigynia; basal leaves wide (0.5–1.5 cm) and shorter than culms; leafy bracts long, their sheaths long, enlarging upward; pistillate spikes far apart, subglobose; pistillate scales from acute to long-awned; perigynia long (1–1.5 cm). A northern species, ranging south in the mountains to North Carolina and Tennessee.

Carex folliculata

35. *Pseudo-Cypereae*

Tall, slender or stout sedges of swamps and shores, or occasionally in shallow water; leaves septate-nodulose; bracts exceeding culms; pistillate spikes dense, elongate, cylindric and 10–15 mm in diameter, erect or drooping; pistillate scales with midvein extending as an awn; perigynia strongly 12–17-ribbed, ovoid-lanceolate, less than 10 mm long, contracted to beak with well developed erect or recurved teeth; achenes 3-angled, style flexuous or bent, stigmas 3. Sedges of temperate and warmer latitudes; represented on all continents.

C. lurida is included in this section by some authors (see 38. *Vesicariae*).

a. Teeth of perigynia outwardly arching, 1–2 mm long; mature perigynia obliquely stipitate, reflexed, in dense spikes, the lower drooping on slender peduncles; leaves 6–15 mm wide ....123. *C. comosa*
aa. Teeth of perigynia erect or nearly so, 1 mm long or less.
    b. Mature perigynia spreading to ascending, nearly round in cross-section, short stipitate ....124. *C. hystricina*
    bb. Mature perigynia reflexed, more or less 2-edged or flattened-triangular in cross-section, obliquely stipitate ....125. *C. pseudo-cyperus*

123. Carex comosa Boott

C. pseudo-cyperus var. *americana* Hochst.

A tall (to 1.5 m), wide-leaved sedge (to 1.5 cm wide) of the eastern half of the United States, and also in the Northwest, often growing in shallow water. The best recognition characters are the backwardly pointing perigynia with long outwardly arching teeth; spikes 3–7 cm long. The Hamilton County record based on Lea's specimens collected in 1837 and 1838 (now in Herbarium ·of Academy of Natural Sciences of Philadelphia).

124. Carex hystricina Muhl.

Wide-ranging swamp sedge, usually in calcareous soils; in Ohio, almost confined to the glaciated area, as it also is in Indiana (Deam, 1943). Perigynia spreading-ascending, the whole spike bristly in appearance; the rough awns of the scales protrude between the perigynia; achenes cuneate-obovoid, the angles thickened. The specific name refers to the bristly character of a porcupine (*Hystrix*, the generic name of the European porcupine). Often confused with *C. lurida* (no. 133) which see for distinctions.

125. Carex pseudo-cyperus L.

A northern species ranging from Newfoundland to Saskatchewan and also in Eurasia. Beak of perigynium shorter than body, the teeth erect to slightly spreading.

36. *Paludosae*

Slender or stout swamp sedges, the sheaths often tinged with purple, leaves more or less septate-nodulose; staminate spikes 2–6, pistillate spikes 2–4, linear- to oblong-cylindric; perigynia firm, glabrous or pubescent, thick in texture, ovoid to slender-ovoid, with short beaks, but long or short teeth. The four Ohio species are similar in aspect.

a.   Teeth of perigynia short, about 0.5 mm long, erect or nearly so; perigynia glabrous, beaks shorter than body; leaf-blades and sheaths glabrous.
  b.   Perigynia with prominent, numerous ribs; old leaf-sheaths becoming fibrillose within; ligule about twice as long as wide; staminate scales obtuse to abruptly awned ....126. *C. lacustris*
  bb. Perigynia faintly ribbed or almost nerveless, nerves impressed; sheaths not breaking into fibers; ligule shorter than to little longer than wide; staminate scales acute or awn-tipped ....127. *C. hyalinolepis*

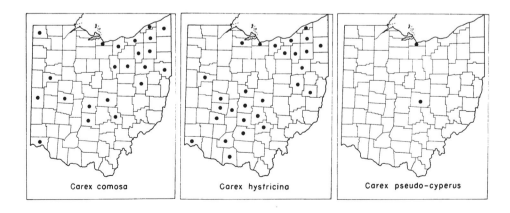

Carex comosa   Carex hystricina   Carex pseudo-cyperus

aa. Teeth of perigynia long (2–3 mm), erect to widely spreading, slender; perigynia glabrous or pubescent, beaks (including teeth) almost as long as body; leaf-blades and sheaths glabrous or pubescent; sheaths purple or brown-purple at mouth.
   b. Perigynia glabrous, the long teeth usually widely spreading; sheaths soft-hairy, brown-purple at mouth; lower leaf-blades hairy on lower side near base ....
                                                         128. *C. atherodes*
   bb. Perigynia pubescent, teeth erect or spreading; sheaths and leaf-blades glabrous ......
                                                         129. *C. trichocarpa*

126. Carex lacustris Willd.

C. lacustris var. *lacustris* (in Gl. and in Gl. & C.); *C. riparia* (in Gray, ed. 7); *C. riparia* Curtis var. *lacustris* (Willd.) Kükenth.

Tall (0.5–1.3 m), usually reddish-purple toward base, lowest sheaths without blades. Widespread in the North, ranging from the Maritime provinces to Manitoba, and south into our area.

127. Carex hyalinolepis Steud.

C. lacustris var. *laxiflora* Dewey; *C. impressa* Mackenz.; *C. riparia* var. *impressa* S. H. Wright

Similar to the last; lower sheaths with blades, not purplish; awns of pistillate scales often equaling or exceeding perigynia. Ranging along the coastal plain from New Jersey to Florida and Texas, and northward in the interior to Ohio, Michigan, Iowa, and Nebraska; in Ohio confined to the western half of the state.

A hybrid, (×*C. subimpressa* Clokey, with hairy perigynia) between this species and *C. lanuginosa* has been found in four northeastern Indiana counties (Hermann, in Deam, 1940).

128. Carex atherodes Spreng.

C. trichocarpa var. *aristata* (R. Br.) Bailey

Basal sheaths purple, without blades, and becoming fibrillose. The strongly ribbed, glabrous perigynia with slender outwardly curving teeth characterize this species. A far-northern species of Eurasia and America, ranging from western New York and Ontario to Oregon and the Yukon; rare in Ohio.

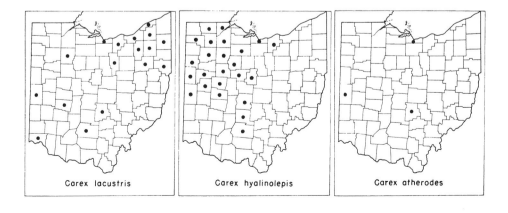

Carex lacustris   Carex hyalinolepis   Carex atherodes

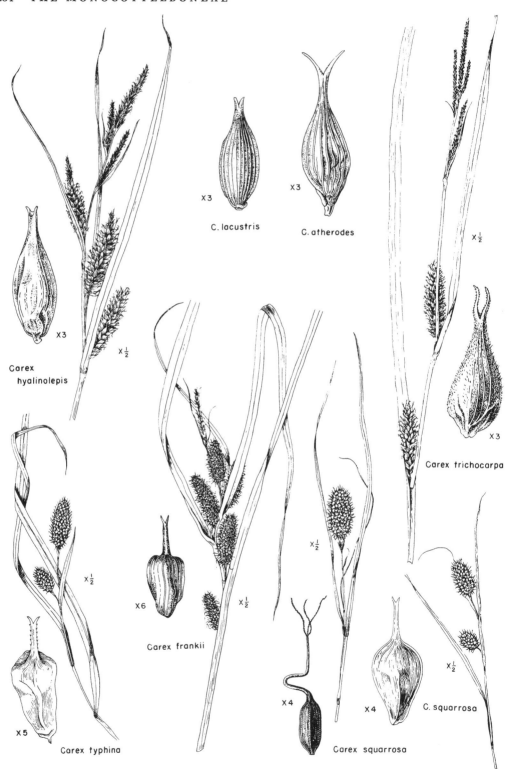

X3
C. lacustris

X3
C. atherodes

X3
Carex
hyalinolepis

X½

X½

X½

X3
Carex trichocarpa

X½

X6
Carex frankii

X½

X½

X4

X5
Carex typhina

X4
Carex squarrosa

X½
C. squarrosa

129. **Carex trichocarpa** Muhl.

Similar to the last, but distinguished from it by the hairy perigynia, and glabrous sheaths and leaf-blades. Mid-northern in range.

Carex trichocarpa

37. *Squarrosae*

Tall sedges of wet woods, swampy creek and lake borders, and roadside ditches; staminate flowers on basal third or less of spikes, or, in *C. frankii* often in terminal all-staminate or largely staminate spike; lateral spikes leafy-bracted, with basal section staminate, or all-pistillate; spikes spherical to cylindric, 8–20 mm thick, much overtopped by the leaf-like bracts; perigynia obconic or broadly obovoid, 3.5–6 mm long, inflated, ribbed, abruptly contracted to beak; achenes 3-angled, style straight or bent; stigmas 3. Almost confined to temperate eastern North America.

a. Scales of pistillate spikes with long rough awns much exceeding perigynia; terminal spike usually staminate; pistillate spikes 2–6, usually crowded ....130.   *C. frankii*
aa. Scales of pistillate spikes (or portions of spikes) shorter than perigynia and obscured by them.
    b. Spikes usually solitary, staminate at base; pistillate portion oblong-oval to sub-globose; beaks of perigynia widely spreading, radiating; style strongly bent or curved near base ....131.   *C. squarrosa*
    bb. Spikes 1–6, the terminal staminate at base; pistillate spikes sub-cylindric; beaks of perigynia more or less appressed and pointing toward apex of spike; style straight below ....132.   *C. typhina*

130. **Carex frankii** Kunth

One of the most easily recognized species of *Carex*, recognized in the field by its crowded, more or less erect pistillate spikes (lower sometimes remote) bristling with the long awns of the scales; staminate spike short-peduncled, often hidden by the pistillate spikes. A widespread and common species of mid and southern latitudes (and also in South America); common in Ohio, but apparently absent from the northeastern part of the state.

131. **Carex squarrosa** L.

Similar in appearance to the next, but usually with only one spike (staminate below) which is shorter in proportion to diameter and (pistillate portion) rounded at both ends, and with perigynia radiating or squarrose; leaves narrower, 2–6 mm wide; style strongly curved near base. Interior in range; in Ohio, more common southward and westward.

132. Carex typhina Michx.

C. typhinoides Schwein.

Distinguished from *C. squarrosa* by its wider leaves (5–10 mm wide), more numerous subcylindric spikes (terminal staminate at base) with beaks of perigynia turned toward apex of spike; style straight or slightly curved near base. Interior in range; in Ohio, common southwestward where it often occurs in depressions in pin oak woods.

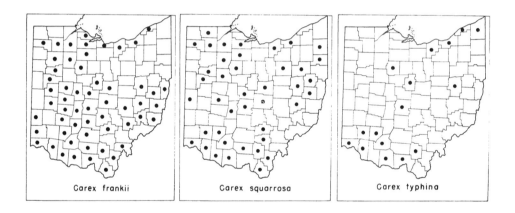

Carex frankii   Carex squarrosa   Carex typhina

38. Vesicariae

Sedges of swampy woods, wet meadows, swamps, or, occasionally, bogs; usually a meter or less in height; leaves from very narrow (in no. 139) to 10 mm in several species, the leaf-like bracts as long as to many times exceeding inflorescence, usually sheathless; staminate spikes 1-several; pistillate spikes 1–8 (usually 2–4), more or less cylindric (except in no. 139); perigynia 4–10 mm long, inflated, ovoid to globose-ovoid, narrowed to beak terminated by 2 sharp teeth (merely notched in no. 139), strongly 7–9 (–10) ribbed; achenes 3-angled (except in one far northern species), style usually sharply bent.

A large section; as interpreted here (following Mackenzie, 1931, Hermann, in Deam 1940, Gleason, 1952, and Gleason and Cronquist, 1962), *C. lurida* and *C. baileyi* (in the *Pseudo-Cypereae* in Fernald, 1950) are included in the *Vesicariae*; these two species have long-awned pistillate scales. The group is well represented in arctic and cold temperate parts of North America; also in Eurasia, northern Africa, and northern South America.

a.   Pistillate spikes cylindric or oblong-cylindric; perigynia numerous, mostly more than 20.
   b.   Pistillate scales with long rough awns.

      c. Pistillate spikes 15–20 mm thick; beak of perigynium half to nearly as long as ovoid body; leaves and bracts 3–7 mm wide ....133.   *C. lurida*

      cc. Pistillate spikes 8–13 mm thick; beak of perigynium equal to or longer than subglobose body; leaves 2–4 mm wide ....134.   *C. baileyi*

  bb. Pistillate scales acute to acuminate, the lower sometimes short-awned.

      c. Bracts several times as long as inflorescence, 4–10 mm wide, sheaths loose; pistillate spikes 3–8, sessile or short-peduncled, all (except lower) close together; perigynia horizontally spreading to reflexed, 6–10 mm long with slender beak half as long as body; teeth long, slender ....135.   *C. retrorsa*

      cc. Bracts from equal to 2–3 times length of inflorescence; perigynia not reflexed.

            d. Achene deeply indented in middle of one angle, about 2× as long as broad; perigynia broadly ovoid, 7–10 mm long, membranous and bladder-like ....136.   *C. tuckermani*

            dd. Achene not indented on one angle.

                  e. Perigynia all ascending, ovoid, narrowed to short (1–2 mm) beak with slender teeth 0.5–1 mm long; basal sheaths becoming strongly filamentous; culms sharply angled, rough ....137.   *C. vesicaria*

                  ee. Perigynia widely spreading, crowded, narrowed to beak about half as long as body; teeth erect to spreading; base of culm with enwrapping sheaths thick and spongy; leaves conspicuously septate-nodulose; culms smooth, except at summit ....138.   *C. rostrata*

aa. Pistillate spikes globose or nearly so; perigynia relatively few (3–15), ovoid, lustrous, beak very short, notched at tip; leaves 1–3 mm wide, involute; culms very slender, smooth ....139.   *C. oligosperma*

133. Carex lurida Wahlenb.

Common and widespread, but apparently absent from the northwestern counties; ranging from Canada to the Gulf and south into Mexico, in wet habitats. Sometimes confused with *C. hystricina*, which has more ribs on the perigynia, and has the ribs "coalesced into a single band on each side of the flattened beak" (Hermann, in litt.), and longer, less strictly erect teeth, and with *C. lupulina*, which has larger perigynia (10–20 mm long), longer teeth (to 2 mm), stigmas radiating from orifice, instead of all protruding from one side as in *C. lurida*, and smooth achene, instead of minutely roughened as in *C. lurida*.

134. Carex baileyi Britt.

An Appalachian and northern species found in Ohio only in Jackson County. Similar to *C. lurida*, but more slender and with narrower leaves.

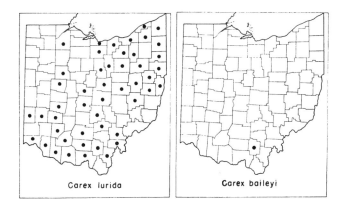

Carex lurida   Carex baileyi

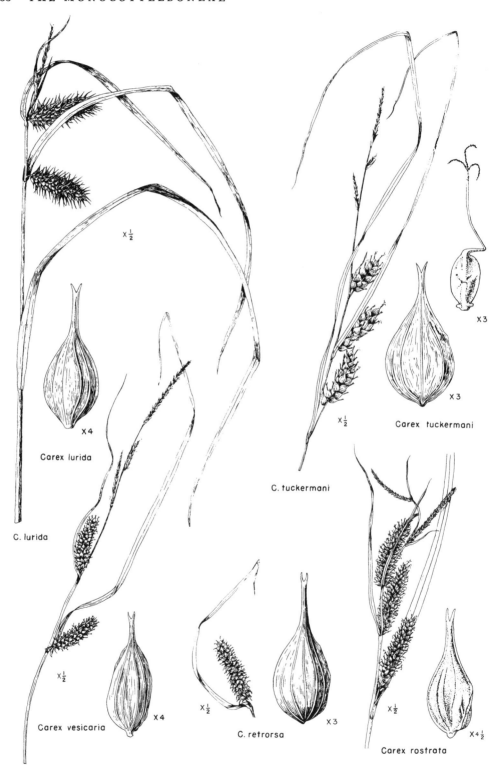

Carex lurida

X½

Carex lurida

X4

C. lurida

Carex vesicaria

X½

X4

C. retrorsa

X½

X3

Carex tuckermani

X3

X3

C. tuckermani

X½

Carex rostrata

X½

X4½

135. Carex retrorsa Schwein.

A northern species ranging across the continent in Canada; known in Ohio only from Lucas County. The closely crowded pistillate spikes, very long (often 5 cm) and wide (4–10 mm) leaf-like bracts with loose sheaths, and the spreading to reflexed perigynia with slender conical beak distinguish this species.

136. Carex tuckermani Boott

A northern species ranging south to New Jersey, Ohio, Indiana, and Iowa. The large (7–10 mm long) thin-walled wide perigynia (more than half as wide as long) and indented achene distinguish this species.

137. Carex vesicaria L.

A highly variable circumboreal species in which a number of poorly marked varieties are sometimes segregated; var. *monile* (Tuckerm.) Fern., *C. monile* Tuckerm., is the only one ranging into Ohio. The filamentous old basal sheaths, sharply angled very rough culms, and comparatively slender pistillate spikes are good characters for field recognition.

138. Carex rostrata Stokes

A northern species of America and Eurasia, in which several varieties have been recognized. Length and shape of pistillate scales vary from south to north; forms within our range can be referred to var. *utriculata* (Boott) Bailey (*C. utriculata* Boott, *C. inflata* var. *utriculata* (Boott) Druce), which has acuminate to awned pistillate scales, wide (to 12 mm) leaves, and crowded perigynia with beak about ¼ as long as body.

139. Carex oligosperma Michx.

A very slender wiry sedge of acid swamps and peat-bogs; very different in appearance from other species of the section because of its filiform-involute leaves, and few-flowered globular pistillate spikes.

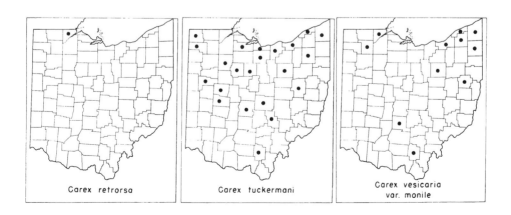

Carex retrorsa   Carex tuckermani   Carex vesicaria
                                                  var. monile

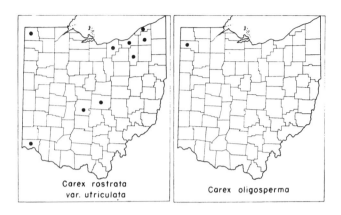

Carex rostrata
var. utriculata

Carex oligosperma

39. *Lupulinae*

Tall sedges of swamps, wet woods, or meadows of eastern North America; staminate spikes sessile or peduncled; pistillate spikes 2 or more (except in depauperate specimens), leafy-bracted, sessile or peduncled, globose to cylindric, 15–35 mm thick and less than 3 times as long; perigynia 10 mm long or longer, much inflated, thin and papery, strongly ribbed (12 or more ribs), ovoid to globose-ovoid, gradually contracted into a beak tipped with 2 stiff sharp teeth; achenes 3-angled, style continuous with achene, straight or bent, bony, not withering; stigmas 3.

a. Pistillate spikes globose or subglobose; styles straight.
   b. Perigynia dull, usually hispidulous, at least toward base, cuneate to base ....
                                                                 140. *C. grayii*
   bb. Perigynia lustrous, glabrous, rounded to base ....141. *C. intumescens*
aa. Pistillate spikes oblong-cylindric; style contorted or abruptly bent near base.
   b. Achenes definitely longer than broad, the sides oval, angles not or very little thickened at bend.
      c. Leaves and larger bracts 2–6 mm wide; perigynia spreading-ascending; pistillate scales blunt to short-acuminate ....142. *C. louisianica*
      cc. Leaves and larger bracts 5–15 mm wide; perigynia appressed-ascending; pistillate scales acuminate to rough-awned ....143. *C. lupulina*
   bb. Achenes as wide as or wider than long, the sides diamond-shaped and concave, angles prominently knobbed ....144. *C. lupuliformis*

140. Carex grayii Carey

C. asa-grayi Bailey

Perigynia 12–18 mm long, radiating in all directions, thus forming large globular heads; perigynia dull, varying from glabrous or almost glabrous (except near base) to prominently hispidulous, in var. *hispidula* Gray, recognized in some manuals.

141. Carex intumescens Rudge

Resembles glabrous perigynium-forms of the last; perigynia slightly smaller (10–16 mm long) and lustrous. More northern plants (var. *fernaldii* Bailey) have perigynia lance-ovoid and achenes broadest near rounded summit. Throughout much of temperate eastern America.

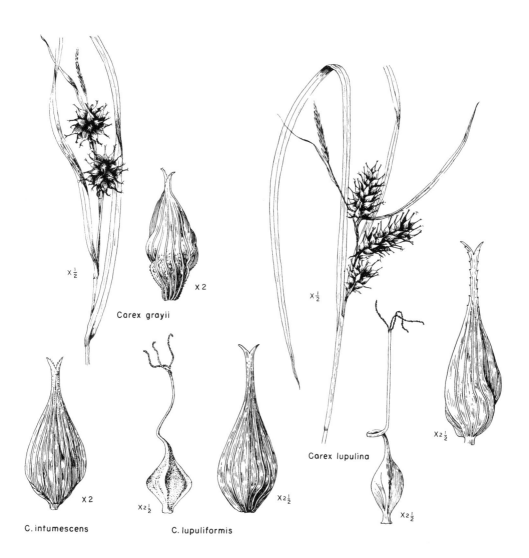

X ½

X 2

Carex grayii

X ½

Carex lupulina

X 2

C. intumescens

X 2½

X 2½

C. lupuliformis

X 2½

X 2½

142. Carex louisianica Bailey

Similar to *C. lupulina*; more slender in appearance, with narrower leaves (3–5 mm wide) and more slender long-peduncled staminate spike (2–3 mm wide instead of 3–5); stems arise singly from slender creeping stolons. A southern species extending north, locally into Ohio. One specimen (Ashtabula County) verified by F. J. Hermann.

143. Carex lupulina Muhl.

The commonest species of the section in Ohio; pistillate spikes sessile or nearly so in the typical variety, peduncled in var. *pedunculata* Gray. In swampy ground throughout much of the Deciduous Forest area of North America.

144. Carex lupuliformis Sartwell

A local but widespread species, distinguished from the preceding by its wide achene with angles knobbed or "nipple-tipped;" perigynia spreading-ascending.

×*C. macounii* Dewey, a hybrid of this species with *C. retrorsa*, is recorded from Lorain County (specimen determined by K. K. Mackenzie).

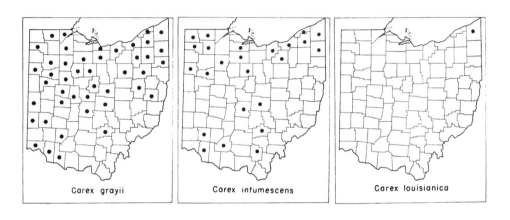

Carex grayii   Carex intumescens   Carex louisianica

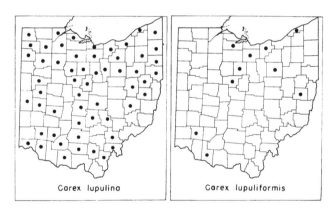

Carex lupulina   Carex lupuliformis

ARACEAE. Arum Family

A very large family of some 2,000 species classified into about 100 genera. Most numerous in humid tropical regions, where many species are high-climbing lianes with cord-like aerial roots, some epiphytic, some arborescent, a few (as *Pistia*, the water-lettuce) floating aquatic. Most of the species of temperate regions are swamp plants, species of *Arisaema* the only eastern American exceptions. A number of the tropical representatives of the family are familiar greenhouse ornamentals; among these, *Anthurium*, a genus with many beautiful species, some with handsome velvety leaves, others with brilliantly colored spathe and spadix, and two root-climbing aroids, *Philodendron* (the largest genus of the family, with well over 100 species) with large deeply lobed leaves; and *Monstera*, whose leaves have large more or less circular perforations; the common calla (not of the genus *Calla* but *Zantedeschia*) from South Africa, frequently grown as a hedge-plant in California, where the plants are hardy; devil's tongue or snake-palm (*Amorphophallus*), with spotted petioles and flower-stalks, and enormous "flowers" sometimes a meter long, the dark red spadix projecting well above the rosy spathe, emitting a very disagreeable odor. The golden club (*Orontium aquaticum* L.), common in swamps of the Atlantic Coastal Plain, is familiar to many travelers who have noted the brilliant yellow of the flowering spadix. Western skunk-cabbage (*Lysichitum*) of swamps in the Pacific Northwest, is easily recognized as a relative of our eastern skunk-cabbage, although it has even larger leaves and yellow or yellow-green spathes. *Pinellia ternata* (Thunb.) Breit., from Japan, with leaves similar to those of Jack-in-the-pulpit (but bulblet-bearing) and inflorescence similar to that of dragon-root, is sometimes adventive around imported Rhododendron plantings.

The corms, tubers, or rhizomes of a number of species are edible, some tropical ones being important sources of food. Temperate zone species were used by the Indians; Jack-in-the-pulpit, arrow-arum, wild Calla, and skunk-cabbage supplied breadstuff. Roots of sweet-flag, candied, yield an aromatic confection.

Ohio species are herbaceous woodland or swamp plants, the flowers closely placed on a spadix, consisting only of stamens and/or pistil, or in *Acorus*, perfect and with perianth of 6 inconspicuous parts; spadix subtended by or partly enwrapped in a specialized bract called a spathe; leaves veiny, simple or compound, broad (and scarcely suggesting monocot foliage) or with linear reed-like leaves; fruit, fleshy closely crowded berries (dry in *Acorus*).

Members of the Araceae, although diverse in appearance, are readily recognized by the characteristic aroid inflorescence—spadix and spathe. All have acrid or pungent juice with abundant needle-like crystals (rhaphides, raphides) of calcium oxalate.

a. Leaves compound; flowers confined to lower part of spadix; spathe well developed ....
                                                                              1. *Arisaema*
aa. Leaves simple; flowers not confined to lower part of spadix, but covering all or most of it.
    b. Leaves usually broad, not linear; spathe well developed, conspicuous, more or less enclosing spadix.
        c. Spathe green, convolute, 1–2 dm long, its margin pale; spadix elongate;

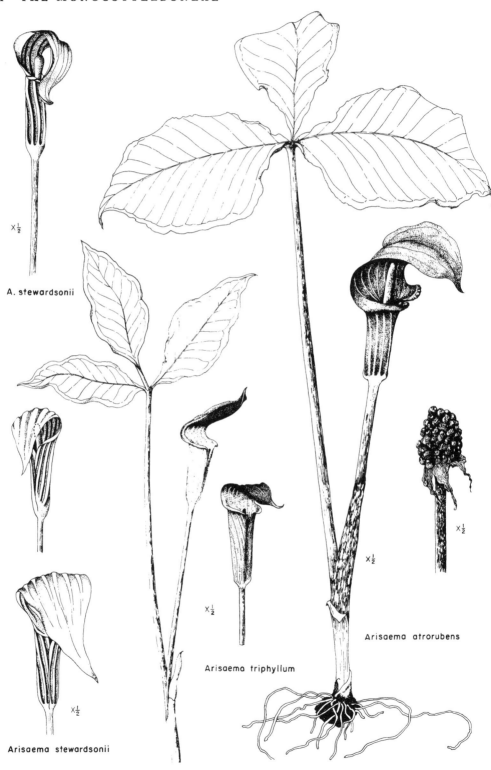

$\times\frac{1}{2}$

A. stewardsonii

$\times\frac{1}{2}$

Arisaema stewardsonii

$\times\frac{1}{2}$

Arisaema triphyllum

$\times\frac{1}{2}$

$\times\frac{1}{2}$

Arisaema atrorubens

leaves broad, triangular to oblong or elliptic, or in aquatic forms narrow but rounded or cordate at base, lateral veins unbranched, but joined near leaf-margin by a scalloped sub-marginal vein; berries green to brown, head of fruit surrounded by persistent base of spathe ....2. *Peltandra*

cc. Spathe white, ovate, flaring, 3–6 cm long, abruptly contracted to linear involute tip; spadix short; leaves ovate with cordate base, lateral veins not branched, curving toward apex; berries red ....3. *Calla*

ccc. Spathe thick and fleshy, variously striped or mottled green and dark purple, 10–15 cm long, widely bulging around subglobose spadix, contracted to inrolled pointed summit; leaves at flowering time scarcely developed, becoming over 0.5 m long, lateral veins branched; berries immersed in spongy spadix ....4. *Symplocarpus*

bb. Leaves linear, 0.5–1 m or more in length, less than 2 cm wide; spathe foliage-like, elongate, spadix thus appearing as a lateral spike of closely placed flowers; fruit dry ....5. *Acorus*

1. ARISAEMA Mart.   JACK-IN-THE-PULPIT.   INDIAN TURNIP

A large genus, most of whose representatives are in temperate and tropical Asia, a few in North America. Plants monoecious or dioecious, the flowers borne on the lower part of the spadix, staminate flowers above pistillate, when both are present; flowers consist merely of stamens (2–5) and/or pistil (a 1-celled ovary), without perianth; fruit a head of red berries. Herbs, with 1 or 2 leaves arising from corm or tuber, their petioles sheathing at base, and enwrapped by a few bladeless sheaths. There is little accord in the treatment of our species, some authors recognizing only two—Jack-in-the-pulpit and dragon-root; others divide the former into 2 or 3 species. Our treatment follows Fernald (1940, 1950).

a. Leaves normally with 3 leaflets (lateral leaflets rarely 2-lobed or divided); spathe much longer than club-shaped or cylindric spadix, arching to horizontal or down-turned summit.

b. Spathe-tube smooth or very shallowly furrowed.

c. Leaves lustrous and green beneath, lateral leaflets tapering to base and apex, not prominently asymmetric; flange of spathe-tube narrow, hood attenuate, narrow-oblong to lance-ovate; spadix slender, little longer than spathe-tube; fruiting head 1.5–2 cm long ....1. *A. triphyllum*

cc. Leaves dull and whitened beneath, lateral leaflets strongly asymmetric, broadly rounded on lower side; flange of spathe-tube wider (2–8 mm), hood abruptly acuminate, broad oblong-ovate; spadix stout, longer than spathe-tube; fruiting head 3–6 cm long ....2. *A. atrorubens*

bb. Spathe-tube deeply furrowed or corrugated; leaves green beneath, leaflets mostly tapering to base and apex, often petiolate; flange of spathe-tube narrow but distinct; hood long-attenuate, curved downward, with apex to one side; spadix stout, noticeably longer than spathe-tube ....3. *A. stewardsonii*

aa. Leaflets 7–13, the central one in line with petiole, the laterals arising successively from 2 widely spreading forks; vertical narrowly convolute spathe much shorter than slender tapering spadix ....4. *A. dracontium*

1. Arisaema triphyllum (L.) Schott.   SMALL JACK-IN-THE-PULPIT

Most Ohio plants, long referred to this species, should be referred to the next. *A. triphyllum* and *A. atrorubens* are connected by intergrading forms (possibly hybrids) where the ranges of the two overlap, but extremes are very distinct, and easily recognized, especially in the field. *A. triphyllum* is usually smaller than the next; the leaves are lustrous and *green* beneath, the lateral leaflets not noticeably asymmetric, the leaflets often contracted to petiole-like base; the spathe narrow and tapering to attenuate apex; spadix slender, 1.5–3 mm thick, little longer than tube of spathe. Hood of spathe varies in color (as it does in the next species), and is the basis for recognition of color forms: solid dark purple in forma *pusillum* (Peck) Fern., clear green in typical *triphyllum*.

This species is southern in range, extending northward along the Atlantic Coastal Plain to Massachusetts, and in the interior, locally to southern Ohio. In southern Ohio, it flowers 2–3 weeks later than the next. Both species are plants of rich woods of ravine slopes and alluvial valley terraces; A. *triphyllum* is also found on rock ledges with an accumulation of humus soil. The seeds of the two species differ slightly in shape, but differences are not enough to identify the species in the absence of other supporting characters.

2. **Arisaema atrorubens** (Ait.) Blume. JACK-IN-THE-PULPIT. INDIAN-TURNIP

The common species through most of interior eastern United States, and most of Ohio. Easily recognized by its strongly asymmetric lateral leaflets, leaves whitened or glaucous beneath (a good field character); spathe-hood broader, abruptly acuminate or acute; and thicker spadix (to 8 mm), often club-shaped. Usually larger and coarser than the last. Named color-forms include forma *zebrina* (Sims) Fern., with striped hood; forma *viride* (Engler) Fern., with all green hood, and the typical full purple hood. Occasional plants having the lateral leaflets divided, thus approaching the doubtfully distinct A. *quinatum* (Nutt.) Schott of the South, are probably but a phase of A. *atrorubens*.

Jack-in-the-pulpit is said to be monoecious or dioecious. Young plants are normally entirely staminate (or with a few rudimentary pistillate flowers at base of spadix) and have only one leaf. Older plants have two leaves, thick stem with sheathing leaf-bases, and both staminate and pistillate flowers, i.e., are fruit-bearing. Staminate plants die down soon after flowering, pistillate stems remain turgid and green until ripening of fruit.

3. **Arisaema stewardsonii** Britt.

A. triphyllum var. *stewardsonii* (Britt.) Stevens

When fresh, the deeply corrugated spathe-tube is conspicuous—ridges white and furrows green or brownish tinged toward flange; within expanded part of

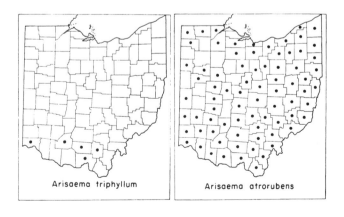

Arisaema triphyllum   Arisaema atrorubens

spathe, the ridges continue as veins and the furrows (now flat) continue as brown shading; spadix to 7 mm thick, often half again as long as spathe-tube; leaves similar to those of *A. triphyllum*. Plants of wet mucky soil, thickets and alder-swamps; Minnesota to Nova Scotia, south to Pennsylvania and northern New Jersey, and in the mountains to North Carolina.

4. **Arisaema dracontium** (L.) Schott. GREEN DRAGON. DRAGON-ROOT

A plant of moist to wet soil of wooded valley flats and seepage spots, almost throughout the Deciduous Forest. Leaf usually solitary, small and with few leaflets in young plants, in older plants elevated on tall petiole (to 5 dm or rarely more), and with more leaflets, a single leaf sometimes 4–5 dm wide, its lanceolate leaflets 1–2 dm long; spadix (the dragon's tongue) yellow or greenish yellow, slender and tapering, to 1.5 dm long and about 3 times length of spathe.

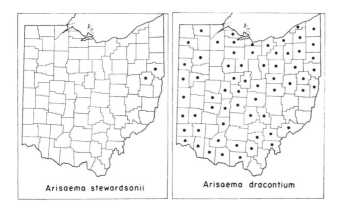

Arisaema stewardsonii   Arisaema dracontium

2. PELTANDRA Raf. ARROW-ARUM

A small genus of temperate eastern North America. Characterized by well developed spathe (in ours convolute) and elongate spadix thickly covered with flowers; staminate above, the anthers sessile and partly embedded, pistillate below, each surrounded by 4–5 white scale-like staminodia. One species in our area.

1. **Peltandra virginica** (L.) Schott & Endl. ARROW-ARUM

In mucky soil or shallow water, borders of ponds or lakes, alluvial swamps, and woodland depressions. Leaves large, 3 or more dm long and half as wide toward base, normally more or less triangular, with basal lobes sometimes almost half as long as leaf, wide or narrow, or in some forms, short or absent, the blade merely rounded or cordate at base. A number of named forms, based on leaf-shape variations. Spathe green, with white or pale margin, the upper part (and staminate part of spadix) decomposing with age, the lower remaining and enclosing the fruit. Largely confined to the northern part of the state, perhaps because of absence of suitable habitats elsewhere. The characteristic venation of the leaf—a prominent midrib and pair of backward-pointing veins extending

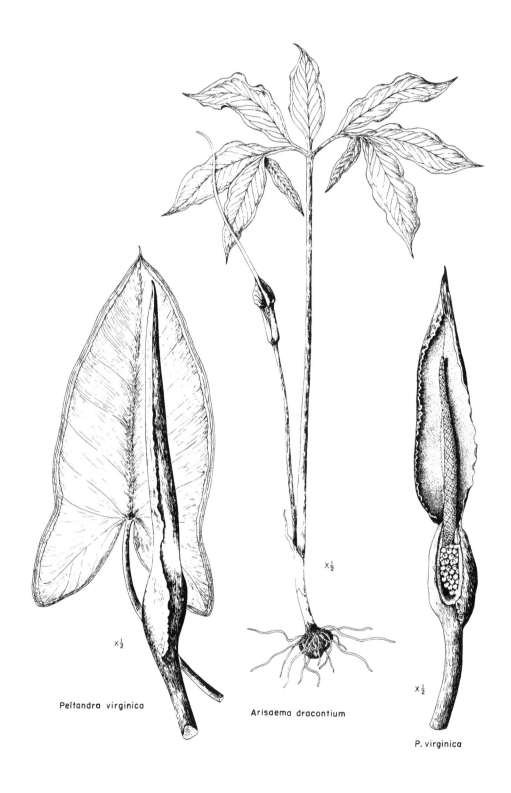

Peltandra virginica

×½

Arisaema dracontium

×½

×½

P. virginica

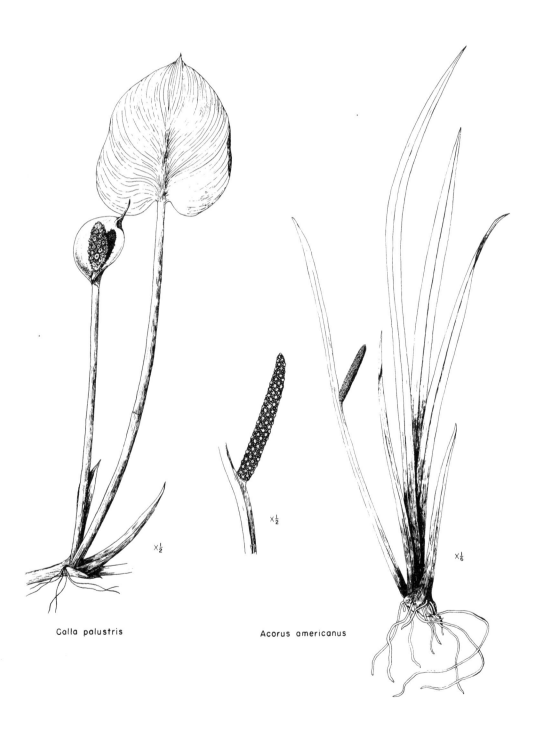

Calla palustris

×½

Acorus americanus

×½

×⅙

almost to tip of basal lobes, very fine veins between the coarser laterals giving the leaf a striate appearance, and the marginal connecting vein indented at each junction with principal laterals—makes recognition easy in the absence of flowers or fruit.

3. CALLA L. WATER-ARUM

A monotypic genus of boreal North America and Eurasia.

1. Calla palustris L. WILD CALLA

Perennial bog herbs with tufts of long petioled cordate leaves and solitary spathe arising from nodes of the elongate horizontal rhizome. Expanded spathe white, broadly oval or elliptic, abruptly contracted into linear tightly inrolled tip 5–10 mm long; spadix short cylindric, covered throughout with perfect flowers; berries red. In Ohio, confined to a few northeastern counties, where an outlier of the Hemlock-White Pine-Northern Hardwoods forest enters the state; almost transcontinental in range in the Northern Coniferous forest.

4. SYMPLOCARPUS Salisb. SKUNK-CABBAGE

Spathyema Raf.

A genus of two species, one common and widespread (in suitable habitats) in the northeastern quarter of temperate North America and northeastern Asia (including Japan), the other (*S. nipponicus* Makino) occupying a limited area in Japan, and discovered only a few decades ago (Li, 1952). Flowers perfect, covering the whole of the globular spadix; stamens 4, style 4-angled, ovary 1-seeded; perianth of 4 inconspicuous arching sepals or segments, their tips meeting.

1. Symplocarpus foetidus (L.) Nutt. SKUNK-CABBAGE
Spathyema foetida (L.) Raf.

Both specific and common name refer to the strong unpleasant odor of flowers (especially as they get old) and crushed or bruised foliage. Often the

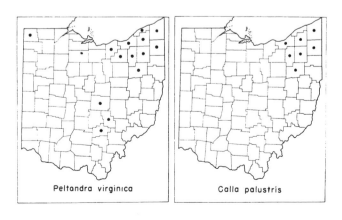

Peltandra virginica   Calla palustris

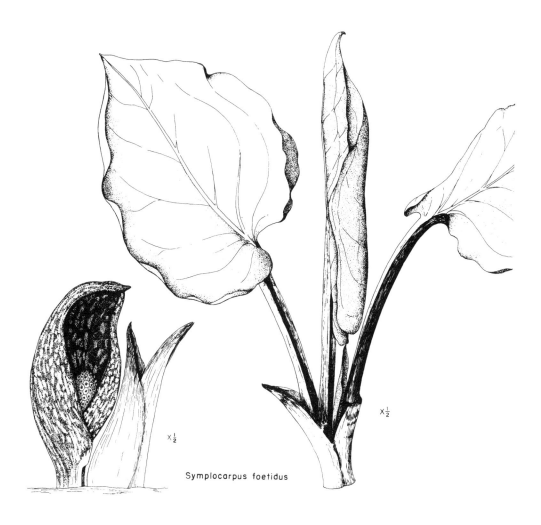

Symplocarpus foetidus

earliest native species to come into bloom, a forerunner of spring. The dark purple, sienna, or green-striped spathes, darker within, push up through mucky soil, sometimes in great numbers. The fleshy spathe is swollen below, open on one side exposing the almost globular spadix, and tapering to a point about equal in length to the swollen lower part, little if at all raised above the ground. Later, the peduncle elongates slightly, its base enveloped in almost leafless sheaths, the young leaves pushing up within. When full grown, the leaves are large (5–6 dm long and over half to nearly as wide), truncate to cordate at base, or sometimes tapering at base, netted veined, on short petioles diverging from crown capping thick but short rhizome from which long very thick roots descend.

5. ACORUS L. Sweet Flag. Calamus

Perennial herbs of swamps, swampy stream-heads, or borders of ponds, with thick creeping aromatic rhizomes, and elongate linear leaves, perfect flowers (6 sepals, 6 stamens, one 2–3-celled ovary) crowded on an elongate spadix, and forming a cylindric slightly tapering spike. North American, European, and Asian species, the number doubtful.

1. **Acorus americanus** Raf. Sweet Flag
   A. calamus L. of Amer. auth.
   The use of Rafinesque's name for this species, instead of the Linnaean name, is based upon recent chromosome studies (Löve & Löve, 1957). *A. calamus* of Linnaeus is a sterile triploid of Europe; our American plant is a diploid, as are also two Asiatic species; a third Asiatic species is a tetraploid. There has long been question as to whether *Acorus* in America is native or introduced from Europe; the chromosome complement points to its indigenous nature.
   Sweet flag is a common species of much of temperate North America, growing in large patches, its leaves about a meter long and bright green. The inflorescence is inconspicuous, born at the top of the leaf-like flowering stem, and subtended by a leaf-like prolongation (a modified spathe) arising to same height as leaves.

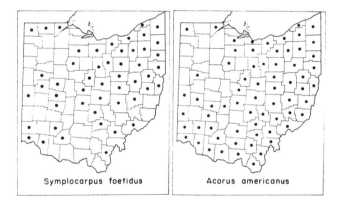

Symplocarpus foetidus   Acorus americanus

LEMNACEAE. Duckweed Family

A family of some 30 species of free-floating aquatic plants, classified into 4 genera; world-wide in distribution. This family includes the smallest flowering vascular plants in the World, species of *Wolffia*. Identification of species is most readily accomplished in the field, as some characters which may be observed in fresh material are poorly preserved in dried specimens. At any time, identification of some species is difficult; the plants are very small, flowers and fruit are rarely produced, plants differ in aspect under different growth conditions, and at different seasons.

"A duckweed plant actually consists of a flattened or attenuated stem with one node and parts of two adjacent inter-nodes fused with any rudiments of the original leaves or petioles. . . . From the node of each plant radiates: (1) the flattened plant body with the nerves or vascular bundles (if developed), (2) the reproductive pouch or pouches through which the new plants, the flowers (if occurring), and the various resting forms grow by budding from the node, and (3) the root or roots (if present)." As a result of budding, 2 or more plants are commonly joined together (but easily separable). Such vegetative multiplication may result, in a single season, in growth of several million individuals from a single plant. In late summer, extensive mats of duckweed may sometimes be seen, composed, if *Spirodela*, of 100,000 to 200,000 plants per square yard of water surface, if *Lemna*, of 300,000 to 800,000 plants, if *Wolffia*, of 1 to 2 million plants. These masses of duckweed furnish food for fish and for ducks and other water-birds; they are also consumed by snails, insects, and spiders. Though small, they play an important part in the balance of life in quiet water. In the fall, resting forms—hibernacula—are produced by some of the species. These are more compact than the vegetative plants of summer with greatly reduced volume of intercellular air spaces, hence are heavier, and sink to the bottom. When the warmth of spring brings about renewed growth and formation of gases, the overwintering plant rises to the surface, and summer forms develop.

Flowers are produced infrequently (are unknown in some species), probably never in shaded situations. Sometimes, drought conditions appear to stimulate flower production. Flowers are borne in a spathe; the staminate consisting merely of a single anther, sessile or with short filament, the pistillate of a single flask-shaped ovary with 1 or a few ovules, the whole appearing as one flower (illustration adapted from Hicks, Pl. 2, 1932). The simple reduced flowers borne in a spathe suggest the Araceae, to which the Lemnaceae is related. (Quotations, measurements, and much of the information in text from papers by L. E. Hicks, 1932, 1937; the following keys are slightly modified from Hicks, 1937.)

Plants with roots and two reproductive pouches from each node.
  Each plant of a group with several roots growing in a fascicle from the node; our
    plants usually reddish below and with a red eye-spot (the node) above; dorsal
    surface of living specimens a glossy green ....1. *Spirodela*
  Each plant of a group with only one root ....2. *Lemna*
Plants without roots and with only one reproductive pouch from each node.
  Plants thick and globular, minute ....3. *Wolffia*
  Plants thin and strap-like, usually submerged and attached in groups shaped like
    rimless wheels ....4. *Wolffiella*

1. SPIRODELA Schleid

A genus of 4 species, ours and 2 others in tropical America, and 1 in tropical Asia.

1. **Spirodela polyrhiza** (L.) Schleid. GREATER DUCKWEED
Plants solitary or in groups of 2–5, glossy green with red eye-spot, 2.2–6.5 mm wide and 2.4–9.5 mm long, with 5–15 nerves radiating from node, and with

usually 4–9 (3–16) roots descending from node (hence the specific name). Overwintering buds (hibernacula) are produced abundantly, and are sometimes washed up in piles on pond margins. Widely distributed in temperate and tropical America, in Europe and Asia.

2. LEMNA L.   DUCKWEED

A cosmopolitan genus of 8 species, 4 or 5 of which are found in Ohio. Plants more or less disk-shaped, with 1 central vein, no or 2–4 lateral veins, and 1 root; 1 pistillate and 2 staminate flowers in each spathe. The maps include stations reported by Hicks, 1937.

a. Plants feather-shaped with basal portions of the long internodes narrowed into petiole-like stems; usually submerged ....1.   *L. trisulca*
aa. Plants oval to oblong, without petiole-like stems, connected plants appearing sessile; usually floating.
   b. Plants symmetric or nearly so.
      c. Plants deep green, thickish convex on both surfaces, margins thick, obscurely 3-veined, cavernous throughout, appearing medium thick when dry ....
              2.   *L. minor*
      cc. Plants usually pale green, lower surface nearly flat, margins thin, obscurely 1-veined, cavernous in middle portions only, membranous when dry ....
              3.   *L. minima*
  bb. Plants asymmetric.
      c. Plant obliquely obovoid, medium thick, usually deep green with some reddish purple, distinctly 3-veined, cavernous throughout; root-sheath with lateral wing-appendages .... 4.   *L. perpusilla*
      cc. Plant long-oblong, thin, pale green, obscurely 1-veined, cavernous in middle portions only; root-sheath not appendaged ....5.   *L. valdiviana*

1.   **Lemna trisulca** L.   SUBMERGED DUCKWEED.   STAR DUCKWEED
Commonly in dense tangled masses just below water-surface, in shaded situations or beneath floating plants. Flowering plants differ in form from submerged vegetative plants, float on surface, and develop stomata. Vegetative plant (expanded portion) 1.4–3 mm wide and 4–10 mm long, with slender petiole-like stem 4–16 mm long; flowering plants 2 mm wide and 3 mm long, convexly curved. No true hibernacula produced, compact forms of the usual submerged plants overwintering. Almost cosmopolitan.

2.   **Lemna minor** L.   LESSER DUCKWEED
The most abundant and most widely distributed duckweed in much of the United States, with a wide range of pH tolerance. Plants floating, round-to elliptic-obovate, 1.5–3 mm wide and 2–4 mm long. Hibernacula produced. Found on all continents; absent far northward.

3.   **Lemna minima** Phillipi.   LEAST DUCKWEED
Rare and local, from Ohio, Minnesota, Kansas, and California southward into South America. "The only Ohio record is from a pond in Paulding County within three or four miles of" the Indiana line.

4.   **Lemna perpusilla** Torr.   MINUTE DUCKWEED
The smallest of the genus, plants 1.2–2.5 mm wide and 2–2.5 mm long.

Winter forms similar to those of *L. minor*. Less extensive in range than the last; very local in our area, a single Ohio record from Mercer County.

5.   **Lemna valdiviana** Phillipi.   PALE DUCKWEED
     L. cyclostasa (Ell.) Chev.
     Usually floating, but sometimes in dense tangled submerged masses; vigorous plants may develop a conical papilla at the node and a row of smaller ones along midvein. Local but wide-ranging, from middle latitudes of North America south into South America; in stagnant water with abundant organic debris.

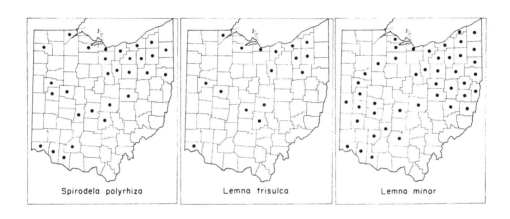

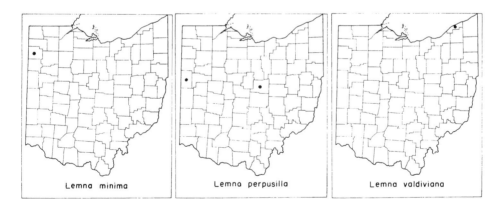

3.   WOLFFIA Horkel.   WATER-MEAL

Members of this genus "resemble floating pinhead dots and are the smallest of all flowering plants." Globular, ellipsoid, or ovoid rootless plants floating at or near the water surface. Actual counts of sample portions of pure cultures of each of our 3 species reveal the enormous numbers of plants in a square yard of water surface: 2,070,000 of *W. columbiana*, 1,630,000 of *W. punctata*, and 1,420,000 of *W. papulifera*. During the drought year of 1933, Buckeye Lake in

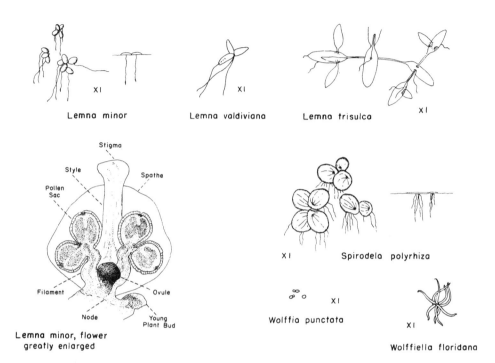

Lemna minor

Lemna valdiviana

Lemna trisulca

Stigma

Style

Spathe

Pollen
Sac

Filament

Ovule

Node

Young
Plant Bud

Lemna minor, flower
greatly enlarged

Spirodela polyrhiza

Wolffia punctata

Wolffiella floridana

central Ohio "had more than one square mile of water surface completely covered with *W. punctata.*" It is estimated that "a single acre of this plant contains a thousand times as many plant individuals as there are human beings in the world." Plants winter as minute hibernacula, which are essentially smaller compact phases of the summer vegetative phase. Eight species, widely distributed in temperate and tropical regions; three in our area.

a.   Plants globose or nearly so, not punctate, loosely cellular; upper surface convex, usually
    with 3 inconspicuous papillae; plants not prominent above water surface ....
                                                         1.   *W. columbiana*
aa. Plants more or less flattened above and gibbous beneath, brown-punctate, more com-
    pactly cellular; plants prominent above water surface.
    b.   Plant round-ovate, strongly gibbous, slightly asymmetric; upper surface with a
        single large conical papilla ....2.   *W. papulifera*
    bb. Plant more or less oblong with upturned acute tip, slightly gibbous, symmetric;
        upper surface with a prominent papilla near center ....3.   *W. punctata*

1.   **Wolffia columbiana** Karst.   COMMON WOLFFIA
    Minute grains, floating at or just beneath water surface, 0.3–0.9 mm wide and 0.32–1 mm long. Plants of greater density, which sink to the bottom, live through the winter or other unfavorable conditions. Flowers rarely produced and difficult to detect because of small size; attached to spadix within spathe are one anther

(the staminate flower) and one ovary containing a single ovule (the pistillate flower). Usually in small permanent ponds but also in protected inlets of larger lakes; widely distributed from northern United States southward through Mexico into South America.

2. **Wolffia papulifera** C. H. Thompson. POINTED WOLFFIA
Plants 1 mm wide and 1–1.5 mm long; upper surface rises from flat margin to a single large papilla on median line, abundantly punctate with brown epidermal pigment cells. In stagnant water rich in organic matter; more restricted in range than our other species: Ohio to Missouri and Kansas, south to Tennessee and Arkansas. Two specimens (Jackson and Portage counties) have been referred to this species.

3. **Wolffia punctata** Griseb. DOTTED WOLFFIA
Plants 0.32–0.57 mm wide and 0.58–0.96 mm long, upper surface flat or slightly convex, rising to the acute apex "like the bow of a boat." Plants floating "with the entire upper surface exposed to the air, profusely brown-dotted throughout with minute pigment cells." Ranging from Connecticut and New York to Minnesota and south to Florida and Texas, and in Jamaica; locally abundant, even on larger lakes.

3. WOLFFIELLA Hegelm.

A small genus with but one representative in the East. Flowers and fruit unknown.

1. **Wolffiella floridana** (J. D. Smith) C. H. Thompson. STAR WOLFFIELLA
Plants strap-shaped, curved, rootless, both surfaces brown-dotted, solitary or coherent in clumps, 0.4–0.7 mm wide and 4–14 mm long. "Restricted to wholly stagnant bodies of water . . . in boggy areas protected from wave action;" sometimes mixed in organic debris such as floating clumps of cattail roots. From Florida, Texas, and Mexico, north locally to Massachusetts, Wisconsin, and Missouri.

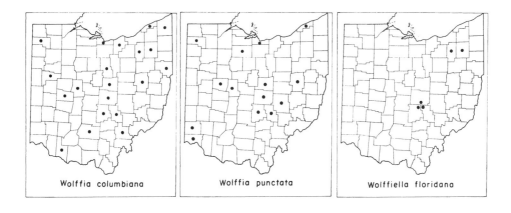

Wolffia columbiana   Wolffia punctata   Wolffiella floridana

X2

X. torta

Xyris caroliniana

Xyris torta

Eriocaulon septangulare

Tradescantia ohiensis

XYRIDACEAE. Yellow-eyed-Grass Family

A family of 2 genera, one of which occurs in Ohio.

1. XYRIS L. YELLOW-EYED-GRASS

Tufted grass-like plants with narrow linear basal leaves, naked scapes, flowers in dense short spikes (ovoid in ours), each flower in axil of subtending spirally arranged closely imbricate bracts, perfect and 3-merous, sepals glumaceous, unequal (2 lateral keeled and boat-shaped), petals yellow. A large genus, chiefly tropical, well represented in the pinelands of the Atlantic and Gulf Coastal plains, local inland.

> Plants hard-based; leaves and scapes spirally twisted; bracts of inflorescence tawny or brown, sometimes with ill-defined green center; keel of lateral sepals finely ciliate or merely pubescent, and with terminal tuft of hair ....1. X. torta
> Plants soft-based; leaves and scapes not twisted (or scarcely so); bracts yellowish or pale brown, with clearly defined green center; keel of lateral sepals erose or jagged in upper half ....2. X. caroliniana

1. **Xyris torta** Sm.
Bogs and wet sandy soil, widely distributed throughout much of eastern United States, but local because of habitat restrictions. Scape 2–5 (–8) dm tall, leaves ⅓–½ as long or more, stiff, 1–2 mm. wide; stems clustered, bulbously thickened at base. Two varieties are recognized, one limited to southeastern Virginia, and the wide-spread typical variety.

2. **Xyris caroliniana** Walt.
Sandy shores, sloughs, swamps and bogs, ranging from Florida and Louisiana north into southern Canada, but local. Scape to 1 m tall, leaves ⅓–½ as long, flaccid, 2–5 mm wide; tufted plants without bulbs, base flattened.

ERIOCAULACEAE. Pipewort Family

A large, chiefly tropical family of aquatic and marsh plants, represented in our area by the genus *Eriocaulon*, a few of whose species are northern.

1. ERIOCAULON L. PIPEWORT

Tufted perennials with narrow somewhat grass-like leaves (all basal), naked scape enclosed toward base in colorless sheath, flowers in terminal subglobose head, white-bearded, as are also the bract-tips; flowers unisexual, in our species 2-merous. Four species in the Manual range, several others in the South and West, one northern species (of North America, Ireland, and Scotland) occurring in Ohio.

1. **Eriocaulon septangulare** With.   WHITE-BUTTONS. DUCKGRASS

Leaves thin, semi-transparent, with rectangular reticulations, awl-shaped, 1–10 (–20) cm long, tapering from colorless base to long very slender tip; scape longer than leaves, usually 3–20 cm long, or much longer in deep water, often 7-striate or angled (hence the specific name); basal sheath ¼–½ length of scape; heads at first hemispheric (and superficially resembling a miniature *Antennaria* head), later subglobose, when bracts of involucre become recurved, 3–8 mm in diam. Growing in sand in shallow water.

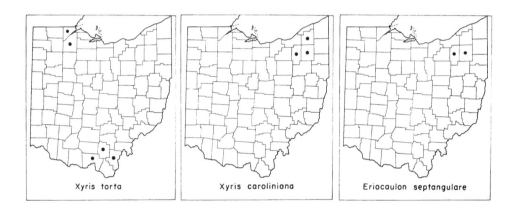

Xyris torta   Xyris caroliniana   Eriocaulon septangulare

COMMELINACEAE.   Spiderwort Family

Annual or perennial herbs, chiefly tropical, with leafy jointed stems, entire parallel-veined leaves, the base dilated and sheathing; flowers perfect, regular or irregular, 3-merous, sepals green (in ours), petals blue (or purple or rosy) or white, ephemeral; stamens 6 in 2 cycles, sometimes of two forms; capsule 3-locular or by abortion 2-locular. A family of several hundred species, classified into about 25 genera. In addition to the following, several other genera are represented by more or less familiar cultivated foliage plants, best known of which is *Zebrina pendula* Schnizl., Wandering Jew, with lavender flowers and beautifully striped leaves, purple beneath, silvery green-white and purplish above. *Tradescantia fluminensis* Vell., also called Wandering Jew, is a greenhouse plant native from Central America to Argentina; it has the leaves red-purple beneath (green in reduced light), ovate acute, and white flowers in sessile terminal clusters subtended by 2 leaf-like bracts. *Rhoea discolor* Hance, an erect-growing member of the family with thick semi-succulent leaves, and small white flowers, is sometimes called Noah's-ark or boat-lily, from the boat-shaped spathe. *Cochliostema*, from Ecuador, epiphytic, with the habit of the better known *Billbergia*, has great axillary panicles of large and beautiful flowers, and is said to be the most beautiful cultivated plant of the family.

Flowers in clusters in folded spathe-like bract, irregular, the two upper petals larger
than lower; leaves 2–4 times as long as wide ....1. *Commelina*
Flowers regular, in 1 or several umbel-like cymes from axils of little-modified upper
foliage leaves; leaves long and narrow, many times longer than wide ....2. *Tradescantia*

1. COMMELINA L. Dayflower

Only a few species of this large tropical genus reach temperate latitudes;
best known is the Asiatic *C. communis*, a rampant grower. The deep sky-blue
flowers (2 upper petals larger than the lower paler blue or white petal), folded
heart-shaped spathe, and Wandering Jew habit of growth readily distinguish
our members of this genus from *Tradescantia*. Also, the stamens are of two
kinds, 3 stamens with fertile anthers, 3 sterile and cruciform. The two Ohio
species are distinguished from other more southern species by a spathe character
—the margins free to the base instead of united near base.

Stems at first erect, later decumbent; leaf-blades ovate-lanceolate, sheaths glabrous on
margin, rarely ciliolate; folded spathes broadly semi-cordate; lower (smaller) petal
white; seeds brownish, rugose ....1. *C. communis*
Stems decumbent, diffusely branched; leaf-blades lanceolate, sheaths long-ciliate;
folded spathes semi-cordate; lower (smaller) petal blue; seeds black, reticulate ....
2. *C. diffusa*

1. Commelina communis L. Dayflower

This widely distributed and abundant naturalized species from Asia is easily
recognized by growth-habit and the two sky-blue upper petals (the lower petal
white or pale). Intensity of color of the upper petals varies, and sheath-margins
may be glabrous, obscurely pubescent, or distinctly ciliolate. On the basis of
these characters, and color of sterile stamens, two intergrading varieties—
communis and *ludens* (Miquel) C. B. Clarke—are sometimes distinguished.

2. **Commelina diffusa** Burm. f.

A southern species ranging north into southern Ohio and southern Indiana;
found in muddy places along streams and on river-banks. More diffusely branched
and decumbent than *C. communis*, and sometimes forming mats.

Commelina diffusa

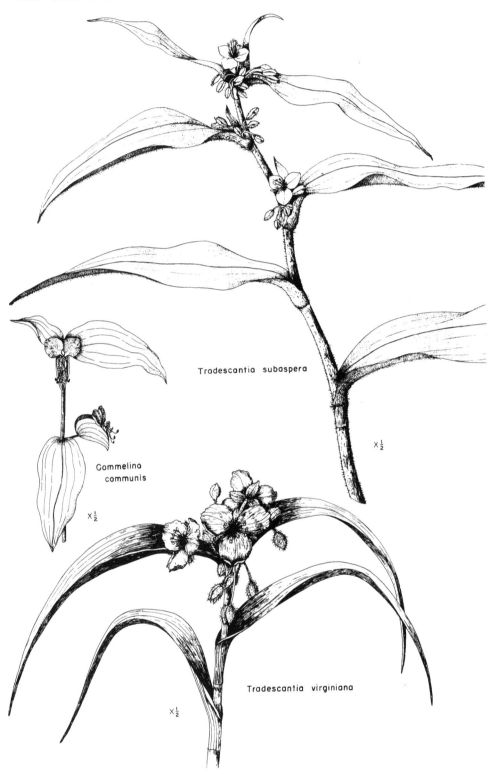

Tradescantia subaspera

X½

Commelina
communis

X½

Tradescantia virginiana

X½

2. TRADESCANTIA L. SPIDERWORT

A genus of tropical and temperate America; distinguished by linear, linear-lanceolate, to elliptic-lanceolate keeled leaves with sheathing base enwrapping stem, regular ephemeral flowers; 3 green sepals, 3 blue (rarely white or rose) petals, 6 stamens with filaments bearded with long hairs (a very distinctive feature); 3-valved capsule which upon ripening dehisces violently. Petals soon shriveling on exposure to sun. The stamen-hairs are so transparent, and their cells so large, that they lend themselves to microscopic examination, and have long been used for laboratory study. Even the germ cells are large and clear and hence well suited to microscope study. Wild species of *Tradescantia* formed the basis for extended field studies involving the genetics of the group and throw light, not only on our native populations, but on garden subjects as well (Anderson, 1952). Although cross-fertile, hybrids are not frequent in the wild state unless two species are growing adjacent to a much disturbed area where hybrids find a suitable habitat. In the garden, they tend to hybridize freely, and have given to the horticultural trade the desirable perennial known as *T. virginiana*, but which actually is true *virginiana* only in part, and contains some *T. subaspera*, which is largely responsible for the later and longer bloom period and persistence of leafy stems. *T. ohiensis* doubtless also played a part in the development of the garden *virginiana*, which has been going on for about 300 years, and so gradually that the changes were scarcely noted. The garden *virginiana* (of mongrel origin) and the wild American *T. virginiana* are not the same.

a. Leaves soft-pubescent, broad, to 5 cm wide below middle, abruptly contracted to petiolar base, then expanded to inflated veiny sheath; cymes usually several, terminal and from upper axils; summer blooming ....1. *T. subaspera*
aa. Leaves glabrous or nearly so, narrower, gradually narrowed toward base, narrower to about as wide as sheath.
    b. Leaves and stem glaucous; sheathing base enlarged and dilated; cymes terminal on main stem and lateral branches ....2. *T. ohiensis*
    bb. Plant not glaucous; leaf-base not prominently enlarged and dilated; cymes solitary, terminal.
        c. Sepals ovate, pubescent with long soft hairs ....3. *T. virginiana*
        cc. Sepals lanceolate; sepals and pedicels densely glandular pubescent ....
                                              4. *T. bracteata*

1. **Tradescantia subaspera** Ker. ZIGZAG SPIDERWORT
    T. pilosa Lehm.
    Coarse summer-blooming perennial, often in clumps, in densely shaded situations. Our specimens referable to the typical variety, with stems more or less zigzag, and upper lateral cymes sessile. Hybrids between this species and *T. ohiensis* occur. Interior United States, mostly west of the mountains from West Virginia and Tennessee to Illinois and Missouri.

2. **Tradescantia ohiensis** Raf. GLAUCOUS SPIDERWORT
    T. canaliculata Raf.; *T. reflexa* Raf.
    A conspicuous plant of prairies of the Middle West, and spreading onto railroad embankments and roadsides throughout its natural range and eastward;

flowering in June. Plants conspicuously glaucous, 4–6 dm tall; flowers usually blue, but sometimes white, purplish, or rose; sepals sometimes with tuft of white hairs at tip.

Both diploid and tetraploid races of *T. ohiensis* are known; these cannot be distinguished except by actual chromosome-counts (12 in diploid, 24 in tetraploid races), gigantism of pollen-grains and of stomata. The tetraploids usually grow in drier situations than the diploids. Intensive studies by Dean (1953, 1959) of occurrence in the Michigan-Indiana-Ohio area give details of distribution in relation to soils and vegetation, and postulate probable time of entrance of the two races, the diploid along post-glacial drainage-lines, the tetraploids during the Xero-thermic Interval. Thus far, Dean has not found tetraploids in Ohio.

3.   **Tradescantia virginiana** L.   SPIDERWORT

A handsome species of open woods, the flowers sometimes 4 cm across, and ranging in color through various shades of rose and purple, rarely white, but most often rich blue; plants low, 1–3 dm tall; leaves from linear and scarcely 1 cm wide, to linear-lanceolate and 2.5 cm wide. Flowering in May. More widespread than the preceding species. Garden forms sold under this name, and flowering into the summer, are of hybrid origin (see under TRADESCANTIA).

4.   **Tradescantia bracteata** Small.

Known from a single collection in 1927; perhaps adventive, rather than indigenous, but growing with other western species. Prairies, from southern Indiana and Michigan westward. Distinguished by its densely glandular villous pedicels and calyx, and lanceolate sepals.

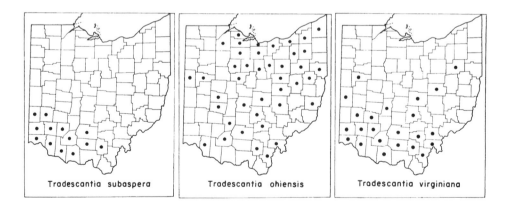

Tradescantia subaspera   Tradescantia ohiensis   Tradescantia virginiana

PONTEDERIACEAE.   Pickerelweed Family

Aquatic or swamp plants, with hypogynous, perfect, regular or somewhat 2-lipped flowers borne in panicles or spikes, or solitary, and emergent from a

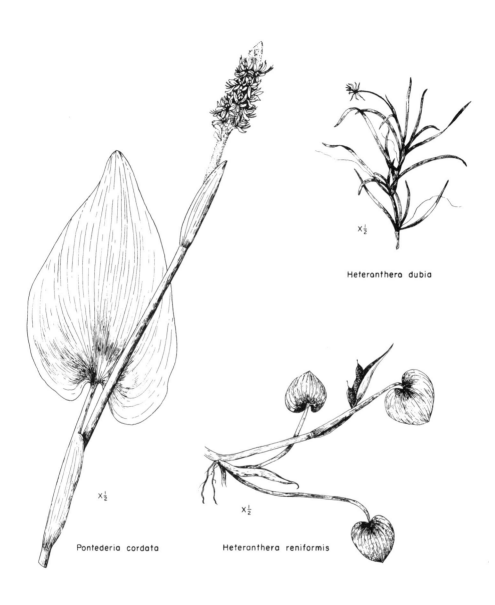

Heteranthera dubia

$\times \frac{1}{2}$

$\times \frac{1}{2}$

Pontederia cordata

$\times \frac{1}{2}$

Heteranthera reniformis

sheath or spathe. Perianth parts 6, petal-like, united below into a tube, scarcely distinguishable into sepals and petals; stamens inserted on perianth-tube. A small family of tropical and warm temperate regions, a very few species northward to cooler latitudes. The water-hyacinth, *Eichornia crassipes* (Mart.) Solms, a floating plant from South America, with large lavender-blue flowers, leaves with round or reniform blades and mid-petiolar globular swelling (merely enlarged in grounded plants) of spongy air-tissue (aerenchyma), is grown as an aquarium plant. In the South, it has become a troublesome pest, blocking waterways with an almost impenetrable growth, the individual rosette-like plants connected by stout floating rhizomes.

> Erect, in swamps or shallow water; leaves basal, large, up to 2 dm long, petioled, usually cordate at base, ovate-triangular to broadly ovate; flowers blue, crowded in spike ....1. *Pontederia*
> Floating or creeping on mud; leaves thin and linear or reniform, to 5 cm wide; flowers white, pale yellow, or pale blue, few (1–8) in loose inflorescence ....2. *Heteranthera*

1. PONTEDERIA L.   PICKERELWEED

Swamp or shallow water plants with erect stem, one stem-leaf similar to basal, and ephemeral flowers in spike, the spike arising from loosely sheathing spathe. Perianth violet-blue, deeply cut and 2-lipped, stamens 6, 3 with long, 3 with short filaments; fruit achene-like, slightly bladdery. A small genus mostly confined to the warmer parts of the Western Hemisphere.

1.   Pontederia cordata L.   PICKERELWEED

Our only species, most frequent in the northern part of the state, where it grows along the mucky borders of lakes and in filled ponds. Widely distributed in eastern United States and adjacent Canada, but local in much of its range because of scarcity of natural ponds. Highly variable in leaf-shape—from narrowly lanceolate (forma *angustifolia* (Pursh) Solms from Lucas County) to broadly triangular-ovate and cordate.

2. HETERANTHERA R. &. P.   MUD-PLANTAIN

Submersed aquatic or floating plants, or creeping on muddy shores. Ephemeral perianth-limb subequally 6-parted; stamens 3; fruit a few to many-seeded capsule. The genus is sometimes divided into *Zosterella*, in which *H. dubia* is placed, and *Heteranthera*, in which *H. reniformis* is placed, these differing in characters here shown in key to species. The generic name refers to the unlike anthers of *Heteranthera* in the limited sense.

> Submersed aquatic with branching stems; leaves linear, translucent, sessile; flowers pale yellow, perianth-tube thread-like; stamens all alike, with sagittate anthers ....
> 1. *H. dubia*
> Floating or creeping on mud; leaves reniform, long-petioled; flowers pale blue or white, perianth-tube slender; stamens dimorphic, 2 with yellow ovate anthers, one with greenish larger anther ....2. *H. reniformis*

1.   Heteranthera dubia (Jacq.) MacM.   WATER STARGRASS
H. graminea Vahl, *Zosterella dubia* (Jacq.) Small
A plant of quiet waters, almost transcontinental in range and southward to tropical America. Easily mistaken for a pondweed except when in flower. Flower arising from a spathe sessile in leaf-axil, the perianth expanding on the water surface.

2. **Heteranthera reniformis** R. & P.

Ranging from tropical America northward on the Gulf and Atlantic coastal plains and in the Missisippi Valley. In Indiana, almost confined to the southern counties (Deam, 1940), and known in Ohio from only one Hamilton County collection of 1837.

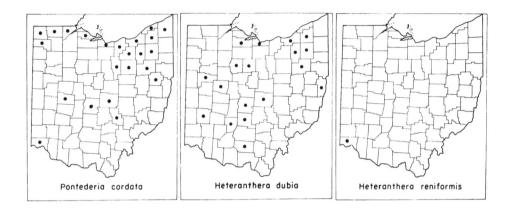

Pontederia cordata   Heteranthera dubia   Heteranthera reniformis

JUNCACEAE. Rush Family

Grass-like or sedge-like plants, distinguished from the grasses and sedges by the regular 3-merous hypogynous flowers; sepals and petals small, chaff-like or scale-like, green or brown or a combination of green and brown, often with colorless membranous margins; stamens 3 or 6, capsule 3-valved. Structurally, these flowers resemble the lilies, to which order the *Juncaceae* belongs. A family of 8 genera, of which 6 are restricted to the southern hemisphere.

Plants glabrous, without hairs; seeds numerous, minute; mostly summer blooming ......
    1. *Juncus*
Plants hairy or cobwebby; seeds 3, larger; spring blooming ....2. *Luzula*

1. JUNCUS L. Rush

A large cosmopolitan genus; species mostly perennial, occurring in a wide variety of habitats—dry or wet soil, swamps, bogs, or shallow water—usually in sunny situations. Stems usually simple (rarely branched), hollow or pithy; leaves basal or cauline or both, or sometimes reduced to bladeless sheaths, flat and grass-like, or terete, then often septate; inflorescence terminal, loosely or compactly branched, subtended by a leaf-like bract which in a few species appears to be a continuation of the stem, thus causing the inflorescence to appear lateral. At the base of each pedicel is a small bract, and in one group of species, a pair of tiny bracteoles (prophylla) also, these closely enwrapping the lower end of perianth.

In distinguishing species, characters of the uninjured sheath-auricle, of the mature capsule, and of the seeds are important; and, in certain groups of species, number of stamens, 3 or 6, is distinctive. The leaf-sheaths may taper to blade, end in a rounded shoulder (auricle), or their margins may extend upward as free prolongations (see sheaths of the common Path Rush, *J. tenuis*). For capsule shape and length in relation to perianth-parts, mature fruit is desirable. The seeds of all species are very small and characters can be seen only with a strong lens; some have dark points at each end; in some, the seed-coat extends beyond the body of the seed at both ends, forming pale "tails" or appendages which may be nearly as long as the body of the seed. The number of stamens, a useful character in identifying some of the species with septate leaves, is more readily determined in fresh than in dried specimens.

The inflorescence in some members of the septate-leaved group is often changed, in whole or in part, to masses of horn-like galls, with the perianth-segments much enlarged and distorted.

Drawings of capsules are after Gleason (1952) by permission of the New York Botanical Garden.

a. Inflorescence appearing lateral, involucral bract appearing like continuation of stem, terete; sheaths at base of stem without blades.
    b. Stems finely striate; basal sheaths papery; inflorescence straw-colored to greenish (reddish in var. *decipiens*) ....1. *J. effusus* var. *solutus*
    bb. Stems not striate, smooth; basal sheaths often lustrous, brownish to chestnut; inflorescence deep brown ....2. *J. balticus* var. *littoralis*
aa. Inflorescence appearing terminal (if appearing somewhat lateral, then bract flat or channeled); stem leaves with blades flat or terete.
    b. Leaves flat or involute, if terete *not* septate.
        c. Annual, with slender fibrous roots; inflorescence one-third as long or longer than stem; flowers scattered along branches of inflorescence ....3. *J. bufonius*
        cc. Perennial, roots thicker, rhizomes often present; inflorescence less than half as long as stem.
            d. Leaves terete or nearly so, channeled, not septate; capsules exserted; each flower with pair of bracteoles.
                e. Capsules and inflorescence greenish; involucral bract stiff, short (1–8 cm) ....4. *J. vaseyi*
                ee. Capsules and inflorescence brown to reddish or chestnut; involucral bracts not stiff, often longer (2–15 cm) ....5. *J. greenei*
            dd. Leaves flat; capsules not exserted (or if so, broad-obtuse), equaling (or slightly longer) to shorter than perianth.
                e. Flowers with pair of bracteoles in addition to bractlet at base of pedicel.
                    f. Flowers strongly secund.
                        g. Involucral bract usually shorter than inflorescence; auricles of leaf-sheath gradually rounded, pale; branches of inflorescence mostly incurved; perianth little longer than capsule
                                      6. *J. secundus*
                        gg. Involucral bract much longer than inflorescence; auricles of leaf-sheath thin, whitish, 2–3 times as long as wide; branches of inflorescence divergent to arched-recurving; perianth much longer than capsule ....
                                      8. *J. tenuis* var. *williamsii*
                ff. Flowers not distinctly secund; involucral bract from shorter than to longer than inflorescence.
                    g. Sheaths of cauline leaves covering half or more of stem; sepals and petals obtuse, perianth scarcely equalling capsule; capsules ovoid ellipsoid, obtuse; plants with elongate horizontal rhizomes ....7. *J. gerardi*

gg. Sheaths of cauline leaves covering only small part of stem (¼ or less); leaves flat but commonly involute; sepals and petals acute to acuminate; rhizomes short.

    h. Auricles appearing as pale to scarious extensions from margin of leaf-sheath; perianth much exceeding capsule.

        i. Auricles thin, whitish, two or more times as long as broad, easily torn ....8. *J. tenuis*

        ii. Auricles about as long as broad, firmer, pale grayish-yellow ....9. *J. platyphyllus*

    hh. Auricles rounded, scarcely extending beyond sheath.

        i. Auricles yellow, darkening with age, hard, glossy; sepals and petals 4–6 mm long, somewhat spreading; perianth longer than capsule ......10. *J. dudleyi*

        ii. Auricles straw-color, drab, or tinged with brown or red, membranaceous; sepals and petals 3–4 mm long, erect; perianth about equalling capsule ......

                11. *J. interior*

ee. Flowers without bracteoles, with only a bractlet at base of pedicel; flowers in heads or glomerules, sepals and petals 2–3.5 mm long, sepals sharply acute, petals obtuse; rhizome thick or knotty.

    f. Stems stout, mostly 6–12 dm tall; principal leaves 4–6 (–7) mm wide, firm, with 5 prominent veins; heads many (20–100) with 2–6 flowers in head; purple anthers often persistent in fruit; seeds orange to rust-color ....12. *J. biflorus*

    ff. Stems slender, mostly 2–5 dm tall; principal leaves 1–3 (–4) mm wide, soft, often with 3 veins more prominent; heads few (2–20) with 5–10 flowers in head; stamens soon withering; seeds brown ....13. *J. marginatus*

bb. Leaves terete, septate.

  c. Flower-heads spherical, dense; sepals and petals very narrow, long-pointed.

    d. Involucral leaf usually longer than inflorescence; capsules tapering to tip or subulate, exserted; stamens 6.

      e. Culms stout, 4–10 dm tall; heads large, averaging more than 10 mm (10–15) in diam.; sheath-auricle pale, scarious, about 3 mm long ......

                14. *J. torreyi*

      ee. Culms slender, mostly 3–4 (–6) dm tall; heads smaller, averaging less than 10 mm (6–12) in diam.; sheath-auricle yellowish, about 1 mm long ....15. *J. nodosus*

    dd. Involucral leaf usually shorter than inflorescence; heads 8–12 mm in diam.; stamens 3.

      e. Capsules tapering to tip, exserted; culms slender, 2–8 dm tall; very local inland ....16. *J. scirpoides*

      ee. Capsules obtuse or abruptly pointed, shorter than perianth and obscured by it; culms slender, 3–8 dm tall; widespread ....

                17. *J. brachycarpus*

cc. Flowers not in spherical heads.

  d. Capsule lance-linear, 4–6 mm long, *much* exserted (to 2 × length of perianth); sepals and petals linear-subulate, 2–3 mm long ....

                18. *J. diffusissimus*

  dd. Capsule prismatic, acute, obtuse, or mucronulate, from about equal to less than twice length of perianth.

    e. Seeds with pale or whitish tail-like appendages.

      f. Sepals and petals soft with scarious margins, obtuse to subacute, shorter than the abruptly short-beaked brown capsule; inflorescence large, heads few-flowered ......19. *J. brachycephalus*

      ff. Sepals and petals stiff, green, greenish, or brownish; capsule slightly exserted.

        g. Stems stiff, erect, tufted, flower with mature capsule 3.5–4.5 (–5) mm long, sepals subulate-tipped; capsule dark straw-color, abruptly narrowed to short beak ....

                20. *J. canadensis*

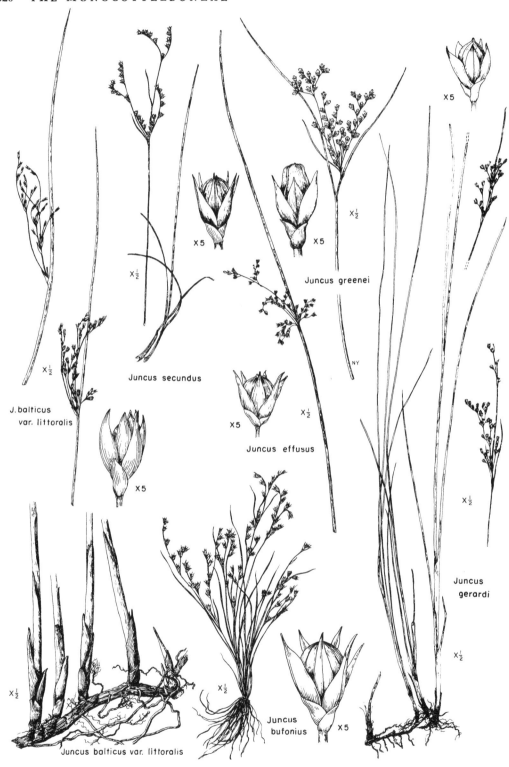

X5

X5

X5

X½

Juncus greenei

X½

X½

Juncus secundus

J. balticus
var. littoralis

X5

X5

X½

Juncus effusus

Juncus
gerardi

X½

X5

X½

Juncus
bufonius

X5

X½

Juncus balticus var. littoralis

gg. Stems loosely ascending, densely tufted; flower with mature capsule 3–4 mm long, sepals lance-linear; capsule brownish, tapering ....21. *J. subcaudatus*
ee. Seeds without whitish appendages, usually dark-tipped.
    f. Capsule about as long as sepals and petals, short-pointed, acute (rarely obtuse) and mucronate; sepals and petals lance-subulate, greenish or straw-color; stamens 3 ....22. *J. acuminatus*
    ff. Capsule distinctly longer than perianth; sepals and petals not subulate, acute or blunt; stamens 6.
        g. Capsule slender-conic, acute, glossy chestnut-brown; sepals acute to acuminate, petals broader; inflorescence short, spreading ....23. *J. articulatus*
        gg. Capsule obtuse to acute, short-pointed; inflorescence spreading-ascending; sepals and petals mostly obtuse ....

24. *J. alpinus*

1. Juncus effusus L.

A highly variable almost cosmopolitan species in which a number of varieties are distinguished; most of our Ohio specimens are referable to the wide-ranging var. *solutus* Fern. & Grisc., a common plant of wet meadows, swamps, and transient pools, growing in dense clumps. One specimen from Wayne County may be referred to the more northern var. *pylaei* (Laharpe) Fern. & Wieg., and a few (from Crawford, Geauga, Tuscarawas counties) appear to be intermediate between this and var. *solutus*. A specimen from Ashland County is best referred to the small-flowered northeastern var. *decipiens* Buchenau, and one from Darke County suggests this variety. These varieties are distinguished as follows:

a. Culms very slender, 1–2 (–3) mm in diam, just above upper sheath; perianth somewhat spreading from base, reddish; sepals soft, linear-lanceolate, 1.5–3 mm long ....var. *decipiens*
aa. Culms to 5 mm thick just above upper sheath; perianth ascending to appressed; sepals firm, 2–4 mm long, lance-attenuate.
    b. Sepals exceeding petals and capsule, slightly spreading; culms grooved below inflorescence ....var. *pylaei*
    bb. Sepals not exceeding petals, usually shorter than to equal to capsule, not spreading; culms finely striate below inflorescence ....var. *solutus*

As these varieties intergrade, and authorities differ in treatment, they are not separated on the map. Specimens of var. *solutus* collected before maturity of capsule have perianth exceeding capsule. *J. effusus* is distinguished from other Ohio species of *Juncus* (except *balticus*) by the apparently lateral inflorescence, and from species of *Scirpus* with comparable lateral-appearing inflorescence by the regular 3-merous flowers with 3-valved capsule.

2. Juncus balticus Willd. var. littoralis Engelm.

A rush of lake shores, ranging from Labrador to British Columbia, and southward into our area; most frequent along the Great Lakes, occasional along railroad tracks inland. Stems tall (to 1 m), usually arising singly at 1–2 cm intervals from elongate straight rhizomes, thus frequently in lines. Sepals and petals dark. Inflorescence dense to open, or, in the more common forma *dissitiflorus* Engelm., open and diffuse.

3. Juncus bufonius L. TOAD RUSH

A low slender annual usually about 10–15 cm tall (sometimes smaller or larger), with fibrous roots, growing in moist clay in open situations. Flowers scattered along branches of the inflorescence, which may make up nine-tenths of the plant. An almost cosmopolitan species, and probably more frequent in Ohio than our map indicates, but overlooked because of its small size. Ohio plants are var. *bufonius.*

4. **Juncus vaseyi** Engelm.

One of our few rushes with terete non-septate leaves. Inflorescence subtended by stiff bract; capsule greenish, exceeding the blunt petals and acute sepals; seeds with long white tails. A northern species.

5. **Juncus greenei** Oakes & Tuckerm.

The somewhat terete leaves deeply channeled, crowded near base of stem; involucral bracts variable in length (2–15 cm long), flexuous; capsule olive-brown to reddish, much exceeding perianth; seeds with minute white tails. Local; a northern species.

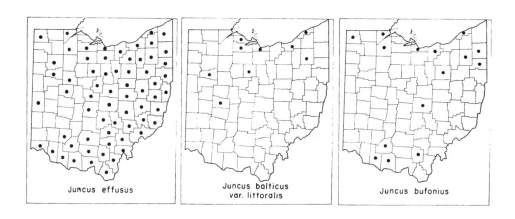

Juncus effusus

Juncus balticus
var. littoralis

Juncus bufonius

Juncus greenei

6. **Juncus secundus** Beauv.

As the specific name suggests, the flowers are arranged along one side of branches of the inflorescence. Stems strictly erect, much exceeding the narrow crowded leaves. A rush of mid-latitudes; very local in Ohio, on sandstone cliffs and sandy soil of clearings.

7. **Juncus gerardi** Loisel. Black-grass

Stems in small tufts, erect, arising from dark, horizontally spreading slender rhizomes; leaf-sheaths covering about half of stem, the upper leaf diverging from stem at about its middle; inflorescence narrow, its branches ascending; sepals and petals obtusely pointed, with faint longitudinal purple-brown stripes; sepals slightly longer than petals, tips incurved; capsule ovoid-ellipsoid, obtuse and mucronate. A plant of salt marshes of the Atlantic and Pacific coastal plains, also in Eurasia and North Africa; inland locally in America, perhaps introduced along railroad tracks. Our plants are var. *gerardi*.

8. **Juncus tenuis** Willd. Path Rush, Yard Rush, Path-grass

This slender wiry rush (especially var. *tenuis*) gets its common names from the fact that it grows abundantly in hard ground along paths; its color, darker than that of grasses to either side, defines the path and makes it look soiled. Three varieties in our area:

Leaves from half as long as stem to longer than stem; stems 1–6 dm tall; branches of inflorescence 1–8 cm long; flowers clustered near tips of branches ....var. *tenuis*
Leaves much shorter than stems; stems 4–9 dm tall; branches of inflorescence longer, to 2 dm; flowers scattered and somewhat secund ....var. *anthelatus* Wieg.
Leaves shorter than stems; stems 1–5 dm tall; branches of inflorescence divergent or arched-recurving, flowers crowded along upper side for entire length .....var. *williamsii*

The typical variety is much more abundant than var. *anthelatus*, is found throughout North America, and is naturalized on all other continents; var. *anthelatus* grows in wetter soil, is not so frequently found along paths, and is more circumscribed in range—Indiana to Maine, south to Missouri, Tennessee, and Georgia; var. *williamsii* is local—specimens from Lorain and Trumbull counties. All three varieties have thin, whitish, easily torn sheath-auricles longer than broad, and very acute or lance-subulate perianth-segments longer than capsule.

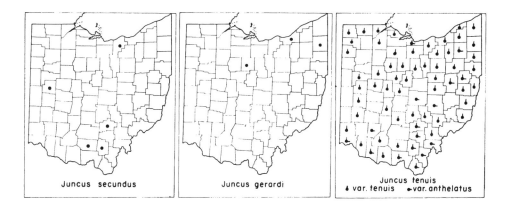

Juncus secundus

Juncus gerardi

Juncus tenuis
▲ var. tenuis ← var. anthelatus

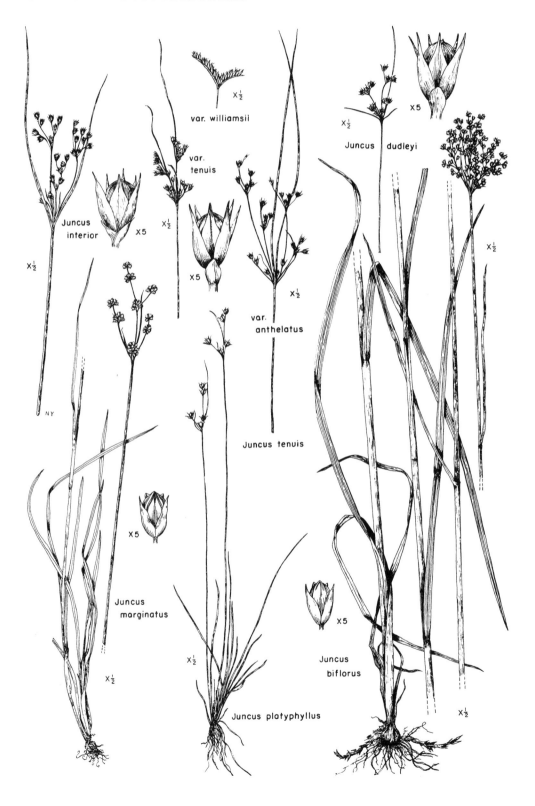

var. williamsii

X½

var. tenuis

Juncus interior

X5

X½

X5

var. anthelatus

X½

X½

Juncus dudleyi

X5

X½

X½

NY

Juncus tenuis

X5

Juncus marginatus

X½

X5

Juncus platyphyllus

X½

X5

Juncus biflorus

X½

9. **Juncus platyphyllus** (Wieg.) Fern.

J. dichotomus Ell. var. *platyphyllus* Wieg.

Although the height of this species is given as 3–10 dm, the three Ohio specimens so determined are smaller, 1.5–3 dm. Plants in tufts, basal sheaths often purplish; leaves flat, involute on drying; sheaths with thin margins extending into truncate or rounded, pale yellowish-gray, firm-membranaceous auricles about as long as broad; inflorescence overtopped by involucral bract, in our specimens 2.5–6 cm high, the flowers on upper side of branches mostly near tips; sepals and petals acuminate, exceeding capsule; bracteoles acute. Plants of "dry or moist acid soils, e. Me. to w. N. Y., s. to Fla." One Ohio specimen (Pike Co., determined by F. J. Hermann) on bank of siliceous gravels.

10. **Juncus dudleyi** Wieg.

Somewhat similar in appearance to *J. tenuis*, but distinguished from it by the hard, glossy yellow rounded auricles of leaf-sheath; leaves half or less than half as long as stems; branches of inflorescence 1–5 cm long, strongly ascending. Common and widely distributed in wet soil.

11. **Juncus interior** Wieg.

Very similar to the preceding; differing from it in the straw-colored to brownish membranous auricles, the acute to acuminate (instead of obtuse to subacute) bracteoles, and shorter, appressed-ascending perianth (3–4 mm long instead of 4–6 mm), about equalling capsule. Ranging in general through the Prairie Region and eastward in the Prairie Peninsula.

12. **Juncus biflorus** Ell.

J. aristulatus of auth., not Michx.

Quite different in aspect from any of the preceding species. Stems stout and tall (to 1 m); leaves 1.5–3 dm long, 4–7 mm wide, firm, with 5 prominent veins; inflorescence large (to 2 dm long) and much branched, flowers 2-several in some 30–100 or more compact clusters or heads; perianth usually slightly shorter than obovoid, blunt or even retuse capsule; seeds orange to rust-color, long apiculate and darkened at each end. Often recognized by the persistent dark purple anthers. Ranging almost throughout the eastern half of the United

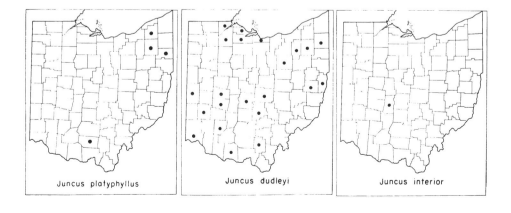

Juncus platyphyllus   Juncus dudleyi   Juncus interior

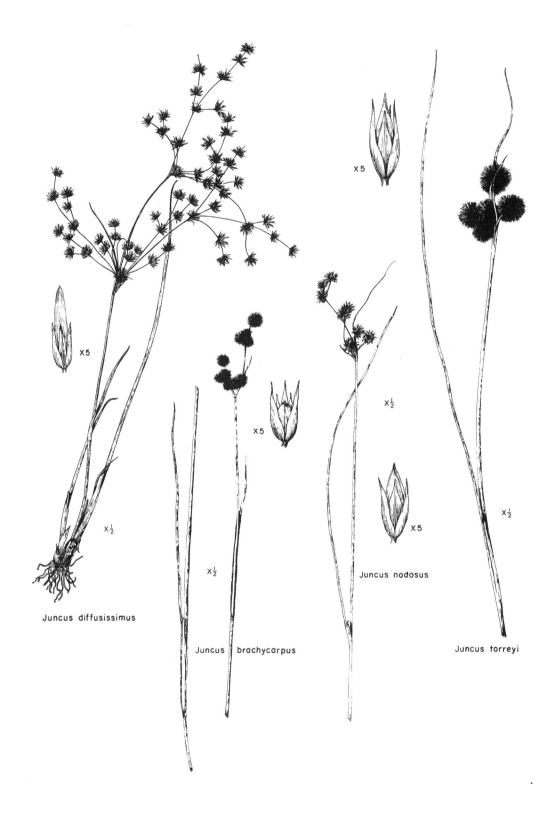

X5

X5

X5

X5

X5

X$\frac{1}{2}$

X$\frac{1}{2}$

X$\frac{1}{2}$

X$\frac{1}{2}$

X$\frac{1}{2}$

Juncus diffusissimus

Juncus brachycarpus

Juncus nodosus

Juncus torreyi

States in wet soil; common in wet meadows in the "white clay" of the Illinoian Till Plain.

13. Juncus marginatus Rostk.

Stems more slender than the last, in tufts, usually less than 5 dm in height; leaves narrow, 1–4 mm wide, soft and grass-like; inflorescence small, rarely 1 dm long, heads 2–20, the subtending bract lance-attenuate; capsule broadly obovoid, the pale brown seeds short-apiculate; stamens soon shriveling, never evident at time of maturity of capsule. Somewhat wider-ranging than the last, but less frequent in Ohio. Larger specimens often resemble *J. biflorus*, from which they can usually be distinguished by number of veins and absence of persistent anthers. A few Jackson County specimens appear to be intermediate.

14. Juncus torreyi Coville

J. nodosus L. var. *megacephalus* Torr.

One of our most easily recognized species, because of the usually few large (1–1.5 cm diam.) globular heads of numerous (30–80 or more) flowers, and rigid linear-subulate sepals and petals about equal in length to the subulate capsule; leaves terete and septate, 1–3 dm long, widely spreading; sheath-auricle thin, white or hyaline, about 3 mm long; stems stout, 4–10 dm tall. In swamps and wet soil, westward in wet prairies; New York and southern Ontario to Kentucky and Tennessee, westward to the Pacific.

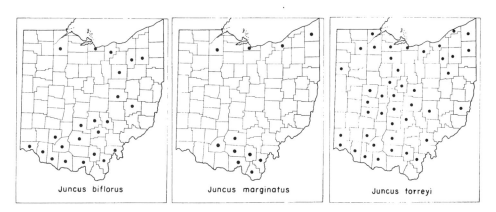

Juncus biflorus   Juncus marginatus   Juncus torreyi

15. Juncus nodosus L.

More slender than the last, and only about half as tall; heads smaller, usually less than 1 cm in diam., with fewer flowers (8–20); leaves shorter, usually less than 1 dm, erect or ascending; sheath-auricle yellowish, about 1 mm long. Galls often replace the normal flower-heads in this species. Almost transcontinental, but more northern than the last; in bogs, swamps, and along wet shores.

16. Juncus scirpoides Lam.

Although occasionally reported, I have seen no specimens referable to this species, which is a coastal plain species ranging in the interior to southern

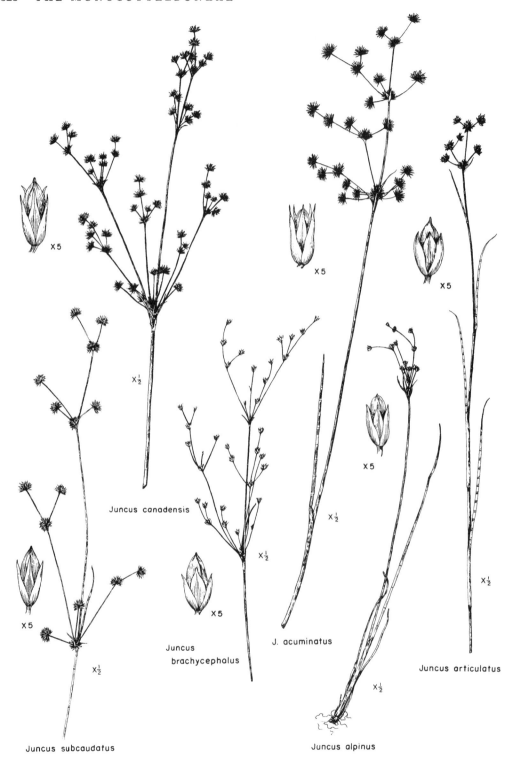

X5

X5

X5

X$\frac{1}{2}$

Juncus canadensis

X5

X$\frac{1}{2}$

Juncus subcaudatus

X5

Juncus
brachycephalus

X$\frac{1}{2}$

J. acuminatus

X$\frac{1}{2}$

X5

X$\frac{1}{2}$

Juncus alpinus

Juncus articulatus

X$\frac{1}{2}$

Missouri and Oklahoma, with disjunct stations at the southern end of Lake Michigan.

17. Juncus brachycarpus Engelm.

A fairly common species of the septate-leaved group of *Juncus* with globular flower-heads. Stems stiff and stout, but more slender than those of *J. torreyi,* and about the same height. Capsules only ½–⅔ as long as perianth, and hidden by it; sepals and petals very slenderly pointed, soft. From northeastern Ohio and southern Ontario west to Missouri and south to Mississippi and Texas; local eastward. Widely distributed in Ohio, abundant in wet meadows on the "white clay" of the Illinoian Till Plain.

18. Juncus diffusissimus Buckl.

A slender species with very diffuse inflorescence, the lower branches long, widely spreading and divaricately branched, the central progressively shorter, middle almost sessile, the whole inflorescence about ⅓ to ½ height of plant. The very slender, elongate capsule, about twice the length of the perianth, is distinctive. A southern species, extending north into southern Ohio and southern Indiana.

19. Juncus brachycephalus (Engelm.) Buchenau

The open or diffuse inflorescence, with many small 2–5-flowered heads and the red-brown abruptly short-beaked capsule about ⅓ longer than the soft scarious-margined sepals and petals characterize this species. In sandy or marly, chiefly calcareous soil of bogs and shores of eastern southern Canada and northern United States.

20. Juncus canadensis J. Gay

Stems in tufts, 4–10 dm tall; leaves prominently septate; inflorescence open to dense, the flowers in heads, these sometimes small with few flowers, sometimes many-flowered, then subglobose; the rigid sepals and petals about equalling or shorter than the round-tipped capsule; seeds with long white tails two-thirds as long as body of seed. Our specimens should be referred to var. *canadensis.* Ranging through much of southern Canada and the U. S. east of the Mississippi River, in wet or marshy, often acid soil.

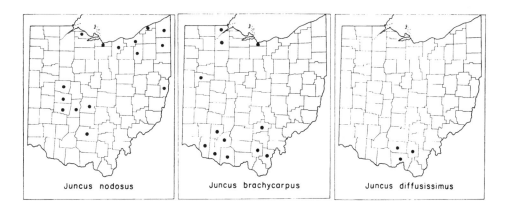

Juncus nodosus   Juncus brachycarpus   Juncus diffusissimus

21. **Juncus subcaudatus** (Engelm.) Coville

J. canadensis var. *subcaudatus* Engelm.

Somewhat similar to smaller specimens of *J. canadensis*, but stems more slender, branches of inflorescence almost filiform, capsule more tapering, and white seed-tails shorter, about one-third as long as body. Inflorescence loose and open; capsule oblong-prismatic and tapering at summit, longer than the lance-linear sepals and petals. Found chiefly on the Coastal Plain from Massachusetts to Georgia, and known also from Ohio and Missouri; a variety has been segregated for plants of spruce bogs of Nova Scotia.

22. **Juncus acuminatus** Michx.

Tufted, erect stems mostly 3–8 dm tall, with usually 2 cauline leaves; inflorescence open, the branches ascending; heads obpyramidal or hemispheric, few to many-flowered, capsule about as long as the green or straw-color lance-subulate sepals and petals, ovoid-prismatic, short-pointed, 3–4 mm long. A common and wide-ranging species of wet meadows and shores, or shallow water, from Maine to Minnesota and south to Florida, Texas, and Mexico, also in the northwest.

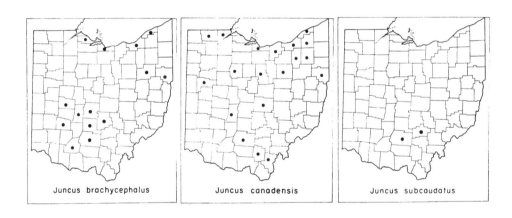

Juncus brachycephalus   Juncus canadensis   Juncus subcaudatus

Juncus acuminatus

23. Juncus articulatus L.

Culms usually clustered, 2–4 (–6) dm tall, slender and erect or arching, with 2–4 slender strongly nodulose leaves; branches of inflorescence ascending to divergent, with few or many 3–11-flowered heads; perianth shorter than the glossy chestnut-brown capsule. Our plants belong to the typical variety. A northern transcontinental species, south to Rhode Island, Ohio, and Indiana.

24. Juncus alpinus Vill.

A northern species, segregated into several varieties; the typical variety does not enter our range. Stems erect or slightly decumbent, 2–5 dm tall, with 1 or 2 slender cauline leaves; inflorescence with few branches. Distinguished by the obtuse sepals and petals with scarious apex, and more or less obtuse, mucronate slightly exserted capsule. Two varieties may be sought in northern Ohio counties:

Heads compact, several- to many-flowered; branches of inflorescence spreading-ascending ....var. *fuscescens* Fern.
Heads loosely few-flowered, usually with one or more of the central flowers on slightly elongated pedicels; branches of inflorescence ascending to erect ....
var. *rariflorus* Hartm. (var. *insignis* Fries; *J. Richardsonianus* Schultes)

Transitional forms between these two varieties occasionally occur. Most of our Ohio specimens are referable to var. *fuscescens*; one from Lake County is referable to var. *rariflorus*, and one from Lucas County approaches this variety.

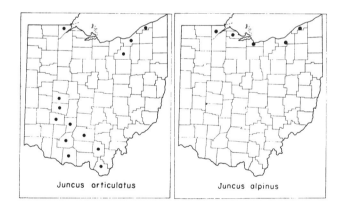

Juncus articulatus   Juncus alpinus

2. LUZULA DC. Woodrush

Juncoides Adams.

A genus of some 65 or more species, many in temperate and colder regions of both hemispheres. Perennials, with narrow grass-like more or less hairy leaves; flowers solitary or in heads in terminal umbellate inflorescence, regular, sepals and petals similar; capsule 1-celled, 3-seeded; seeds with pale caruncle.

Authors do not agree in interpretation of species. The correct specific name for *L. carolinae* S. Wats. (*L. acuminata* Raf.; *L. saltuensis* Fern.) is doubtful.

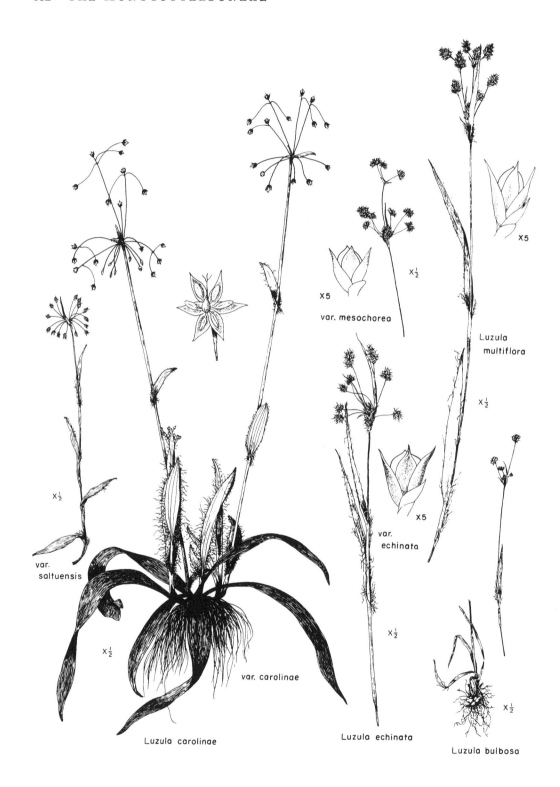

var.
saltuensis

X½

var. carolinae

var. mesochorea

X5

var.
echinata

X5

Luzula
multiflora

X5

X½

Luzula carolinae

Luzula echinata

Luzula bulbosa

Following G. N. Jones (Rhodora, 1951) and F. J. Hermann (in correspondence, 1962), specimens are determined as *L. carolinae.* The two varieties often distinguished (var. *carolinae* and var. *saltuensis*) intergrade in Ohio and are not geographically separable. The characters usually used to separate these so-called varieties are given in the key.

Luzula campestris (L.) DC., which we have long thought of as our common species, is, by Fernald, considered to be Eurasian, and Ohio woodrushes related to it are given specific rank: *multiflora, bulbosa, echinata.* These, Gleason believes, grade into one another and should be retained under *L. campestris.*

Most of our Ohio specimens of *Luzula* have been examined by F. J. Hermann, and his determinations are here followed. In Ohio, there are many transitional forms which cannot be placed in a species-pigeonhole. Immature specimens of this complex cannot be determined satisfactorily.

a. Flowers solitary at tips of spreading or drooping filiform rays of inflorescence.
    b. Rays usually unbranched, with a solitary terminal flower; perianth pale brown to straw-color; filaments ¼–⅓ as long as anthers .... 1b. *L. carolinae* var. *saltuensis*
    bb. Rays often branched, 1–5-flowered; perianth often dark chestnut-brown; filaments as long as anthers .... 1a. *L. carolinae* var. *carolinae*
aa. Flowers in compact heads or spike-like clusters on long or short rays of inflorescence (or some sessile).
    b. Rays of mature umbel ascending to erect; heads mostly cylindric, longer than broad; stems noticeably leafy.
        c. Heads or spikes pale, sepals and petals lanceolate, with soft tips; cauline leaves relatively large, ciliate especially toward base; perianth 3–4 (–5) mm long, usually exceeding capsule .... 2. *L. multiflora*
        cc. Heads or spikes brown, sepals and petals ovate-lanceolate with firm tips; leaves smaller; perianth 2–3 mm long, shorter than capsule; plant usually with bulbs .... 3. *L. bulbosa*
    bb. Rays of mature umbel strongly divergent to reflexed (few ascending), very slender; heads subglobose; cauline leaves few, small.
        c. Sepals and petals 3–4 mm long, tips attenuate, soft, much exceeding capsule; filaments ½ length of anthers or less .... 4a. *L. echinata* var. *echinata*
        cc. Sepals and petals 2.5 mm long, tips acute to acuminate; shorter than to slightly exceeding capsule; filaments more than ½ length of anthers ....
                                       4b. *L. echinata* var. *mesochorea*

1. **Luzula carolinae** S. Wats. EVERGREEN WOODRUSH
    L. saltuensis Fern.; *L. acuminata* Raf.; *L. carolinae* var. *saltuensis* (Fern.) Fern.; *L. acuminata* var. *carolinae* (Wats.) Fern.; *Juncoides carolinae* (S. Wats.) Kuntze.; *J. pilosum* of some Amer. authors; *L. vernalis* of early manuals.

Distinguished from other Ohio species by the filiform, mostly unbranched rays of the inflorescence, each with a solitary terminal flower, and numerous wide (to 1.5 cm) evergreen basal leaves. Forms with some of the primary rays of inflorescence again branched, and with usually chestnut-colored perianth may be referred to the southern var. *carolinae*; those with rays mostly unbranched and with paler perianth may be referred to the reputedly more northern var. *saltuensis.* Because of the many intermediates and lack of geographical separation in Ohio, all representatives of the species are referred to *L. carolinae* by F. J. Hermann (1962) and all are included in one distribution map.

2.   **Luzula multiflora** (Retz.) Lejeune, var. *multiflora*

This could well be called Pale Woodrush, from the very pale green to buff heads of the inflorescence. Our largest species, usually 3–5 (–9) dm tall, very leafy, leaves 9–14 cm long and 4–6 mm wide; branches of inflorescence strongly ascending, heads often more than 1 cm long. Ohio specimens referable to var. *multiflora* (*L. campestris* var. *multiflora* (Retz.) Celak.)

3.   **Luzula bulbosa** (Wood) Rydb.   BULBOUS WOODRUSH

L. campestre var. *bulbosa* Wood; *L. multiflora* var. *bulbosa* (Wood) Hermann.

Although this is a widespread eastern American species, it is apparently rare in Ohio. Distinguished in part by the hard whitish bulb-like tubers commonly hidden by dried leaf-bases. The small brown perianth (averaging 2.5 mm long) shorter than capsule, the erect or ascending rays of inflorescence, and the small cauline leaves further characterize this species. A number of Ohio specimens are intermediate between this and *L. echinata.*

4.   **Luzula echinata** (Small) F. J. Hermann

Distinguished by lax umbel in which some or all of the rays are divergent or reflexed, very slender to hair-like rays, and subglobose heads. Two intergrading varieties in our area:

4a.   **L. echinata** var. **echinata**

L. campestris var. *echinata* (Small) Fern. & Wieg.

Heads 6–9 mm thick; sepals and petals soft-tipped, lanceolate, 3–4 mm long, much longer than capsule.

4b.   **L. echinata** var. **mesochorea** F. J. Hermann

Heads smaller, 5–7 mm thick, sepals and petals shorter, 2.5 mm long, shorter than capsule; often confused with *L. bulbosa,* which also has perianth segments· shorter than capsule, and heads about the same size, but rays of inflorescence are erect or ascending instead of divergent.

To a composite map of the distribution of the *L. multiflora, L. bulbosa, L. echinata* complex as shown by the accompanying maps can be added two

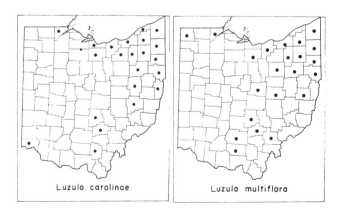

Luzula carolinae   Luzula multiflora

additional counties from which immature or intermediate specimens have been examined: Lake, Richland.

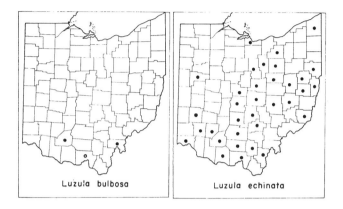

Luzula bulbosa

Luzula echinata

LILIACEAE. Lily Family

A very large family chiefly of herbaceous plants, especially abundant in warm regions, both humid and arid. Flowers, with few exceptions, regular, perianth of 6 distinct or united parts (4 in *Maianthemum*) in two series, all usually petaloid and similar (unlike in *Trillium*), almost always free from ovary which is then superior (partly inferior in *Aletris*); stamens 6 (rarely 4) and placed opposite the perianth parts, usually hypogynous but in some genera (as *Polygonatum*) adnate to perianth; pistil one, usually 3-locular, style 1, divided or trifid or absent, stigmas 3 or 1 with 3 lobes, terminal or lateral on the inner side of divisions of style; fruit a capsule, either loculicidal or septicidal, or a berry. In a few genera, flowers are unisexual (or some unisexual), species then monoecious or dioecious (*Chamaelirium, Smilax*). Inflorescence various—paniculate, racemose, umbellate (terminal or axillary), or of one or a few flowers (either terminal or axillary). Leaves are generally parallel-veined or (as in *Trillium*) venation is parallel-reticulate. Plants with bulbs, corms, rhizomes, tubers, or fibrous roots; stems (except in *Smilax*) mostly unbranched, or, in *Uvularia, Disporum*, and *Streptopus*, forking.

Because of the marked variation in flower and inflorescence characters, the Lily Family is sometimes divided into a number of tribes, subfamilies, or families. Some characters resemble those of the Amarylldaceae, which has resulted in suggestion to transfer some genera to that family. Usually, members of the Liliaceae may be distinguished by the wholly superior ovary, but the ovary of *Aletris* is partly inferior, suggesting Amaryllidaceae. Floral structure is similar to that of the Juncaceae, but perianth parts are distinctly petaloid, not chaffy or glumaceous.

The family contains many native species with large showy flowers (as lilies) or large handsome inflorescences (as in *Stenanthium*), as well as numerous garden plants: tulips, hyacinths, grape hyacinth (*Muscari*), lily-of-the-valley (*Convallaria*), *Scilla*, *Yucca*, *Hosta*, etc. Occasional reports of some of these as escapes do not warrant their inclusion in the Ohio flora.

Of the many (about 240) genera, 22 are represented in Ohio by native species, and others by naturalized or adventive species. Three genera (*Allium, Asparagus, Smilax*), each poorly represented in Ohio's flora, comprise about one-fourth of all the species of this family of nearly 4000 species.

a. Stems leafy, but with leaves usually crowded near base, reduced in size upward; inflorescence a terminal raceme or panicle; fruit a capsule.
    b. Inflorescence a raceme, the flowers placed singly or in groups of 2–3.
        c. Leaves and stems smooth, glabrous.
            d. Lower leaves oblanceolate or obovate, forming a loose rosette; cauline leaves reduced in size upward; dioecious, staminate raceme crowded, white, pistillate raceme greenish ....1. *Chamaelirium*
            dd. Lower leaves narrowly oblanceolate to almost linear, crowded; cauline leaves reduced to small widely spaced bracts; flowers perfect, perianth granular-roughened ....26. *Aletris*
        cc. Leaves smooth and glabrous, longitudinally folded, 2-ranked, linear and grass-like; stem glutinous at least above; flowers perfect, white, in raceme ....
            2. *Tofieldia*
    bb. Inflorescence a panicle.
        c. Leaves linear or linear lanceolate, many times longer than broad.
            d. Stems and inflorescence glabrous.
                e. Sepals and petals white, without glands; flowers of terminal raceme of panicle perfect, of branches staminate ....3. *Stenanthium*
                ee. Sepals and petals white within, greenish on back, with large ob-cordate or 2-lobed gland below middle; flowers all perfect ....
                    4. *Zygadenus*
            dd. Stems, at least above, and rachis of inflorescence pubescent; sepals and petals whitish to greenish or purplish, clawed, and with a pair of glands at base of blade; plants with perfect and imperfect flowers ....
                5. *Melanthium*
        cc. Leaves, at least lower, only 2–4 times as long as broad; stems pubescent; sepals and petals green or dark purple, without glands; plants with perfect and imperfect flowers ....6. *Veratrum*
aa. Without this combination of characters.
    b. Leaves all basal or appearing basal, if 1–2 on stem, these cylindric and hollow, or much reduced (absent at flowering time in 1 species of no. 8).
        c. Flowers solitary on slender peduncle; leaves 2, usually mottled ....
            12. *Erythronium*
        cc. Flowers several to many, variously arranged.
            d. Flowers in terminal umbels, umbel-like clusters or corymbs.
                e. Flowers large, orange or yellow, perianth segments about 1 dm long; leaves linear, numerous ....10. *Hemerocallis*
                ee. Flowers smaller.
                    f. Leaves linear or terete and hollow; fruit a capsule.
                        g. Leaves with silvery or whitish midrib; inflorescence corymbose or short-racemose ....14. *Ornithogalum*
                        gg. Leaves green, without whitish midrib.
                            h. Plants with odor of onion or garlic; umbels dense, many-flowered ....8. *Allium*
                            hh. Plants without onion- or garlic-odor; umbels loose, few-flowered ....9. *Nothoscordum*
                    ff. Leaves large, oblong or elliptic.
                        g. Leaves withering before flowering-time; fruit a capsule; plants with strong onion-odor ....8. *Allium tricoccum*
                        gg. Leaves present at flowering time; fruit a berry ....
                            18. *Clintonia*

dd. Flowers not in umbels or umbel-like clusters.
 e. Flowers racemose.
  f. Perianth of 6 distinct segments.
   g. Raceme elongate; leaves 2–4 dm long, 1–3 cm wide ....
                       13. *Camassia*
   gg. Raceme short, bracted; leaves narrow, with silvery or whitish midrib ....14. *Ornithogalum*
  ff. Perianth-segments united.
   g. Flowers white, outer surface granulose-mealy; leaves narrowly oblanceolate to linear ....26. *Aletris*
   gg. Flowers blue (white in albinos), outer surface of perianth not mealy; leaves linear ....15. *Muscari*
 ee. Flowers in large panicle, creamy-white; leaves linear, evergreen, fibrous on margin, arising from short woody (hidden) stem ....
                      16. *Yucca*
bb. Leaves variously disposed on stem, not basal, and not reduced in size upward.
 c. Flowers in axillary umbels; stems herbaceous or woody ....27. *Smilax*
 cc. Flowers not in axillary umbels; stems herbaceous.
  d. Plants finely branched; leaves reduced to scales, functionally replaced by short capillary modified stems (cladophylls) ....17. *Asparagus*
  dd. Plants with simple, forking, or once- (or twice-) branched stems; leaves developed.
   e. Flowers large, erect or nodding, perianth orange to red, spotted within ....11. *Lilium*
   ee. Flowers smaller, perianth-segments less than 5 cm long.
    f. Inflorescence a terminal raceme or panicle.
     g. Flowers small, perianth-segments less than 5 mm long; fruit a berry; stems 1–8 dm tall.
      h. Flowers 4-merous ....20. *Maianthemum*
      hh. Flowers 6-merous ....19. *Smilacina*
     gg. Flowers larger, perianth-segments 6–15 mm long; fruit a capsule; stems 8–20 dm tall ....6. *Veratrum*
    ff. Inflorescence not a terminal raceme or panicle; fruit a berry or capsule.
     g. Leaves in 1 or 2 whorls at or near summit of stem; fruit a berry.
      h. Leaves 3 in terminal whorl; flowers large, solitary, terminal ....25. *Trillium*
      hh. Leaves (on flowering plants) in 2 whorls, lower of 5–9 leaves, upper of 3 (–4) leaves; flowers greenish yellow, few, in sessile terminal umbel ....24. *Medeola*
     gg. Leaves alternate.
      h. Flower axillary; fruit a berry.
       i. Stems simple; perianth-segments united; berries blue ....23. *Polygonatum*
       ii. Stems branched or forking; perianth-segments distinct; berries red ....22. *Streptopus*
      hh. Flowers terminal (appearing axillary in no. 7); stems branched or forking; fruit a capsule or berry.
       i. Stems pubescent at least when young; flowers and fruit obviously terminal, 1–4 in cluster ....
                   21. *Disporum*
       ii. Stems smooth and glabrous; flowers (except on very young stems) and fruit appearing lateral because of elongation of axillary branch, usually solitary ....7. *Uvularia*

1. CHAMAELIRIUM Willd.   Devil's-bit. Blazing Star

Dioecious; stems mostly 3–6 dm tall, with leaves crowded at base, gradually

Stenanthium gramineum

Chamaelirium luteum

decreasing in size and more widely spaced upward, the pistillate taller than staminate, sometimes to 1 m tall, terminated by a crowded raceme; perianth of 6 narrow segments, white in staminate, greenish in pistillate flowers; stamens 6, anthers white; ovary ellipsoid, styles 3, stigmatic on inner face. A genus of one (possibly two) species of eastern North America.

1. **Chamaelirium luteum** (L.) Gray. FAIRY-WAND

Easily recognized by its simple wand-like stem, and, on staminate plants, white compact spike-like raceme, and by its oblanceolate to obovate basal leaves crowded into an irregular rosette, and oblanceolate to lanceolate smaller stem-leaves. Pistillate inflorescence less conspicuous, but often longer, the 3-locular capsules on slender pedicels. Throughout much of the Deciduous Forest region, generally confined to acid soils.

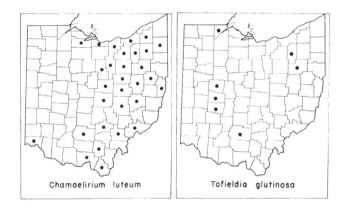

Chamaelirium luteum   Tofieldia glutinosa

2. TOFIELDIA Huds. FALSE ASPHODEL

Slender herbs with leaves near base of stem, 2-ranked, linear, grass-like, folded lengthwise, their bases enfolding stem and base of leaf next above. Flowers in terminal dense raceme, perfect, small, with 6 perianth-segments, 6 stamens, and 3-locular ovary with 3 subulate styles. A genus of about 20 species, mostly in cool-temperate latitudes of North America and in the Andes of South America. Only one species in Ohio.

1. **Tofieldia glutinosa** (Michx.) Pers.
Triantha glutinosa (Michx.) Baker

Plants of marly bogs and wet calcareous ledges and shores, northern in range. The specific name refers to the dark sessile glands on stem and rachis of inflorescence. Flowers usually 2 or 3 together at nodes of the rachis, each group subtended by small greenish to colorless bracts, each flower subtended by thin whitish ovate bracts which look like a tiny cup at immediate base of perianth; perianth-segments about 4 mm long; pedicels glutinous with short hair-like glands.

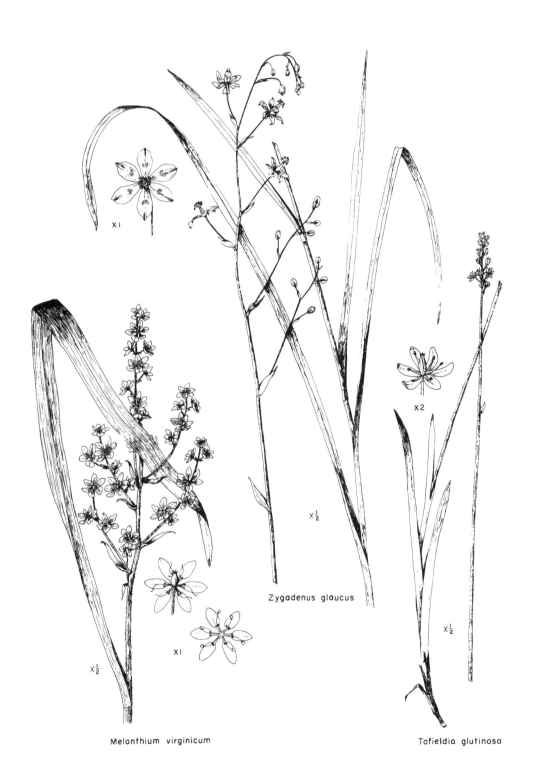

X1

X½

X2

Zygadenus glaucus

X½

X1

X½

Melanthium virginicum

Tofieldia glutinosa

3. STENANTHIUM Gray FEATHER-BELLS

Tall perennials with long grass-like keeled leaves and paniculate inflorescence of many white flowers; flowers of the branches of panicle staminate, those of terminal unbranched axis perfect; perianth-segments linear-lanceolate, attached at base to ovary; stamens much shorter than perianth-segments; ovary 3-lobed, each carpel prolonged into an outwardly curving style, and developing into a 3-beaked short-cylindric capsule.

Genus usually interpreted as containing but a single variable species.

1. **Stenanthium gramineum** (Ker) Morong. FEATHER-BELLS
Three varieties may be distinguished, the last two of which occur in Ohio:

a. Flowers small, perianth-segments 3–4.5 mm long; panicle lacy, open; stems slender, less than 5 mm thick near base; capsules spreading to reflexed ......var. *micranthum* Fern.
aa. Flowers larger, perianth segments 5–10 mm long; stems usually more than 5 mm thick at base.
   b. Panicle lax; capsules spreading to reflexed, leaves narrow, averaging about 6–8 mm wide ....var. *gramineum*
   bb. Panicle dense, up to 1 m in length; capsules spreading to ascending; leaves wider, mostly over 1 cm wide ....var. *robustum* (S. Wats.) Fern.

Although these varieties intergrade, extremes are distinct; var. *robustum* is a handsome plant, attaining a height of 2.5 m, with panicle about ⅓ total height; var. *gramineum* is much more delicate, its leaves narrower, its panicle more open. Furthermore, var. *gramineum* blooms 2 weeks to a month sooner than var. *robustum* in Ohio; the two do not occur in the same patch. More field study (and perhaps cytogenetic studies) are necessary before the status of the varieties can be established. On the distribution map, the occurrence of var. *gramineum* is indicated by an upwardly-pointing appendage to the dot, var. *robustum* by a downwardly-pointing appendage; dots for specimens too incomplete for satisfactory determination have no appendages.

4. ZYGADENUS Michx.

A genus of about 20 species, sometimes divided into several genera distinguished by flower, capsule, and inflorescence differences; all the species are North American except one in eastern Asia. All species are probably poisonous; the plants (bulbs and foliage) contain a violently poisonous alkaloid, zygadenine. Of the inclusive genus, only one species occurs in Ohio.

1. **Zygadenus glaucus** Nutt. WAND-LILY. WHITE CAMASS
Z. chloranthus of Gray ed. 7; *Anticlea elegans* (Pursh) Rydb. in part; *Anticlea glauca* (Nutt.) Kunth

Bulbous perennial with narrow grass-like leaves mostly crowded near base of stem; stem and leaves conspicuously glaucous, blue-green, glabrous; inflorescence an open panicle, its upper part and branches loosely racemose, each branch and flower subtended by ovate green bracts; flowers white or greenish, perianth-segments 7–12 mm long, adnate to base of ovary, each bearing a conspicuous obcordate dark gland below middle; stamens 6; capsules 3-lobed, with recurved beaks, little exceeding the persistent withered perianth. Plants of wet calcareous cliffs and shores, and marly bogs, northern in range.

5. MELANTHIUM L. BUNCHFLOWER

Stems tall, 8–15 dm, stiff, pubescent above, leafy in lower half; leaves (in the Ohio species) linear, elongate; panicle large; petals and sepals clawed, with pair of glands at base of blade; flowers of two kinds, those of terminal raceme of panicle and lower part of its lateral branches perfect, the others staminate or with abortive pistil only. A small genus, all its species North American.

1. Melanthium virginicum L. BUNCHFLOWER

Distinguished from other more or less similar species of related genera by its pubescent stem and inflorescence, and gland-bearing distinctly clawed sepals and petals free from ovary. Principal leaves 3–4 dm long, widest near middle, tapering to base and attenuate apex; flowers creamy or greenish-yellow, turning brown or purplish, in large panicles whose lateral branches are simple racemes; stamens 6, their filaments adnate to claw of perianth-segments; capsules 3-lobed, 10–18 mm long, the carpels rounded on back and beaked. Rare and local in Ohio; long extinct in Hamilton County, the record based on specimens collected before 1852.

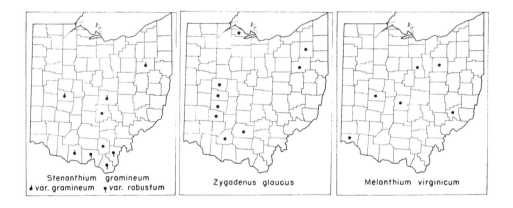

Stenanthium gramineum
▲ var. gramineum ▼ var. robustum

Zygadenus glaucus

Melanthium virginicum

6. VERATRUM L. FALSE HELLEBORE

Coarse somewhat pubescent plants with broad 3-ranked strongly veined somewhat corrugated leaves and large terminal panicle of (in eastern species) dull green to brownish or dark purple flowers; sepals and petals not clawed (but narrowed to base), without glands; capsules (developing from perfect flowers of terminal raceme) ovoid, about 2 cm long. A genus of some 40 or 50 species of north-temperate latitudes.

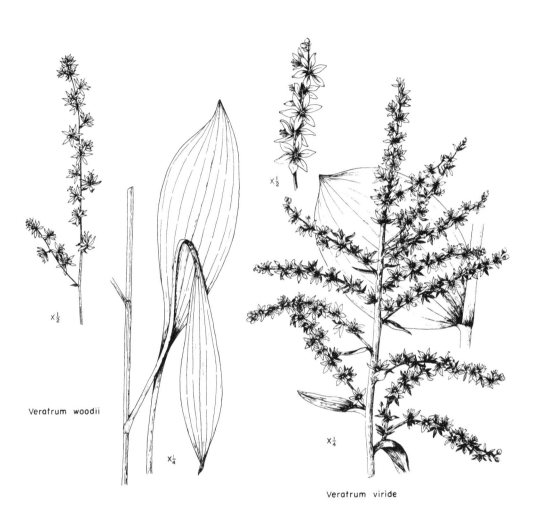

Veratrum woodii

Veratrum viride

$\times\frac{1}{2}$

$\times\frac{1}{4}$

$\times\frac{1}{2}$

$\times\frac{1}{4}$

Leaves broad, oval or elliptic, sessile and clasping at base; stems leafy to inflorescence, but leaves reduced in size upward; flowers yellowish-green ....1. *V. viride*
Leaves broadly oblanceolate, tapering to base; stems sparingly leafy, the upper leaves linear, bract-like; flowers dark purple ....2. *V. woodii*

1. **Veratrum viride** Ait. WHITE HELLEBORE. INDIAN POKE

A plant of northern range (and southward in the mountains), found in Ohio only in the northeastern counties, in wet meadows or low woods where it may be associated with Skunk-Cabbage. Before the flower-stalk starts to elongate it is sometimes mistaken for it; White Hellebore is violently poisonous, and the gatherer of Skunk-Cabbage greens must learn to distinguish it, chiefly by its strongly pleated sessile leaves. At flowering time, the tall leafy stem, broad

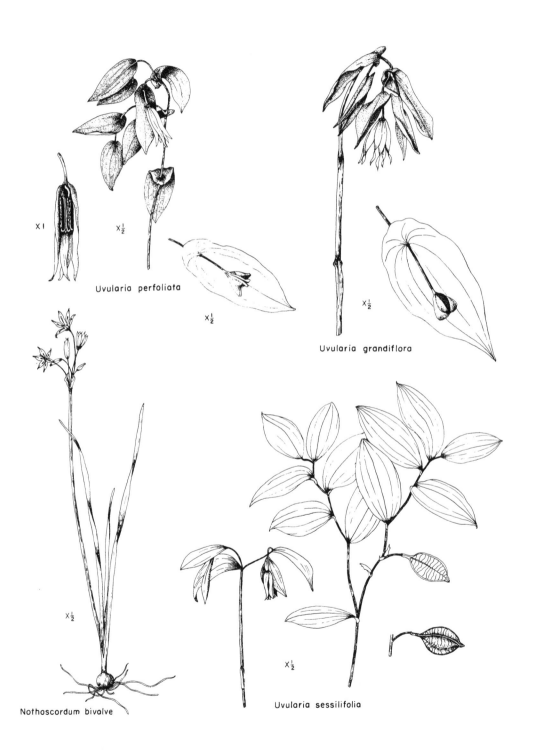

X I

X$\frac{1}{2}$

Uvularia perfoliata

X$\frac{1}{2}$

X$\frac{1}{2}$

Uvularia grandiflora

X$\frac{1}{2}$

Nothoscordum bivalve

X$\frac{1}{2}$

Uvularia sessilifolia

fluted leaves, and 6-parted liliaceous flowers in large panicle, distinguish this species.

2. Veratrum woodii Robbins

A plant of west-central range, from western Ohio and west-central Kentucky westward to Iowa, Missouri, and Oklahoma, found in rich soil in hillside woods. Apparently, plants flower about every 4 or 5 years (Deam, 1940), and when not flowering, must be recognized by the large basal leaves. Leaves narrowed and greatly reduced upward. The only Ohio specimens seen were collected in 1898.

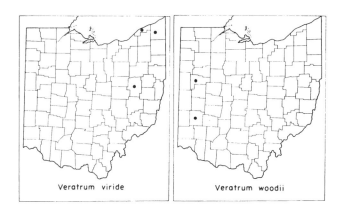

Veratrum viride   Veratrum woodii

7. UVULARIA L. BELLWORT

Plants with slender glabrous stems forking (in flowering plants) above the middle, and slender long or short rhizomes; leaves oblong to oval, perfoliate or sessile, reduced to bladeless sheaths on lower half or more of stem; flowers yellow, pendulous, lily-like, of 6 similar and distinct perianth parts, 6 stamens with elongate-linear anthers and a 3-carpel ovary developing into a 3-angled loculicidal capsule; styles 3, separate or united to about middle. The flowers are terminal on the forks of the stem, but appear lateral and axillary because of growth of a branch from axil of uppermost leaf. The systematic position of *Uvularia* is in doubt, some authors classifying it with the genera preceding in the arrangement here followed (that of Fernald, 1950), others placing it near *Disporum* and *Streptopus* which it more nearly resembles.

a. Leaves perfoliate.
    b. Flowers 2–3.5 cm long; perianth-segments conspicuously glandular-roughened within; capsules truncate, each lobe 2-beaked at summit and with 2 dorsal ridges; leaves glabrous beneath ....1. *U. perfoliata*
    bb. Flowers larger, 2.5–5 cm long; perianth-segments not glandular roughened within; capsule-lobes without dorsal ridges, truncate, pointed, or rarely acuminate at tip; leaves more or less pubescent beneath ....2. *U. grandiflora*
aa. Leaves sessile, not perfoliate, glaucous beneath; flowers 1–2.5 cm long; capsules narrowed to base and summit, stalked, sharply 3-angled ....3. *U. sessilifolia*

1. **Uvularia perfoliata** L.   BELLWORT. STRAW-BELL

A plant of acid or circum-neutral soil, almost confined, in Ohio, to the Allegheny Plateau and adjacent Lake Plains, with outlying stations in acid soils of the Till Plains. Smaller than the next and less freely flowering. Readily distinguished at flowering time by the glandular-papillose inner surface of perianth, later, by capsule characters and orientation of capsule with respect to branches— upper branches spread horizontally at about capsule-level, capsule not hidden by foliage. Eastern in range, rarely occurring west of the Appalachian provinces.

2. **Uvularia grandiflora** Sm.   BELLWORT

A wide-ranging species of rich calcareous soil of woods. Larger than the preceding; flowers often deeper yellow, rarely orange-yellow with orange stripe down middle of each perianth-segment, these slightly spirally curved; capsules without the 2 dorsal ridges of the preceding, usually truncate but (rarely) the lobes prolonged to acuminate tips; upper branches ascending above capsules and tending to hide them. Although there is some overlap in size-range of all parts of these 2 species, there is little difficulty in distinguishing them at any season. For further discussion of similarities and differences, see Anderson and Whitaker (1934).

3. **Uvularia sessilifolia** L.   MERRY-BELLS

Oakesia sessilifolia (L.) S. Wats.; *Oakesiella sessilifolia* (L.) Small

The sessile leaves, paler and glaucous beneath, and stipitate, sharply 3-angled capsule distinguish this species from the two preceding. The similar but more southern *U. pudica* (Walt.) Fern. (*U. puberula* Michx.) is distinguished by the bright green shining leaves and absence of capsule-stalk above receptacle. Usually in moist woods; somewhat more northern in general range than the other species.

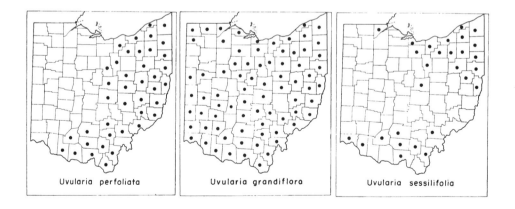

Uvularia perfoliata   Uvularia grandiflora   Uvularia sessilifolia

8.   ALLIUM L.   ONION. GARLIC

One of the largest genera of the Liliaceae, containing about 300 species distributed throughout the north temperate zone; many species in the West, and in Asia. Bulbous, scapose plants with strong onion or garlic odor; bulb-coats membranous or fibrous-reticulate; leaves, in most species, narrow, flattened or terete and hollow, basal or confined to basal part of stem; flowers in terminal umbel subtended by 1–3 bracts, greenish, white, pink, or purple, rarely yellow, the 6 sepals and petals similar, withering and persistent below capsule; stamens attached at base of perianth; capsule 3-locular, bearing the persistent filiform style; seeds 1–2 in each locule.

Several old world species of *Allium* are grown for food, especially *A. cepa* L., the common onion, *A. porrum* L., leek, *A. sativum* L., garlic, and *A. schoenoprasum* L., chives; any one of these may occasionally be adventive in waste places or near gardens; *A. sativum* has been reported from 5, *A. schoenoprasum* from 4 Ohio counties. Several species are showy garden plants; a few, especially *A. vineale* L., field garlic, are troublesome weeds.

a. Leaves linear, narrow and long, present at flowering time.
　　b. Umbel erect, some or all of the flowers replaced by bulblets.
　　　　c. Leaves flat, 2–4 mm wide, subbasal; outer coat of bulb fibrous reticulate ....
　　　　　　　　　　　　　　　　　　　　　　　　　　　1. *A. canadense*
　　　　cc. Leaves terete, hollow (easily flattened), cauline and subbasal; outer coat of
　　　　　　bulb membranous ....2. *A. vineale*
　　bb. Umbel nodding because of bend in stem near base of umbel; umbel not bulblet
　　　　bearing; leaves flat, 2–4 (–6) mm wide, basal; outer coat of bulb membranous ......
　　　　　　　　　　　　　　　　　　　　　　　　　　　3. *A. cernuum*
aa. Leaves lanceolate, lance-elliptic, to elliptic, 2–8 cm wide, narrowed to petioles, present
　　in early spring and dying before flowering time; umbel born on stiff erect stem, pedicels
　　stiff, ascending; perianth white or greenish-white; seeds black, shiny, 1 in each lobe of
　　capsule, persisting after opening of capsule ....4. *A. tricoccum*

1. **Allium canadense** L.　WILD GARLIC
A common and widespread species, usually in moist soil on roadsides and woodland paths, and commonly in floodplain woods. Umbel composed of sessile bulblets and slender pedicelled pink flowers; buds and developing bulblets are at first enclosed in bracts making a slender-pointed conical terminal mass. Capsules rarely developed, and reproduction is almost entirely from bulblets.

2. ALLIUM VINEALE L.　FIELD GARLIC
A common and troublesome weed in lawns, fields, and pastures, roadsides, and gardens; reproducing freely both by multiplication of bulbs and from the many bulblets which often entirely replace flowers in the umbel. Easily recognized by its slender terete and hollow leaves part way up the stem. Flowers, when produced, greenish to purplish, the inner stamens (those opposite petals) with broad filaments cut at tip into 3 awn-like processes, the anther on center projection.

3. **Allium cernuum** Roth　NODDING WILD ONION
The most attractive of the native Ohio species, with pale pink to (usually) deep rose-colored flowers in umbel turned to the side or slightly downward because of bend just below tip of stem; this character distinguishes the species. Frequent in dry usually calcareous soil on gravel banks, rock ledges, roadsides, and in prairie-patches where it often grows abundantly. Ranges farther west than our other species, to British Columbia, Arizona, and Texas.

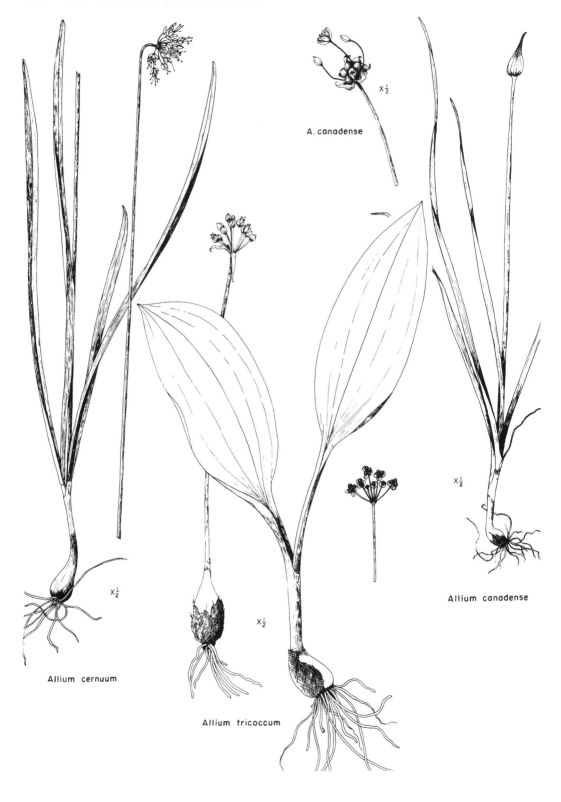

A. canadense

X½

Allium cernuum

Allium tricoccum

X½

Allium canadense

4. **Allium tricoccum** Ait. WILD LEEK. RAMP

Leaves large, flat, bluish-green, variable in shape and size, sometimes narrow, sometimes broad-elliptic, petioled, usually 2–3 from a bulb with finely fibrous-reticulate outer coat, and crowning a rhizome-like base from which the fleshy roots descend. Leaves appear in early spring and wither by early summer at which time the flowering stem starts growth. The fruiting umbels are conspicuous in late summer and early fall when one shiny black bead-like seed protrudes from each open locule of capsule. Common in rich mixed deciduous and beech or beech-maple woods. Ranges from New Brunswick and New England westward to Minnesota and southward to Maryland, Georgia, and Tennessee.

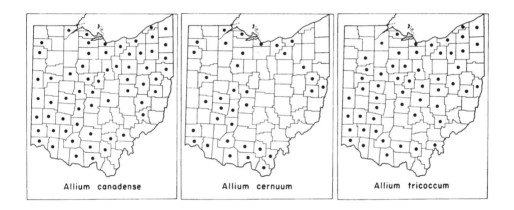

Allium canadense   Allium cernuum   Allium tricoccum

9. NOTHOSCORDUM Kunth. FALSE GARLIC

Similar to *Allium* and distinguished from that genus by absence of onion-odor, and by the greater number of seeds, 6–10, in each locule of capsule. A genus of about 30 species, mostly in the southern hemisphere (South America and Africa); 1 in China, 1 in the southwestern states, and 1 in the East. The common name of the genus is a translation from the Greek.

1. **Nothoscordum bivalve** (L.) Britt.

Bulbous perennial with flower stems (in Ohio specimens) mostly less than 2 dm tall, slender, with loose few-flowered terminal umbel subtended by lance-ovate bracts about 1 cm long; flowers (in May and rarely in Oct.) on slender pedicels of unequal length; perianth-segments creamy white, the mid-nerve prominent, 10–12 mm long; leaves from shorter than to slightly exceeding flowering stem, 1–3 mm wide; capsules subglobose, blunt. In Ohio and in all the Kentucky localities where I have seen it, *Nothoscordum* is confined to dry limestone slopes and cliffs, and prairie patches; Deam (1940) gives for its habitat (in southwestern Indiana) "alluvial bottoms . . . and in low ground in the post oak flats." This is a species whose general range is far to the south and southwest, to Florida, Texas, western Oklahoma, and Mexico to Central and South America. Habitats vary with geographic location; in Oklahoma it sometimes whitens lawns as spring beauty does in Ohio. Now known in Ohio in a limited area of southern Adams County; specimens from Champaign Co. (Univ. of Mich. herbarium) were collected in 1838 in "warm, sandy soil." Another specimen, from Clark County "near Springfield," is a Sullivan collection of 1840 and is now in the Gray Herbarium (data from R. L. Stuckey who examined the Gray Herbarium specimen).

Nothoscordum bivalve

10. HEMEROCALLIS L.   Day-lily

A Eurasian genus of about 15 species and many horticultural hybrids; one species abundantly naturalized in Ohio. Flowers large, lily-like, but the perianth parts united below the middle into a tube, the segments widely spreading or recurved, in few-flowered terminal often branched clusters on a leafless scape, tube of perianth enclosing ovary, stamens inserted at throat of tube, with versatile anthers (as in *Lilium*); leaves all basal (at base of scape), linear, keeled; roots fleshy-fibrous or with tubers. The name, Day-lily, is also applied to species of *Hosta* (*Funkia*). In addition to the following, the Yellow Day-lily, *H. lilio-asphodelus* L., commonly known as *H. flava*, occasionally spreads from cultivation (nomenclature from Dress, in Baileya, 1955).

1.   HEMEROCALLIS FULVA L.   ORANGE DAY-LILY
This well-known day-lily is common along roadsides almost throughout Ohio, and a beautiful sight when in bloom, if not ruined by poison spray. Since its early introduction as a garden plant it has spread throughout and beyond the "manual range." The extent and rapidity of spread is remarkable, as seeds are rarely if ever produced in this country; spread has been by accidental carriage of tubers from place to place. Its pollen has been used in the production of many showy and brilliantly colored hybrids.

11.   LILIUM L.   LILY

A genus of about 80 species, 7 or 8 eastern and about 10 western American, the remainder in Eurasia, mostly eastern Asia. Bulbs scaly, stems erect, leafy, the leaves in whorls or scattered, usually narrow and sessile; flowers 1-several, at summit of stem. Flowers large and showy, of 6 distinct and deciduous perianth-parts, bell-shaped or funnelform, yellow to orange or red and dark-spotted in eastern American species; stamens 6, filaments long, anthers versatile; style elongate, stigma 3-lobed; capsules ovoid to short-cylindric, loculicidal, each locule containing many closely packed, roughly circular, flat seeds.

All of our native species are rather local in distribution, occurring singly or sometimes in patches. A number of exotic species (and hybrids) are in cultivation, most familiar of which are the tiger lily, *L. tigrinum* L., orange with many dark spots and axillary purple-black bulblets; the white lily of gardens, *L. candidum* L.; the Easter lily, *L. longiflorum* Thunb.; *L. elegans* Thunb., with erect orange-red flowers; and the showy lily, *L. speciosum* Thunb., with red-spotted white flowers suffused with rose-pink.

a.   Flowers erect; sepals and petals distinctly clawed; leaves whorled or alternate ....
                                             1.   *L. philadelphicum*
   b.   Leaves lanceolate, mostly whorled; capsules 2.5–3.5 cm long ....var. *philadelphicum*
   bb. Leaves linear to narrowly lanceolate, mostly scattered or upper and lowest in
       whorls; capsules 4–7 cm long ....var. *andinum*
aa. Flowers nodding, sepals and petals not clawed, sessile; leaves mostly whorled; capsules
    erect.
   b.   Sepals and petals recurved almost from base, each with a distinct green area
       (1–1.5 cm long) at base, sepals with 2 sharp dorsal ridges; margins of leaves and
       principal veins on lower surface smooth or rarely with low rounded papillae ....
                                        2.   *L. superbum*
   bb. Sepals and petals recurved from near middle, or only outwardly curved; green
       zone shorter (less than 1 cm long), not well-marked; sepals rounded along midrib,
       without dorsal ridges; margins of leaves and principal veins beneath roughened
       with distinct, sometimes very small, spicules.
      c.   Sepals and petals recurved from below middle, their tips often reaching or
         curving beyond base of tube; filaments strongly outwardly arching ....
                                 3.   *L. michiganense*
      cc. Sepals and petals recurved from above middle or only horizontally curved;
         filaments nearly parallel to style ....4.   *L. canadense*

1.   **Lilium philadelphicum** L.   WOOD LILY

The upright orange to red flowers distinguish this lily from other Ohio species; the pine lily, *L. catesbaei* Walt. of the Southeast, also has upright flowers. The typical variety is Appalachian in range; growing in open woods and thickets; its leaves are narrow oval-lanceolate, tapering to each end, mostly in whorls, often with a few scattered alternate leaves between whorls; capsules obovoid-oval, 2.5–3.5 cm long. The var. *andinum* (Nutt.) Ker (*L. umbellatum* Pursh), Western Red Lily or Prairie Lily, is interior in range, from Ohio westward and northwestward to Minnesota, British Columbia, Colorado, and New Mexico, growing in prairies, open woods, meadows and glades; its leaves are linear, occasionally linear-lanceolate, long-tapering to apex, mostly irregularly alternate, usually a whorl at uppermost node, and often at lowest 1–3 nodes also; capsules oblong, 4–7 cm long.

2.   **Lilium superbum** L.   TURK'S-CAP LILY

The tallest and largest flowered Ohio lily; flowers orange, dark-spotted, sepals and petals strongly recurved from near base, resulting in more open basal

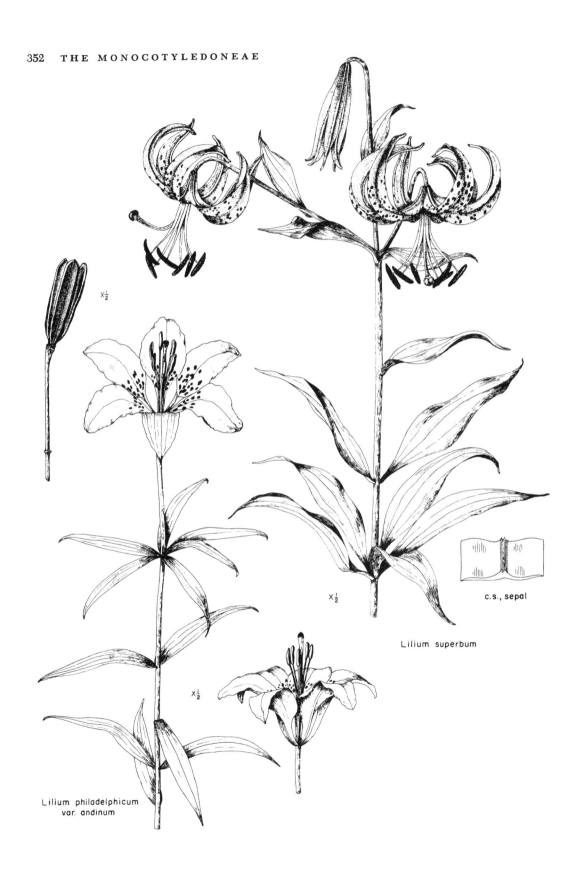

$X\frac{1}{2}$

$X\frac{1}{2}$

$X\frac{1}{2}$

c.s., sepal

Lilium superbum

Lilium philadelphicum
var. andinum

cup of perianth, thus exposing the sharply delimited green basal zone; flowers rarely 1–2, usually 3-many, umbellate or alternately arranged; sepals and petals attenuate to apex, the sepals with 2 sharp dorsal ridges at mid-nerve, a distinctive character and one which results in marked angularity of buds; anthers linear, 15–25 mm long, longer and more slender than those of the next 2 species; leaves elliptic-lanceolate to almost linear, 8–30 mm wide, long attenuate at apex, margins and veins beneath smooth, rarely with almost microscopic low rounded papillae. Appalachian and southern in range, in acidic soils of meadows and mountain woods; rare in Ohio.

3. **Lilium michiganense** Farw. MICHIGAN LILY

This and the next are closely related and, in Ohio, where their general ranges meet and overlap, there seem to be intermediate forms. In its most typical aspect, sepals and petals are strongly recurved from below middle, their tips meeting at or extending slightly beyond base of tube, basal cup not open as in preceding species, flower more funnelform; filaments strongly outcurving, anthers oblong, 7–15 mm long, shorter and more slender than those of preceding; leaves lanceolate, acute to acuminate, leaf-margins and veins beneath roughened with numerous spicules (inconspicuous on shade-plants). Although this species has been confused with *L. superbum* (and is included in it by Gleason, 1952) it is readily separated from it by characters given in key and text, some of which are comparative and overlapping but not the sepal-midvein character— 2 sharp dorsal ridges in *L. superbum*, low rounded strip in *L. michiganense* and *L. canadense* (see figures, adapted from Wherry, 1942). Interior in range, in wet prairies, stream flats, and mucky soil along roads and on shores.

4. **Lilium canadense** L. CANADA LILY

Although this species often goes by the common name, Wild Yellow Lily, and, in popular books on wildflowers (originating in the East) is shown with yellow flowers, this would be a misnomer in Ohio where flowers are orange to red (these referable to forma *rubrum* Britt.). Leaves vary greatly in width; median leaves of the typical variety are narrow (5–10 times as long as wide), those of var. *editorum* Fern. are broader (2–5 times as long as wide); these are said to differ also in thickness of perianth-tube. Such variations overlap in Ohio plants, and varieties are not here distinguished. Sepals and petals outwardly curved or recurved from about the middle or beyond, and filaments parallel to style, two of the more tangible characters separating *L. canadense* from *L. michiganense*. Essentially Appalachian in range, the most frequent wild lily through much of eastern Ohio; usually in woods, or in woodland borders. Often, but not always about two weeks later than *L. michiganense* when growing in the same latitude. Like many Appalachian species, *L. canadense* ranges westward locally in the area of Illinoian glaciation of southwestern Ohio and southeastern Indiana, and also occurs in the hilly unglaciated southern section (the Knobs) of Indiana. The range of *L. michiganense* extends westward from the Allegheny Plateau through the area of Wisconsin glaciation. The two species look quite different in the field.

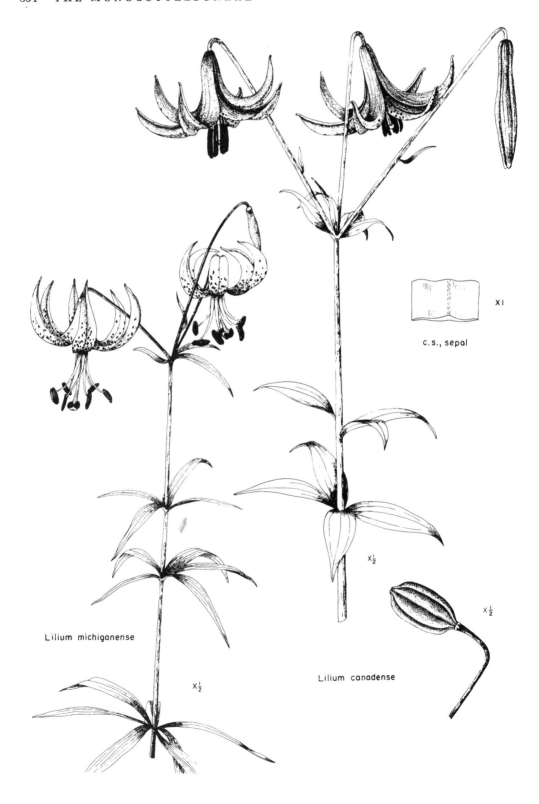

X 1

c.s., sepal

Lilium michiganense

X ½

Lilium canadense

X ½

X ½

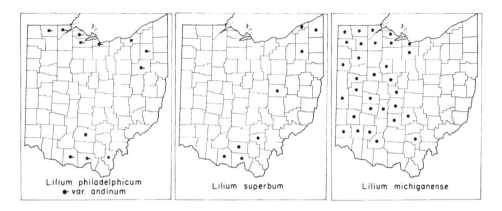

12.   ERYTHRONIUM L.   FAWN-LILY.   DOG'S-TOOTH-LILY

A genus of about 25 species, over half in the western states. The European *E. dens-canis* L. is the basis for one common name, usually written as dog's-tooth-violet, but as the Erythroniums are in no way related to violets, the name should be dropped; locally, dog's-tooth-lily has long been in use. The flowers, nodding and borne singly (in eastern species) on a scapose stem, are similar to those of *Lilium*, with 6 separate and deciduous lanceolate perianth parts, 6 stamens with linear but not versatile anthers, 3-locular ovary, style with 3 stigmas, and loculicidal capsule with plump seeds; leaves on flowering plants 2, usually mottled, appearing basal but actually borne near middle of stem, about half of which is underground, and ascending from a deep-seated bulb-like corm. In current manuals, the capsule is described as "obovoid, contracted at base" and as "obovoid to oblong." To either of these must be added: or ellipsoid and tapering at apex to pronounced beak, in order to include *E. rostratum*. The sepals and petals are very sensitive to temperature changes, spreading or recurving when warmed (usually by the sun, but the same movement can be induced

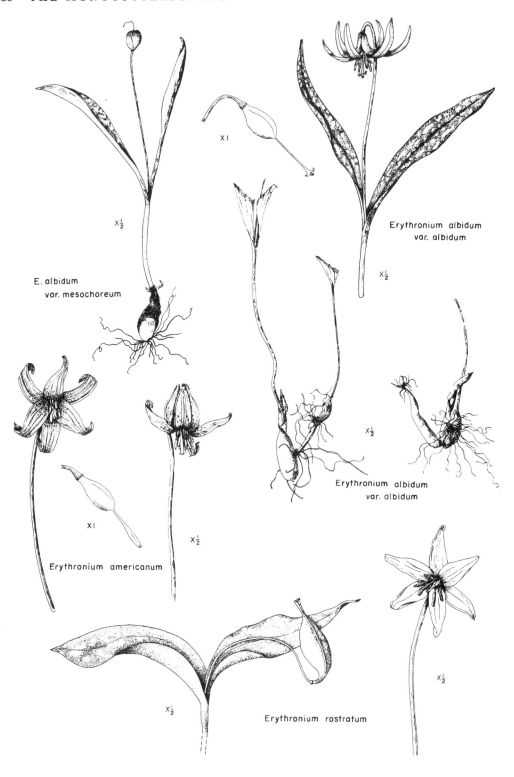

E. albidum
var. mesochoreum

X½

Erythronium albidum
var. albidum

X½

Erythronium albidum
var. albidum

X½

X1

Erythronium americanum

X½

X1

X½

Erythronium rostratum

X½

by bringing the flowers from a cool place to a warm room) and closing or partly closing in response to a downward temperature change.

a. Flowers white, the outer surface of sepals and often the whole perianth suffused with dull blue or pink; style linear below, enlarged upward (clavate); stigmas 3, spreading or recurved ....1. *E. albidum*
aa. Flowers yellow, more or less brown-tinged on outer perianth parts (sepals); petals more or less auricled at base; stigmas erect.
    b. Perianth parts often brown-dotted on inner surface; petal-auricles inconspicuous, not encircling a filament; style slender, pale, withering and soon deciduous; capsule obovoid, truncate ....2. *E. americanum*
    bb. Perianth parts not dotted with brown on inner surface; petal-auricles well developed, encircling a filament; style thick, green, persistent; capsule ellipsoid to ellipsoid-obovoid, tapering to stout, long beak which is part of persistent style ....
                                             3. *E. rostratum*

1. **Erythronium albidum** Nutt. WHITE FAWN-LILY

When in flower in early spring, this species can be recognized at once by flower-color—white, more or less suffused with dull blue or pink. A species of the interior, ranging from southern Ontario and Minnesota southward to Kentucky, Georgia, Missouri, and Oklahoma; "rare eastward to Pa. and D. C."; widespread in Ohio in mesic or dry woods, and sometimes so abundant as to whiten whole hillsides. In southwestern Ohio, many populations indicate introgression from the Ozarkean var. *mesochoreum* (Knerr) Rickett; these bloom earlier than typical *E. albidum*, and grow in drier sites. Vegetative propagation in typical *albidum* is by lateral offshoots connected with the principal bulb by slender threads, and easily detached; few specimens show this feature. In var. *mesochoreum*, offshoots are not produced and the new bulbs develop at base of old one, and are generally included in the loose old wrappings. Plants showing this character occur in the Adams County prairies and on dry slopes; otherwise they are similar to, but smaller than typical *albidum*.

2. **Erythronium americanum** Ker. YELLOW FAWN-LILY

Flowers yellow, variable in amount of brown pigment and anther-color, yellow or reddish-brown, and perhaps consisting of a number of distinct races. The general range extends much farther eastward than the preceding, through New England to Nova Scotia; widely distributed in Ohio where it is the commoner species in the northeastern counties. After the sepals and petals have fallen, this can be distinguished from *E. albidum* by style and stigma characters, as illustrated, and from *E. rostratum* by ovary or capsule and style differences.

3. **Erythronium rostratum** Wolf. GOLDEN-STAR. BEAKED YELLOW FAWN-LILY

A yellow-flowered species of southern range—"northern and central Alabama, and Tennessee. Flowering specimens from Arkansas, eastern Oklahoma and southeastern Kansas are tentatively (without fruit) identified as this species" (Parks and Hardin, 1963). This species is very easily recognized in flower and fruit; its flowers are bright golden yellow, sepals purplish brown on back, petals with a few fine brown lines, all widely spreading (not recurving) in sunlight; its ellipsoid to ellipsoid-obovoid capsule with long tapering beak and persistent style is held erect because of an upward curve of the apical 1–2 cm of the other-

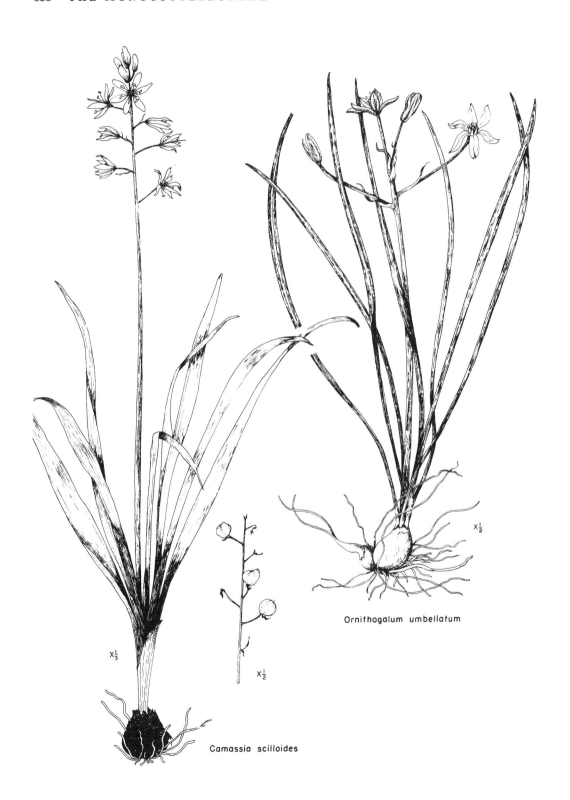

Camassia scilloides

X⅓

X½

Ornithogalum umbellatum

X½

wise downwardly arching peduncle (see figure). The wide-open flower is flat and star-like, and held in a more or less vertical or upturned plane. Known in Ohio only from mesic slopes and flats of Rocky Fork in Brush Creek Township, Scioto County, where I first found it in fruit on April 30, 1963, and in flower April 11, 1964 (Braun, 1964).

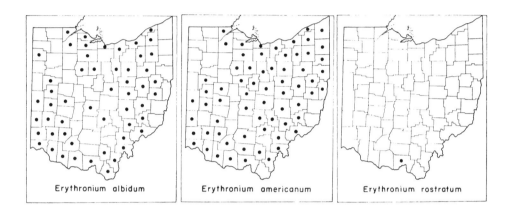

Erythronium albidum   Erythronium americanum   Erythronium rostratum

13. CAMASSIA Lindl.

A small genus of bulbous plants with one eastern species, the others western. Flowers in a loose terminal raceme on leafless stem, white, blue, or purple; perianth of 6 spreading distinct parts; stamens 6, with short versatile anthers; style filiform, stigma 3-lobed; capsule about as long as broad, triangular in cross-section, 3-locular; leaves linear, basal.

1. **Camassia scilloides** (Raf.) Cory.   WILD HYACINTH
   C. esculenta Robins. not Lindl.; *Quamasia hyacintha* (Raf.) Britt.
   The pale blue flowers in long open raceme, or later, the triangular-globose capsule, and linear basal leaves 2–5 dm long and 5–20 mm wide distinguish this from other spring flowers of Ohio. Perianth parts 10–15 mm long, 3-nerved near

Camassia scilloides

center, pedicels slender, 1–2 cm long, subtended by narrowly linear bracts about as long; base of plant wrapped in a membranous sheathing extension of the bulb-coats. Through much of Ohio, wild hyacinths grow in floodplain woods and along streams; in the southwestern counties, the usual habitat is steep rocky calcareous wooded slopes; in the Erie Islands, it is "especially common in hackberry woods" (Core, 1948). A similar contrast of habitats less than 50 miles apart has been noted near St. Louis (Erickson, 1941).

14. ORNITHOGALUM L.

A large old world genus of about 100 species, many in Africa. Perianth of 6 separate and spreading, mostly white, segments; stamens 6, the filaments broad, anthers oblong, versatile; style 3-sided, stigma 3-lobed; capsules loculicidal, seeds few. Bulbous plants with narrow basal leaves and erect flowering scape. One species naturalized in Ohio.

1.   ORNITHOGALUM UMBELLATUM L.   STAR-OF-BETHLEHEM
Flowers white, widely expanded and star-like in sunlight, the segments green on back, in a bracted short raceme or corymb, the lower pedicels much longer than upper, the flower cluster flat-topped, filaments hyaline, flat, anther attached to subulate tip of filament; leaves 3–4 dm long, longer than scape, 2–4 mm wide, silvery white along midrib. A native of Europe, abundantly naturalized and often a troublesome weed difficult to eradicate; reported from a third of the Ohio counties.

Another species, *O. nutans* L., from western Asia, is reported as growing in a cultivated field in Athens County; it has ascending to nodding, short-pedicelled flowers in an elongate raceme, and wider (4–8 mm) leaves.

15.   MUSCARI Mill.   GRAPE-HYACINTH

A genus of the Mediterranean region and southwest Asia, a few species in cultivation and occasionally reported as escapes. Early spring bulbous perennials with small globular or ovoid tubular, usually blue flowers, 6-toothed at summit, in dense raceme; leaves linear.

M. botryoides (L.) Mill., the common grape-hyacinth, with small globular flowers on short pedicels, and flat but channeled leaves, is reported from 11 counties. *M. comosum* Mill., larger in all respects, and with pedicels several times as long as the tubular perianth, is reported from Scioto County.

16   YUCCA L.

A genus of warm temperate and tropical latitudes of North America, best developed in arid and semi-arid regions of Mexico and southwestern United States. Stems woody, short (2–4 dm) and hidden by the long narrow leaves, or taller (in some southern and southwestern species) and simple or branched. Flowers large, white or whitish; perianth of 6 separate oval to oblong parts, petals

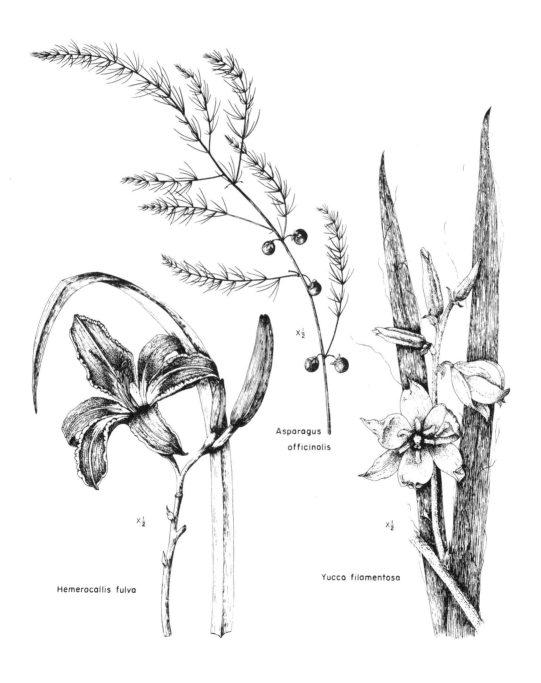

Hemerocallis fulva

X½

Asparagus
officinalis

X½

Yucca filamentosa

X½

longer than sepals; stamens 6, shorter than petals, their filaments outwardly curved; stigmas 3; capsules dry or fleshy, 3-locular or imperfectly 6-locular; seeds flat, numerous.

The pollination story of Yucca is an interesting example of symbiosis in which two organisms, one plant and one insect, are mutually benefited. The flowers are visited by a small moth (*Tegeticula, Pronuba*), which collects pollen from freshly opened anthers, pressing it into a small ball. When enough pollen has been gathered to form a ball about 1 mm in diameter, the moth inserts eggs into the ovary, usually of another flower, then runs up the style and pushes the ball of pollen grains into the cavity between the 3 lobes of the stigma. The pollen-grains germinate, the pollen tubes grow downward and finally reach the ovules. Egg cells in the ovules are fertilized by sperm cells from the pollen tube, and growth of seeds begins. The eggs laid by the Pronuba moth hatch into tiny larvae which feed on the growing seeds, consuming some of them, but by no means all. At maturity of larva, a small hole is bored in the capsule-wall, through which the larva crawls, drops to the ground and pupates, to emerge the following summer as a moth. Without pollination by the moth, the Yucca would produce no seeds, without seeds, the moth could not reproduce.

Wherever growing, Yuccas are striking plants. Most often noticed by the traveler are the Spanish bayonet, *Y. gloriosa* L., of coastal dunes; *Y. elata* Engelm. of desert-grasslands of the Southwest, with usually unbranched trunk, relatively narrow leaves, and large panicle of flowers or capsules elevated on a long stalk; and the grotesquely branched Joshua tree, *Y. brevifolia* Engelm. (*Y. arborescens* Trel.) of the Mohave Desert, each branch terminating in a compact cluster of relatively short stiff leaves, the panicle essentially sessile.

1.   YUCCA FILAMENTOSA L.

Woody stem short, hidden by the numerous long stiff linear leaves (4–8 dm long and 3–7 cm wide), fibrous on margins; peduncle long, panicle large, the combined height up to 3 m. Often planted and occasionally escaped and established on gravelly banks and along roads.

Fernald (1944c, 1950) distinguished two southern species of *Yucca*, both of which are cultivated northward and found as escapes: *Y. filamentosa* and *Y. smalliana* Fern. In his key, he states differences in leaves and inflorescence. Most Ohio specimens have some characters of one, some of the other species. Following Gleason and Cronquist, 1963, both are here included in *Y. filamentosa*.

ASPARAGUS L.

One of the largest genera of the Liliaceae with about 300 species, all native of the Old World, ranging from Siberia to the Cape of Good Hope. Flowers perfect or unisexual, of 6 separate or partly united segments; fruit a berry; stems much branched, herbaceous or woody, some climbing (with stems to 15 m long); leaves much reduced, scale-like, functionally replaced by cladophylls (modified stems arising from the axils of the bract-like leaves), these in fascicles or rarely single. The asparagus-ferns, *A. plumosus* Baker and *A. sprengeri* Regel, are often grown in greenhouses for florists' use; the greenhouse Smilax is *A. medeoloides* Thunb., with leaf-like cladophylls solitary in the bract-axils. One species, the garden asparagus, is used for food.

1. ASPARAGUS OFFICINALIS L.   GARDEN ASPARAGUS.

Dioecious; stems at first succulent, later much branched, from thick matted rootstalks; ultimate branchlets filiform, in fascicles from axils of minute scarious bracts which are the true leaves; flowers greenish, narrowly bell-shaped, on slender jointed pedicels in groups of 1–3 from bract-axils; fruit a red, few-seeded berry. Escaped from gardens, often established along roadsides and in waste places.

18. CLINTONIA Raf.

Perennial herbs with creeping rhizome, 2–4 large obscurely ciliate elliptic basal leaves, these sheathing base of flower-stem; stem 2–4 dm tall, with terminal umbel or umbellate cluster of perfect flowers; pedicels pubescent; perianth-segments 6, distinct; stamens 6, filaments long, very slender; ovary 3-locular, fruit a berry. A genus of 6 species, 2 in eastern Asia, 2 western American, and 2 eastern, all woodland plants.

Flowers more than 1 cm long, greenish yellow; umbel loose or flowers scattered near
 tip of stem; berries blue ....1.   *C. borealis*
Flowers less than 1 cm long, white speckled with purple; umbel regular and compact;
 berries black ....2.   *C. umbellulata*

1. **Clintonia borealis** (Ait.) Raf.   YELLOW CLINTONIA.   BLUEBEAD.

A plant of the Northeast, from Labrador to Manitoba, south to Newfoundland, New England, Michigan and Minnesota, and south in the higher mountains to North Carolina and Tennessee, often in spruce-fir forests and wooded bogs; in Ohio, known only from the northeastern corner, an area included in the Hemlock-White Pine–Northern Hardwoods forest region. Distinguished by its greenish yellow nodding bell-shaped flowers, with perianth-segments often 2 cm long, in loose terminal cluster, and blue, short-ovoid to spherical berries.

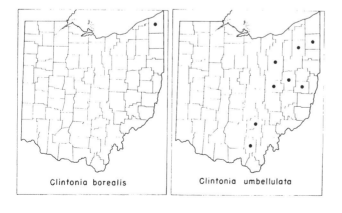

Clintonia borealis   Clintonia umbellulata

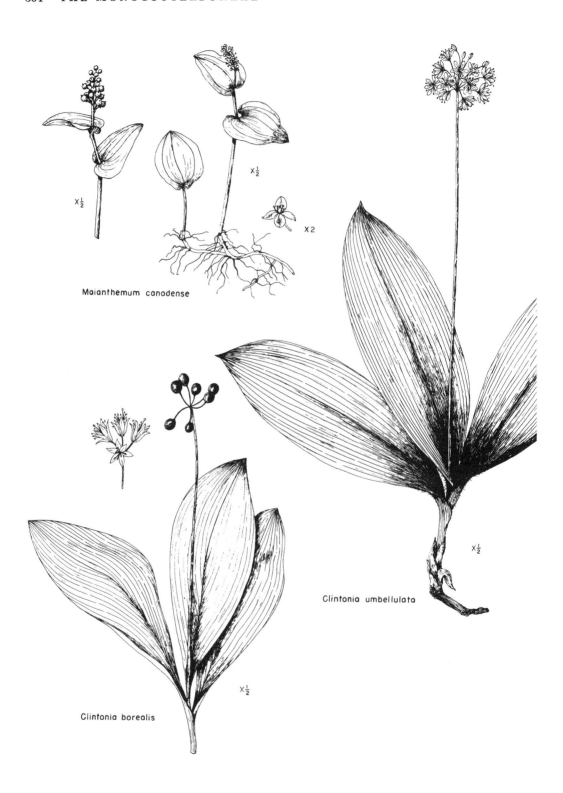

Maianthemum canadense

Clintonia umbellulata

Clintonia borealis

2. **Clintonia umbellulata** (Michx.) Morong.   WHITE CLINTONIA

An Appalachian species, ranging from northern New Jersey, New York, and eastern Ohio southward in the mountains (at lower elevations than the preceding) to Kentucky, Tennessee, and North Carolina, in rich deciduous woods. Distinguished by its speckled white flowers in compact umbel, with perianth segments shorter and blunter (5–8 mm long), and black berries. Local in Ohio, in the Allegheny Plateau section.

19.   SMILACINA Desf.

Vagnera Adans.

Leafy-stemmed plants with creeping rhizomes; flowers small, white, in racemes or panicles, perianth segments 6, stamens 6, the filaments dilated at base, slender above; anthers short; ovary 3-locular, 2 ovules in each locule; fruit a globular 1–2 seeded berry. A genus of North America and eastern Asia.

a. Flowers on short pedicels in terminal panicle, its branches (sometimes much shortened) simple racemes; stems arched ascending; berries green more or less densely mottled with red ....1. *S. racemosa*
aa. Flowers on long pedicels in terminal raceme; stems erect (weak in no. 3).
    b. Raceme sessile or nearly so; stems erect; berries green striped with black, or dark
                                          2. *S. stellata*
    bb. Raceme long-peduncled; stems weak, erect; berries dark red ....3. *S. trifolia*

1. **Smilacina racemosa** (L.) Desf.   SOLOMON'S PLUME.

Stems very leafy, slightly zigzag, arched-ascending, never vertical, finely pubescent; leaves oblong to elliptic, 1–2.5 dm long and to 1 dm wide, abruptly narrowed at base to short petiole, and at apex to acuminate tip, bright green; perianth small, 1.5–3 mm long; berries irregularly spotted or mottled with dark red. Two varieties may be distinguished: var. *racemosa*, with sessile or short peduncled (to 7 cm) ovoid panicle, its longer racemes 2–6 cm long; var. *cylindrata* Fern., with long-peduncled cylindric panicle, its few-flowered racemes 0.5–2.5 cm long. The latter is more southern than the typical variety, and in the area of overlap, the two intergrade. Most of the Ohio specimens seen belong to var. *racemosa*; two are var. *cylindrata* in its extreme form, shown by a drawing of part of a Clinton County specimen; a Hamilton County flowering specimen has the lateral branches of the panicle so shortened that flower-clusters are spherical. Plants of mesic, rich woods.

2. **Smilacina stellata** (L.) Desf.

Stems leafy, upright, usually glabrous; leaves glaucous, lanceolate to oblong-lanceolate, sessile and slightly clasping, gradually narrowed to acute apex; perianth larger, 4–5 mm long; raceme short-peduncled; berries with longitudinal stripes. More northern than the last, ranging across Canada and far southward in the uplands in East and West, in moist sandy or boggy woods, and on shores and bluffs.

3. **Smilacina trifolia** (L.) Desf.

A small far-northern species of bogs and boggy woods, ranging almost

X2

var.
cylindrata

X½

X½

S. racemosa

Smilacina racemosa

X½

Smilacina stellata

X½

Smilacina trifolia

X½

across the continent in the North; also in Siberia. Stems slender, weak, 0.5–2 dm tall; leaves glabrous, 1–4, usually 3, oblanceolate to elliptic, tapering to narrow or petiolar base, acute to blunt at apex, 6–12 cm long; raceme long-peduncled, few-flowered, loose, pedicels long; berries dark red.

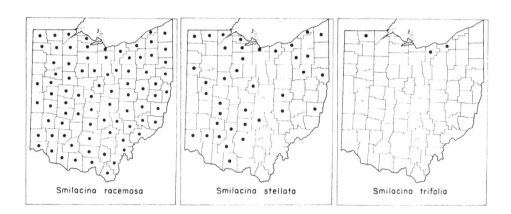

Smilacina racemosa   Smilacina stellata   Smilacina trifolia

Maianthemum canadense

20.   MAIANTHEMUM Weber

Similar to *Smilacina*, and by early botanists referred to that genus; differing in its 4-merous flowers—4 perianth-segments, 4 stamens, 2-locular ovary, and 2-lobed stigma. A genus of only 3 species, 1 Eurasian, 1 in the Pacific Northwest, and 1 eastern.

1.   **Maianthemum canadense** Desf.
Unifolium canadense (Desf.) Greene.
This dainty little woodland plant is often called "Wild Lily-of-the-Valley," a name which should be discarded because there is a true wild Lily-of-the-Valley, *Convallaria montana* Raf., in the Southern Appalachians, and because *Maianthemum* does not resemble a *Convallaria*. The name, False Lily-of-the-

Valley, has also been used. Flowering stems usually about 1 dm tall (0.5–2.5), with 1–3, usually 2 sessile cordate ovate leaves, and numerous long-petioled single leaves arising from the extensively creeping slender rhizomes; flowers small, 3–5 mm wide; berry pale red. Northeastern in range, and southward in the mountains; in woods and second-growth thickets. A variety with leaves distinctly ciliate and more or less pubescent beneath, var. *interius* Fern., is more western, ranging from Ohio northwestward to Alberta and northern British Columbia; most of the western Ohio specimens belong to this variety.

21. DISPORUM Salisb. FAIRYBELLS

Plants with a mass of heavy fibrous roots from a short rhizome; stems pubescent (at least when young), leafy, forking or more irregularly branched, branches diverging at angle of about 60°; leaves below first branch reduced to loose bladeless sheaths; leaves sessile, ovate, oval, or oblong, acuminate, ciliolate on margin, pubescent beneath especially on veins, surface nearly glabrous when old; flowers terminal, on slender pedicels, but at first partly enwrapped in unfolding leaves; perianth of 6 similar distinct parts; stamens 6; ovary 3-locular; fruit a berry, fleshy in one eastern species, dry and capsule-like in appearance in the other.

The eastern species of the genus, to which the above description applies, belong to a small section (5 species) of this American-Asiatic genus. Chromosome numbers are $2n = 12$ in *D. maculatum*, 16 in one western species, 18 in *D. lanuginosum* and others, and 22 in *D. trachycarpum* which approaches the limits of the "manual range" in Manitoba (Jones, 1951; Ownbey, 1953). Chromosomes of *D. maculatum* differ from those of all other North American species in the low basic number ($n = 6$) and in the absence of a pair of satellited chromosomes, having instead a thin terminal segment on one pair. "The chromosome situation [in this section of the genus] suggests that they represent but a remnant of a one-time large and complicated assemblage" (Ownbey, 1953). It would appear, from the low chromosome number of *D. maculatum*, that this is the oldest species of the section; it also has the most limited and local distribution.

> Flowers yellowish green, perianth segments long-attenuate, strongly outwardly curving above middle, much longer than stamens; fruit a fleshy red berry ....
> 1. *D. lanuginosum*
> Flowers white, with lavender or purple spots; perianth segments clawed, expanded blade ovate to ovate-lanceolate, acute, abruptly turned outward at summit of claw, shorter than or about equal to stamens; fruit a dry, straw-colored, 3-lobed, papillose and hairy capsule-like berry ....2. *D. maculatum*

1. **Disporum lanuginosum** (Michx.) Nicholson.

Somewhat suggestive of *Uvularia*, but easily distinguished by the long-attenuate outwardly arching perianth-segments and pubescent leaves, or in fruit by the ellipsoid red berry. Flowers yellowish green, in terminal clusters of 1–3 (–4), nodding; leaves pubescent beneath, pubescence on veins dense, whitish.

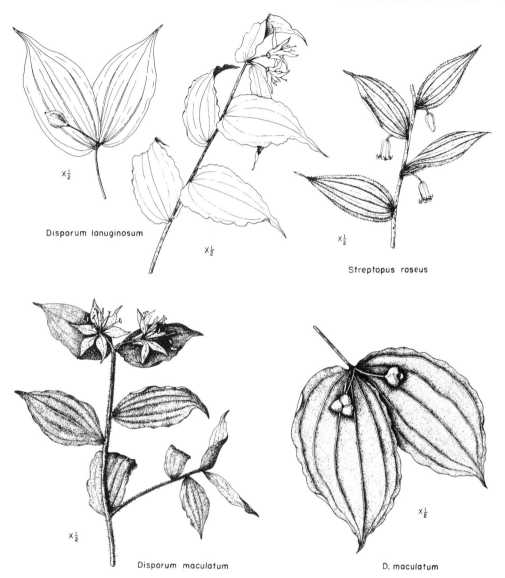

Disporum lanuginosum

Streptopus roseus

Disporum maculatum

D. maculatum

mostly directed apically, margin ciliolate, the cilia directed apically; leaves usually slightly narrower than the next, more gradually acuminate. Rich woods, almost confined to the Appalachian upland from New York and Ohio southward to South Carolina and Alabama.

2.   **Disporum maculatum** (Buckl.) Britt.

One of Ohio's most beautiful spring wildflowers, very local in rich woodland, usually on lower northerly slopes; can be recognized at once by its similar white sepals and petals speckled with purple. Flowers on slender pedicels, 1–4 in terminal cluster. The expanded flowers for the most part stand with perianth in a vertical plane, so that one looks directly at the wide-open flower; the clawed portion of the segments forms a small cup, the widened blade turned outward about at right-angles. In summer, this species can be recognized by its 3-lobed triangular fruit, pale straw-color when mature, coarsely papillose, with a long hair and a number of shorter ones from apex of each papilla, entire surface pubescent and glandular dotted; fruit ripens in early August. In vegetative condition, *D. maculatum* may be distinguished by its more irregular marginal cilia and vein-pubescence.

This species is so local in distribution that few botanists, except in the Southern Appalachians, have seen it in bloom or in fruit—which doubtless explains the errors in current manuals. Local in northern Alabama (Sand Mt.), in the mountains of northern Georgia, of North Carolina and Tennessee, formerly abundant in the rich (now destroyed) forests of Black Mountain and Pine Mountain, Kentucky, and very local at the western margin of the Cumberland Plateau of Kentucky, local in the western part of the unglaciated Allegheny Plateau of southern Ohio; two southern Michigan stations, based on two herbarium specimens of 1884 and 1922 (Jones, 1951) seem strangely out of place for an ancient species otherwise confined to vegetationally old areas of the Appalachian upland; the first of these (from Ann Arbor) is almost certainly an error caused by shifting of collector's labels.

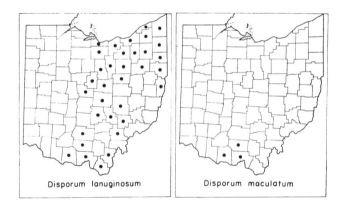

Disporum lanuginosum

Disporum maculatum

22. STREPTOPUS Michx.   Twisted-stalk.

A genus of 7 species of the cooler latitudes of the Northern Hemisphere. The common name is a translation from the Greek *streptos*, twisted, and *pous*, stalk or foot, and refers to the abruptly bent peduncle. Stems comparatively stout, forking or less regularly branched; leaves sessile or clasping, ovate, acuminate; flowers axillary, but the peduncle fused to the stem-internode above its origin, becoming free opposite the next leaf; perianth of 6 distinct segments, bell-shaped at base, spreading to recurved above; stamens 6, filaments short and flattened; ovary 3-locular, style slender; fruit a subglobose to ellipsoid many-seeded red berry.

1. **Streptopus roseus** Michx.

The only Ohio species of the Liliaceae with axillary rose-pink to purple flowers and many-seeded red berries. Three varieties are sometimes distinguished, differing chiefly in length of rhizome-internodes, number of sepal-nerves, and pubescence. The one Ohio specimen seen is referable to var. *perspectus* Fassett, which ranges from Labrador to Ontario and southward in the mountains to Georgia and Kentucky.

Streptopus roseus

23. POLYGONATUM Mill. Solomon's-seal.

Plants with arching to nearly erect unbranched stems arising from horizontal knotty rhizomes, a character to which the generic name, meaning *many* and *knee*, refers. As long ago as 1731 (P. Miller) the origin of the common name was explained thus: "It is also call'd *Solomon's-Seal*, because the Knots of the Root somewhat resemble a Seal." These seals are the scars left by the annual stems. Stems naked below, above with numerous alternate sessile, short-petioled, or somewhat clasping 2-ranked leaves; flowers axillary, greenish (in eastern American species), pendent, the peduncles with 1–15 (rarely more) flowers; pedicels jointed near flower; perianth tubular, with 6 small somewhat outwardly curved lobes; stamens inserted on perianth-tube; ovary 3-locular, style slender, stigma capitate; berries dark blue or black, several seeded.

A genus of about 20 species of Europe, Asia, and North America from the Atlantic to the Black Hills, and from southern Canada to Florida and northeast Mexico, with an outlying species in western New Mexico and adjacent Arizona.

The Ohio species of *Polygonatum* are easily recognized in the field when several individuals may be observed; the overlap in measurements in the descriptions which follow will not there be confusing. Additional characters could be enumerated, among them length and breadth of the small terminal joint of pedicels. When compared, at the same stage of flower or fruit development, differences in relation of length to diameter can be discerned. Fernald's discussion of specific distinctions (Rhodora, 1944a) is of interest in any detailed study of *Polygonatum*.

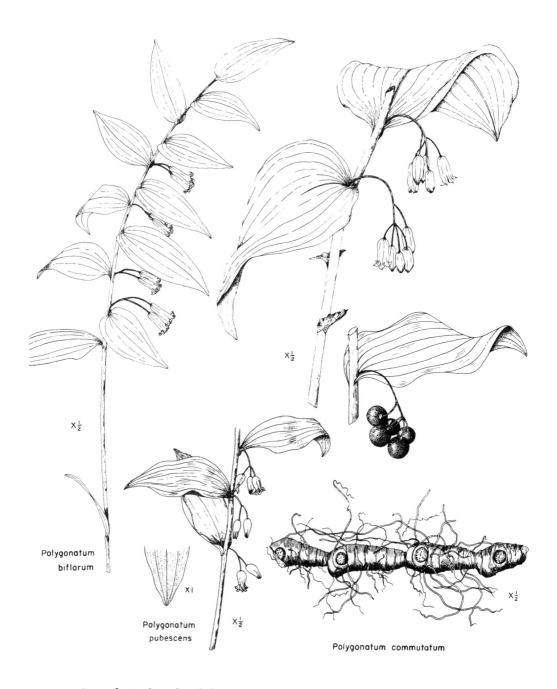

Polygonatum
biflorum

Polygonatum
pubescens

Polygonatum commutatum

A cytological study of the American species of *Polygonatum* (R. P. Ownbey, 1944) reveals that *P. pubescens* and *P. biflorum* are diploids with chromosome number, $n = 10$; and that *P. commutatum* (*P. canaliculatum*) is a tetraploid with chromosome number $n = 20$. In common with most tetraploids, "*P. com-*

mutatum is usually distinguished by the greater vigor . . . ," by such characters as "general robustness, a stouter stem, a larger number of leaves and flowers, broader and thicker leaves, coarser peduncles, larger flowers and fruits." Ownbey further states that "the only certain way to distinguish small *biflorum*-like plants of *P. commutatum* growing within the distributional range of *P. biflorum* is by chromosome number." Such doubtful specimens are omitted from the Ohio distribution maps.

a. Leaves pubescent on veins beneath; lowest peduncle from 1st or 2nd (rarely 3rd) axil, 1–2 (–4) flowered ....1. *P. pubescens*
aa. Leaves glabrous on both surfaces; lowest peduncle usually from 3rd or 4th axil.
    b. Peduncles terete; lowest peduncle usually from 3rd (or 4th) axil; 1–3 (–5) flowered; leaves sessile, flat, with 3–5 (–7) prominent veins; stems slender, 2–7 dm tall, leafy portion shorter than lower leafless portion ....2. *P. biflorum*
    bb. Peduncles flattened; lowest peduncle usually from 4th or 5th axil, 1–15 (rarely more) flowered; leaves slightly clasping, corrugated toward base, with 9–15 (–19) prominent veins; stems stout, 6–20 dm tall, leafy portion longer than lower leafless portion ....3. *P. commutatum*

1. **Polygonatum pubescens** (Willd.) Pursh.   HAIRY SOLOMON'S-SEAL.
    P. biflorum, in part, of older manuals.
Easily distinguished from other species by pubescence on veins of the lower leaf-surface, and, usually, by position of lowest peduncle—from 1st or 2nd axil. Leaves elliptic to broadly oval, about half as wide as long (length 6–12, width 2–5 cm), contracted to petiolar base, with 3–9 prominent veins; flowers earlier than other species, perianth less slender-tubular than in *P. biflorum*, 8–12 mm long, stamens inserted high on perianth-tube. A widespread woodland species, occurring almost throughout the Deciduous Forest from Kentucky, Indiana, and Minnesota eastward.

2. **Polygonatum biflorum** (Walt.) Ell.   SMOOTH SOLOMON'S-SEAL.
    P. commutatum in part.
A comparatively small, slender species, with leaves glabrous, paler and glaucous beneath, and lowest peduncle usually from 3rd axil. Leaves lanceolate, averaging nearly 4 times as long as wide (length 5–10, width 1.5–3 cm.), sessile,

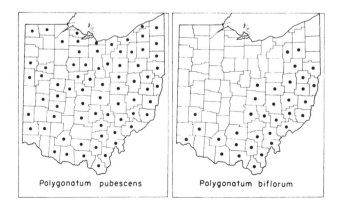

Polygonatum pubescens   Polygonatum biflorum

usually with 3–5 prominent veins; perianth slender cylindric, 10–18 mm long, lobes small; stamens inserted near middle of perianth-tube. More southern than the last, ranging from Florida and Texas northward to the limits of Wisconsin glaciation, only locally beyond this line; usually in dry woods in circumneutral soils, or, if in wet woods, then in acid soil. Ownbey (1944) distinguishes, as var. *melleum* (Farw.) R. Ownbey, plants with upper leafy and lower naked portions of stem about equal, many-nerved leaves (nerves all fading above middle), and 2–4 flowered peduncles. This is "restricted to the region about Lake St. Clair . . . in Michigan and Ontario"; a Sandusky County specimen can be referred to this variety, and an Erie County specimen has some of its characters.

3.   **Polygonatum commutatum** (Schultes f.) A. Dietr.   LARGE SOLOMON'S-SEAL
     P. canaliculatum, misapplied; incl. in *P. biflorum* in Gl. & C.
     A tall stout species with leaves glabrous, slightly paler and glaucous beneath, and lowest peduncle usually from 3rd or 4th axil, often from 5th or 6th. Leaves elliptic to ovate-oval, about 2.5 times as long as wide (length 7–15, width 3–7 cm), puckered or plaited into subpetiolar clasping base, usually with 9–15 (–19) prominent veins; peduncle flattened; perianth thick-cylindric, 15–20 mm long, lobes larger, 5–6 mm long; stamens inserted at or above middle of perianth-tube. Mid-latitudinal in range, over and beyond most of the area of *P. biflorum*, in rich woods of ravine-slopes, alluvial flats, and often on roadsides.
     Although the number of pedicels branching from one peduncle is given as 1–15 (usually 3–8), in rare individuals it is much higher. One Clermont County specimen (a winter collection without leaves) has branched peduncles showing leaf-scars, the secondary branches with long approximate to widely spaced pedicels; the lowest peduncle with its branches is 25 cm long, and bore 28 flowers. This variation in *P. commutatum* is comparable to forma *ramosum* (McGivney) Fern. of *P. biflorum*.
     Because *P. commutatum* is a tetraploid, doubtless derived from the diploid *P. biflorum*, Cronquist, 1963, does not distinguish it as a species, but includes it in *P. biflorum*.

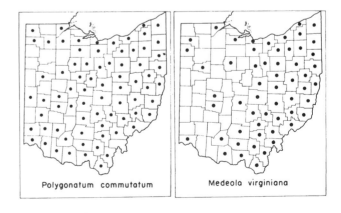

Polygonatum commutatum   Medeola virginiana

24. MEDEOLA L. INDIAN CUCUMBER-ROOT

A monotypic genus of eastern North America.

1. Medeola virginiana L.

Plants with horizontal white tuber, which has the taste of cucumber, hence the common name; stems slender, when young with loose coat of soft woolly hairs, remnants of which persist about the nodes, bearing 2 whorls of leaves on flowering stems, 1 terminal whorl on younger plants; leaves of principal whorl 5–9, oblanceolate, 7–15 cm long and one-third to one-fourth as wide, tapering to base, acuminate at apex; those of upper whorl shorter and relatively broader, usually 3, the lower fifth purple at fruiting time; flowers nodding below leaves, greenish yellow, in sessile terminal umbel on pedicels about 1 cm long, elongating in fruit; perianth of 6 similar distinct recurved parts; stamens 6; ovary globular, 3-locular, styles 3, united only at base, widely divergent and recurved, extending beyond perianth segments, stigmatic along inner side, often purple; fruit a purple-black, few-seeded berry. The sterile stems, bearing only 1 whorl of leaves, superficially resemble *Isotria*, but are easily distinguished by hairs on stems and lack of succulence. Widespread but local in much of the eastern half of the United States and southern Canada; in woods, usually in slightly acid soils.

25. TRILLIUM L. TRILLIUM

One of the most easily recognized genera of vascular plants, comprising about 25 species of temperate North America and eastern Asia. Stems from a short rhizome, erect, bearing a whorl of 3 leaves at summit, and a terminal sessile or peduncled flower; perianth of 3 green sepals and 3 white or colored petals; stamens 6, filaments short, anthers linear, adnate; ovary 3-lobed or 3-angled, style short or absent, stigmas 3, divergent; fruit a berry. In many of the species, aberrant forms occur, with leaves and floral parts in twos or fours, or with floral parts replaced by leaves. Seedling plants have only one leaf on stem-like petiole. The flower of each peduncled species has a distinctive profile-aspect, as shown in the illustrations; this character, usually obscure in herbarium specimens, should be noted from fresh flowers.

Most of the species are attractive spring wildflowers, of which the showiest is the Large White Trillium, *T. grandiflorum;* the Snow Trillium, *T. nivale,* is the earliest to bloom. All species vary considerably, but, with the exception of *T. flexipes* (*T. gleasoni*) are readily recognized as to species. Some of the color-variants have been given form names; these are not included in the present treatment for they have little if any taxonomic value, and only some of the color-forms have received names. The Ohio distribution patterns display interesting correlations with physical features.

a. Flowers sessile.
    b. Leaves sessile; sepals spreading or ascending; petals ascending, narrowed to sessile base ....1. *T. sessile*
    bb. Leaves petioled; sepals strongly reflexed; petals erect but bulging out at middle, distinctly clawed at base ....2. *T. recurvatum*

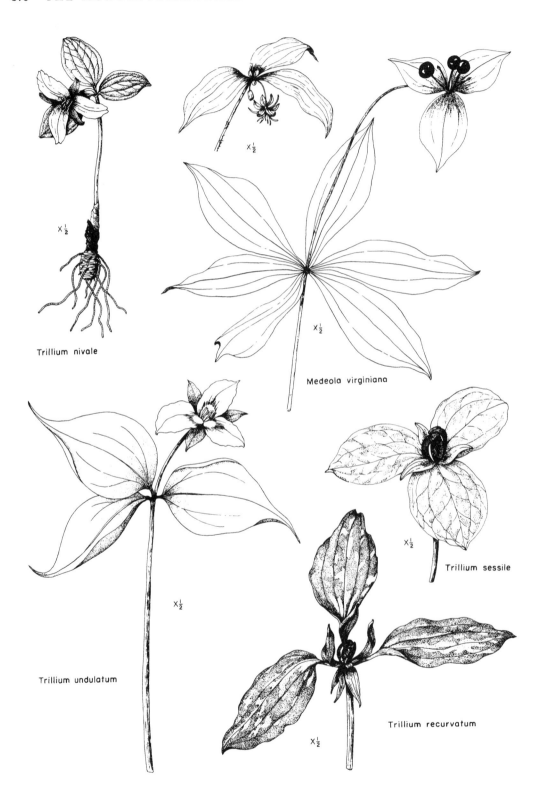

Trillium nivale

$\times \frac{1}{2}$

$\times \frac{1}{2}$

$\times \frac{1}{2}$

Medeola virginiana

$\times \frac{1}{2}$

Trillium sessile

$\times \frac{1}{2}$

Trillium undulatum

Trillium recurvatum

$\times \frac{1}{2}$

aa. Flowers peduncled.
 b. Leaves petioled, ovary and fruit subglobose to obtusely 3-angled, not winged.
  c. Leaves oval to ovate, usually obtuse; petals white .... 3. *T. nivale*
  cc. Leaves ovate, acuminate; petals white, conspicuously veined with red toward base ....4. *T. undulatum*
 bb. Leaves sessile, rarely contracted to petiole-like base, but not definitely petioled; ovary and fruit 6-angled or 6-winged.
  c. Petals erect or strongly ascending at base, flaring above, white; filaments and anthers 15–25 mm long; stigmas slender, straight; ovary small, hidden by the erect petals ....5. *T. grandiflorum*
  cc. Petals not erect at base, petals and sepals more or less similar in curvature, varying from white to dark purple-red; stigmas stout, recurved; ovary not hidden by petals.
   d. Ovary very dark purple (except in albino forms); petals spreading from base ....6. *T. erectum*
   dd. Ovary white or only tinged with purple or pink.
    e. Anthers creamy-white; filaments short, about a third as long as anthers, not exposed in fresh flower ....7. *T. flexipes*
    ee. Anthers pink or purple; filaments ⅔ as long as anthers or longer, plainly visible in fresh flower ....8. *T. cernuum*

1. Trillium sessile L. WAKE-ROBIN

The sessile flowers with ascending to spreading sepals distinguish this species. Petals brown-red to greenish-red, or occasionally green or yellowish green, oblanceolate to elliptic, ascending; sepals green, often tinged with purple, at first ascending, soon spreading; leaves usually mottled. Interior in range; widespread in Ohio, most frequent in the calcareous soils of western Ohio, in dry to mesic woods; early spring.

2. Trillium recurvatum Beck. PRAIRIE WAKE-ROBIN

The petioled leaves, clawed petals, and strongly recurved sepals distinguish this more western species. Flowers red-brown or maroon, richer in color than the last; petals widest above middle, acute to acuminate at apex, narrowed below to a slender (clawed) base, bulging outward at middle to widely spreading; sepals (alternate with leaves) recurved and lying against stem. Although a woodland species, the common name is not too inappropriate, because it suggests the geographic range. Very rare, or perhaps extinct in Ohio, no records since 1897; found throughout Indiana; abundant in woods but a few

Trillium sessile

Trillium recurvatum

miles west of the Ohio state line. The drawing is of an Indiana plant growing 5 miles from the line.

3.   **Trillium nivale** Riddell.   SNOW TRILLIUM

Dainty little plants, mostly 10–20 cm tall; one of the earliest spring flowers, and the earliest of all Trilliums; a covering of fresh snow does not harm the plants, flowers will be wide open with snow all around. Leaves ovate to elliptic, petioled, bluish green, obtuse to subacute; flowers on ascending peduncles; petals white, varying greatly in size and shape, from narrowly oblong (5 mm wide, 20 mm long) to broadly elliptic (1.5 cm wide, 4 cm long), erect at base, flaring above (flower-shape much as in *T. grandiflorum*); after pollination, the peduncle curves down, the fruit hidden by the leaves, and on small plants, almost in contact with the ground. Range, as stated in manuals, "W. Pa. and W. Va. to Minn., Neb. and Mo.," suggests a more extensive range than detailed distribution maps disclose; the plants are confined to areas where calcareous rocks outcrop or are close to the surface, as in southern Ohio, one known Kentucky station, a southeast-northwest band in Indiana, a few counties in the Missouri Ozarks, etc. On the whole, the species is rare and local, although sometimes fairly abundant where found; in rich woods on talus slopes and on shaded calcareous ledges. Riddell's type was collected along the Scioto River in Franklin County about 1832.

4.   **Trillium undulatum** Willd.   PAINTED TRILLIUM

A species of northeastern range, extending south, at rather high elevations to Georgia and Tennessee; known in Ohio only from Ashtabula County. The red color at base of white petals and conspicuous red veining distinguishes this species in flower. Leaves ovate, petioled, acuminate; petals wavy-margined, narrowly or broadly lanceolate, more or less spreading at base; peduncles remaining erect in fruit; berry bright red.

5.   **Trillium grandiflorum** (Michx.) Salisb.   LARGE WHITE TRILLIUM

Different in flower-shape from other species (except *T. nivale*), the sepals spreading, the petals strongly ascending at base (base therefore tube-like), flaring above. Leaves sessile or contracted to very short petiolar base, ovate,

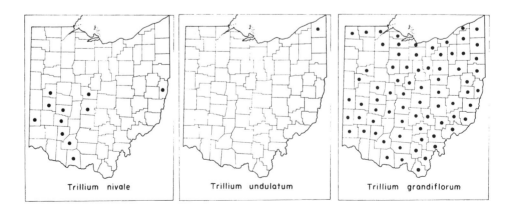

Trillium nivale   Trillium undulatum   Trillium grandiflorum

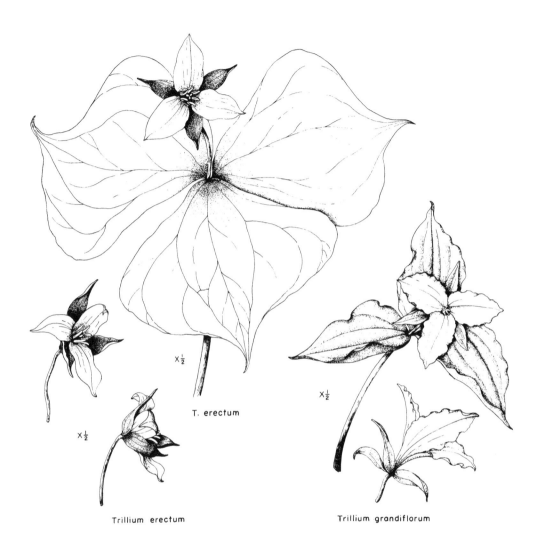

$X\frac{1}{2}$

T. erectum

$X\frac{1}{2}$

$X\frac{1}{2}$

Trillium erectum

$X\frac{1}{2}$

Trillium grandiflorum

roundish, or rhombic, acuminate, large; peduncle erect or ascending; petals white changing to pink after pollination, varying greatly in width; filaments and anthers together about 2 cm long, erect; ovary hidden by stamens and erect petal-bases; berry dark, red-black, about 2 cm in diam. Ranging from Maine and Quebec westward to Minnesota, and southward to Georgia and Tennessee (in the mountains), and Kentucky; abundant and widespread in Ohio, but strangely absent from the southwestern corner of the state south of the Wisconsin glacial border but abundant in Butler, Warren, and Clinton counties immediately to the north of the boundary; absent also in adjacent Indiana and in the Bluegrass section of Kentucky though found both to east and west. This distribution pattern (in Ohio and Indiana) suggests post-Wisconsin migration from the

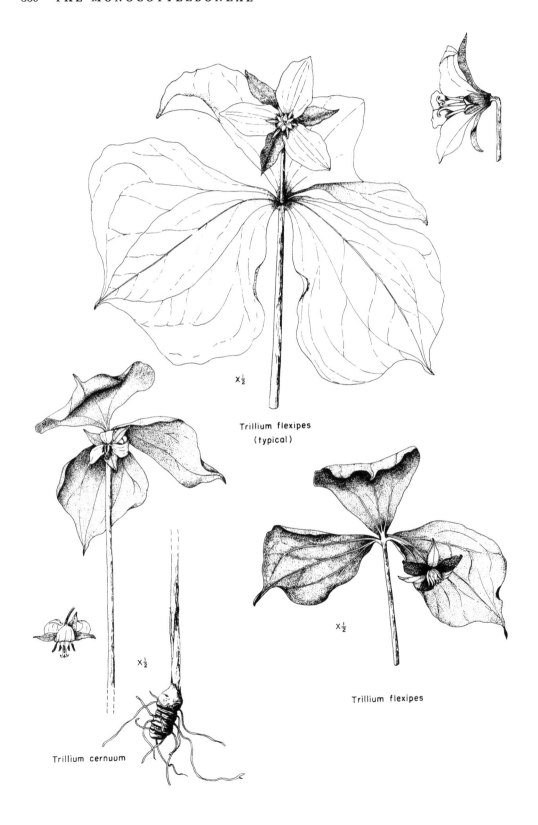

X½

Trillium flexipes
(typical)

X½

Trillium cernuum

X½

Trillium flexipes

Appalachian upland. The species is abundant in Adams County (and the Kentucky counties to the south) as far west as the hilly Silurian area; it does not extend westward onto the unglaciated Interior Lowland or the Illinoian glacial area.

6. Trillium erectum L.

A large and showy species with bright maroon-red, white, or rarely yellow petals and small, very dark ovary. Because the sepals and petals spread from the base, the small dark ovary is plainly visible and should make recognition of this species easy in the field; the peduncle may be erect, divergent, or even declined, and should not be used as a recognition-character. Leaves broadly rhombic, their margins often meeting, abruptly acuminate; petals from narrow to broadly ovate, the open flowers often 8–10 cm across; stamens exceeding stigmas, filaments slender, more than half as long as anthers; berry 6-angled, red, 2–2.5 cm in diam. An Appalachian species ranging through much of New England and the Maritime provinces of Canada southward in the mountains to Georgia and Tennessee, westward to the margin of the Cumberland and Allegheny Plateaus, and in the Lake region, westward to Michigan. In Ohio, not found west of the Allegheny Plateau (except near Lake Erie); however, there is a Miami County specimen (O.S.U. herb.) without collection data which probably originated from a nearby nursery which long has offered this species for sale.

7. Trillium flexipes Raf.

T. gleasoni Fern., *T. declinatum* (Gray) Gleason; *T. erectum* var. *declinatum* Gray.

The most confusing of the peduncled Trilliums; extremely variable as to color, flower size, peduncle-length, and angle of peduncle—upright to turned under the leaves. The only fairly constant characters are: ovary white or only tinged with purple, sharply angled with decurrent ridges from base to each margin of each recurved stigma (6 ridges); and short, broad filaments less than half length of the creamy white anthers. Plants with nodding flowers have often been erroneously referred to *T. cernuum*, those with long-peduncled erect flowers to *T. erectum*. Distinguishing characters of the 3 species are given in the key.

Nomenclature, also, is confused. Fernald (1944b) briefly discussed this problem, and quoting an old description by Rafinesque, states "The inclined peduncle about equaling (in early anthesis) the sessile, acuminate, obovate leaves, the acuminate, lanceolate sepals and the equal, oblong white petals are all good characters of *Trillium Gleasoni*," and that "there is no reasonable doubt about *T. flexipes*," which name therefore has priority over *T. gleasoni*. Rafinesque's description was based on plants from "West Kentucky and Tennessee"; Missouri plants from "Hidden Valley," about 1 mile east of St. Albans (sent to me by Robert L. Dressler, Missouri Botanical Garden) agree with this description; so also do a few southwestern Ohio specimens. Most differ in color, peduncle length and angle. There do not appear to be definite geographic races; rather, appearance suggests some contamination or isolation at some time in the Pleistocene.

Only detailed field observations throughout the range of *T. flexipes*, accompanied by statistical studies, can resolve the complex, which may have some admixture of *T. cernuum* or of *T. erectum*. Such studies may make it desirable to recognize two entities (species or perhaps varieties), the typical *T. flexipes* and *T. gleasoni*. Cronquist and Gleason (1963) question the synonymy of these two names. The map shows the distribution of the complex in Ohio.

8. **Trillium cernuum** L.   Nodding Trillium

A rather small-flowered species—petals 1.5–2.5 cm long—with short peduncle (1–4 cm) recurved below the leaves, and pink or purplish anthers (3–7 mm long) on filaments two-thirds to as long as anther. Two varieties are recognized, var. *cernuum* of the Northeast, and var. *macranthum* Wieg. of the North, based chiefly on size differences, especially width of petals and length of anthers. Known in Ohio from one specimen from Lake County; often erroneously reported because of frequent occurrence of nodding, rather short-peduncled forms of *T. flexipes*. The drawing is of a Michigan plant sent to me by Dr. E. G. Voss.

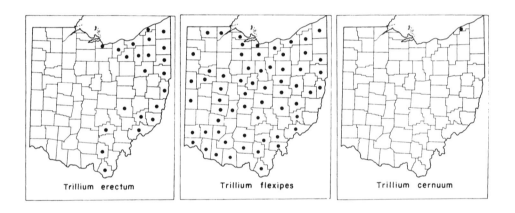

Trillium erectum   Trillium flexipes   Trillium cernuum

26.   ALETRIS L.   Stargrass. Colic-root.

A small genus of eastern United States and eastern Asia; besides the following, 3 more occur in the Southeast. Basal leaves in a rosette, narrow, strongly ascending to spreading; stems scape-like, stem-leaves reduced to bracts; inflorescence a spike-like raceme; perianth tubular, granular-roughened, 6-cleft at summit, its base adhering to ovary; stamens 6, inserted at summit of perianth-tube; ovary partly inferior, 3-locular, style slightly 3-lobed; capsule small, enclosed in the withered but persistent perianth. The genus is sometimes placed in the Amaryllidaceae because of the inferior ovary.

1. **Aletris farinosa** L.

The specific name refers to the mealy outer surface of the white perianth. Leaves firm, dull pale green, narrowly oblanceolate to almost linear, 6–20 cm long; stems slender, 4–8 dm tall, bracts small and widely spaced. In dry or moist, usually sandy soil; ranging from Maine to Minnesota and south to Florida and east Texas, but local, and absent from extensive areas within its range.

Aletris farinosa

27. SMILAX L. Greenbrier. Catbrier.

One of the largest genera of the Liliaceae, with about 300 species, many of them tropical. Eastern American species are divisible into 2 distinct groups classed as subgenera or sometimes as genera: *Nemexia*, herbaceous, with annual stems, and *Eusmilax* or *Smilax*, with woody stems. Leafy-stemmed plants, lower leaves reduced to bracts, upper usually broad, with several strong longitudinal ribs, interspaces netted veined, stipulate, many species with tendrils (modified stipules); dioecious, flowers greenish yellow, in umbels on axillary peduncles, perianth of 6 distinct segments; staminate flowers consist of perianth and 6 stamens, attached at base of perianth segments; pistillate flowers of perianth, 1–6 abortive stamens, and pistil with 3-locular ovary and 1–3 recurved stigmas; fruit a small, few-seeded dark blue or black berry (red in some southern species).

The woody species (mostly spiny or prickly) often form dense tangled thickets, the shoots arising from much-branched rootstocks, which Indians and early settlers used in the preparation of a jelly; the herbaceous species occur singly or in small groups, the young succulent shoots, except for tendrils, somewhat resembling asparagus, and occasionally used as a vegetable. The Smilax of florists is a species of *Asparagus, A. medeoloides* Thunb. (*A. asparagoides* Wight.), with leaf-like cladophylls.

a. Stems annual, herbaceous.
    b. Stem erect, 2–6 dm tall, usually without tendrils (weak tendrils sometimes on upper petioles if stem elongate); leaves mostly clustered near tip of stem, on taller stems scattered except toward tip; umbels from axils of bracts below leafy part of stem, sometimes also from lowest leaf ....1. *S. ecirrhata*
    bb. Stem leaning or climbing by tendrils, tall, simple to branched, evenly leafy above lower leafless but bracted stem-section; umbels from axils of normal leaves.
        c. Leaves green beneath; umbel hemispheric; berries black, not glaucous ....
            2. *S. pulverulenta*

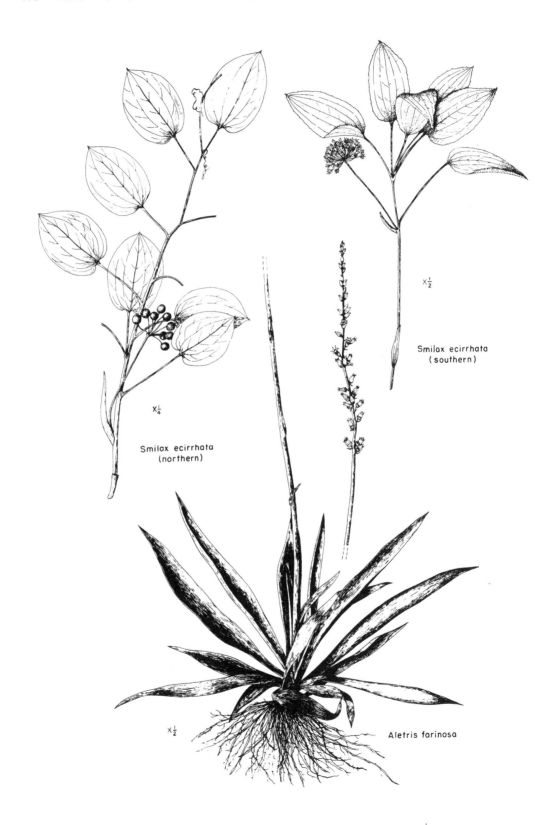

Smilax ecirrhata
(southern)

$x\frac{1}{2}$

$x\frac{1}{4}$

Smilax ecirrhata
(northern)

$x\frac{1}{2}$

Aletris farinosa

cc. Leaves pale or glaucous beneath; umbels crowded, becoming globose; berries glaucous.
    d. Leaves glabrous beneath; petioles 1–4.5 cm long; bladeless bracts appressed-ascending ....3. *S. herbacea*
    dd. Leaves minutely pubescent beneath; petioles 2.5–9 cm long; bladeless bracts spreading-ascending ....4. *S. lasioneuron*
aa. Stems perennial, woody; petioles tendril-bearing.
    b. Leaves whitened beneath, ovate or rounded; stems slender, terete, irregularly armed with slender prickles; berries blue-black, glaucous ....5. *S. glauca*
    bb. Leaves green on both sides.
        c. Stems terete, usually more or less densely covered, except on young branchlets, with sharp but slender terete black bristles; peduncles longer than petioles; berries black, without bloom ....6. *S. hispida*
        cc. Stems angled (or sometimes terete), armed with stout green spine-like prickles somewhat flattened toward base.
            d. Stems 4-angled or terete, not striate, green, rigid, with stout flattened prickles; leaves broadly ovate to suborbicular, sometimes narrowly ovate; leaf-margins more or less denticulate or erose (under a lens), not conspicuously thickened; peduncles about as long as petiole or shorter ....
                                                  7. *S. rotundifolia*
            dd. Stems rounded or with one prominent angle, scurfy at base; leaves from ovate to ovate-hastate, with much constricted sides; margins colorless but noticeably thickened, entire, or on young shoots denticulate to erose; peduncles longer than petioles ....8. *S. bona-nox*

1. **Smilax ecirrhata** (Engelm.) S. Wats.
   Nemexia ecirrhata (Engelm.) Small

Distinguished by erect stem with leaves more or less whorled at summit with sometimes a few below whorl, or in most northern and central Ohio specimens, scattered or in pairs along elongate stem; and by position of slender, long-peduncled umbels, 1 or 2 from axils of 2–3 cm long bladeless bracts, occasionally one from axil of lowest leaf. Leaves finely white-pubescent on veins and veinlets beneath, irregularly finely erose or denticulate on margins (a character easily felt in the field), various in shape, narrow- or broad-oval or oblong, cordate or subcordate at base, enlarging after flowering time, in most Ohio specimens from as wide as long to ¾ as wide, very blunt and apiculate at apex. Young plants of *S. herbacea* or *S. lasioneuron* are sometimes mistaken for this species.

Two distinct growth-forms occur: one with short (1.5–3 dm) erect stems with 1–3, or no leaves below a terminal whorl, and another with more elongate (often zigzag) stems with scattered and irregularly spaced leaves, with or without terminal group of 2 or 3, and often with petiolar tendrils on upper leaves. The first of these is more southern in range (chiefly south of the Wisconsin glacial border) and is the only form I have seen in many localities in Kentucky and in southern Ohio (shown in × ½ drawing, and by figure in the first edition of Britton and Brown (1896); the other is shown in the × ¼ drawing (from Erie County specimen), and is suggested by Gleason's figure (1952); this occurs (in Ohio) only north of the limits of Wisconsin glaciation. The Ohio distribution of the two growth-forms is shown in the map—dots with northward-pointing appendage for the northern form, with southward pointing ones for the southern form. In rich, often moist woods, widespread but infrequent.

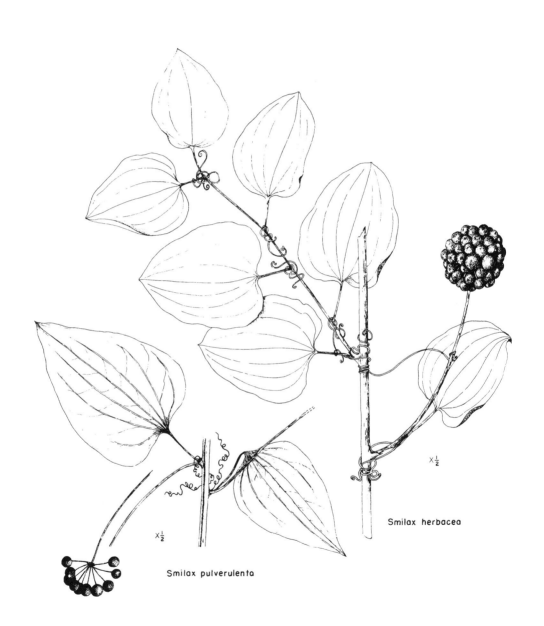

X½

Smilax herbacea

X½

Smilax pulverulenta

2. **Smilax pulverulenta** Michx.

S. herbacea L., var. *pulverulenta* (Michx.) Gray; *Nemexia pulverulenta* (Michx.) Small

Similar to the next 2 species; distinguished from both by green under-surface of leaves, abrupt short-acuminate tip, and black berries in hemispheric umbel (umbel of 10–35 flowers); from the next, by its long (3–9 cm) petioles and fine colorless pubescence on veins beneath. Infrequent in woods, often in calcareous soil.

3. **Smilax herbacea** L.   CARRION-FLOWER

Nemexia herbacea (L.) Small

The most abundant of the 3 similar species, all tall (to 2 m), branched, and supported by tendrils; distinguished from the preceding by the glabrous, pale or glaucous lower leaf surface, blunt or cuspidate leaf-tips, dense many-flowered (20–100) umbel, and dark blue glaucous berries; from the following, with which it intergrades, by the absence of pubescence on veins beneath, and the usually shorter (1–4.5 cm) petioles. Most luxuriant along woodland borders or roadsides; flowers ill-scented.

4. **Smilax lasioneuron** Hook.

S. herbacea var. *lasioneuron* (Hook.) A. DC.; *Nemexia lasioneuron* (Hook.) Rydb.

Probably best considered a variety of the last, with which it intergrades; similar in leaf-shape to the last, but petioles often longer (2.5–9 cm), and lower leaf-surface minutely pubescent on veins, characters which vary and hardly suffice to distinguish species. More western in range than the two preceding species.

5. **Smilax glauca** Walt.   GREENBRIER. SAWBRIER.

Stems slender, wiry, much branched, forming dense tangled masses climbing over bushes and small trees, especially in second-growth or logged forests; prickles irregularly spaced, slender; leaves narrowly to broadly ovate to sub-rotund, whitened beneath, the lower surface densely but finely papillose or pulverulent beneath in var. *glauca*, smooth and glabrous beneath in var.

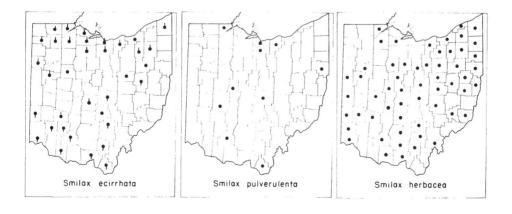

Smilax ecirrhata   Smilax pulverulenta   Smilax herbacea

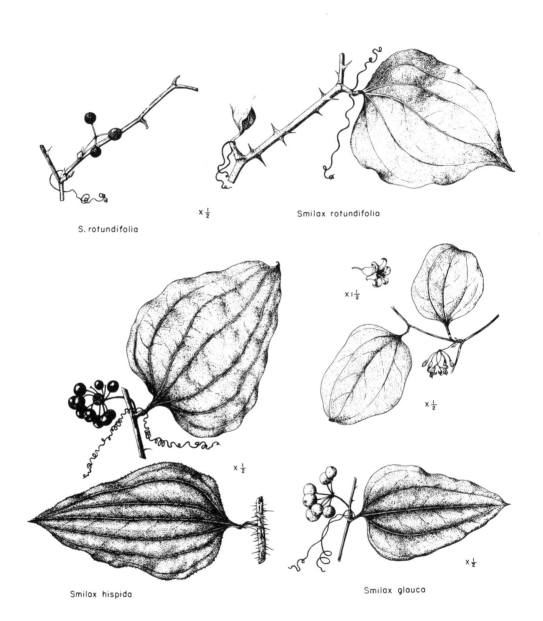

S. rotundifolia

x ½

Smilax rotundifolia

x 1½

x ½

x ½

Smilax hispida

Smilax glauca

leurophylla Blake. Leaves turning red in autumn, long-persistent. Southern in range, extending north into Ohio, Indiana, and Illinois in acid soils.

6. **Smilax hispida** Muhl. BRISTLY GREENBRIER

   S. tamnoides L., var. *hispida* (Muhl.) Fern.

Stems more densely prickly than other species, the prickles black, slender, and terete; ultimate branchlets often without prickles; leaves bright green, persistent in mild winters, ovate to elliptic or rounded; peduncles much longer than petioles of subtending leaves; berries black, without bloom. High-climbing, but seldom forming thickets; ranging farther west than our other species; found throughout Ohio.

7. **Smilax rotundifolia** L. GREENBRIER. SAWBRIER.

The strongly armed green but rigid terete or quadrangular stems distinguish this *Smilax* from the 2 preceding species; from the following, by the usually thinner leaves without hard colorless margin and absence of stellate scurf near base of stems. Plants with branchlets 4-angled were formerly segregated as var. *quadrangularis* (Muhl.) Wood. Leaves narrowly to broadly ovate, rounded to base, occasionally somewhat fiddle-shaped. A common and widespread species.

Smilax lasioneura

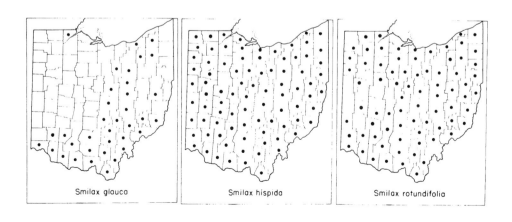

Smilax glauca

Smilax hispida

Smilax rotundifolia

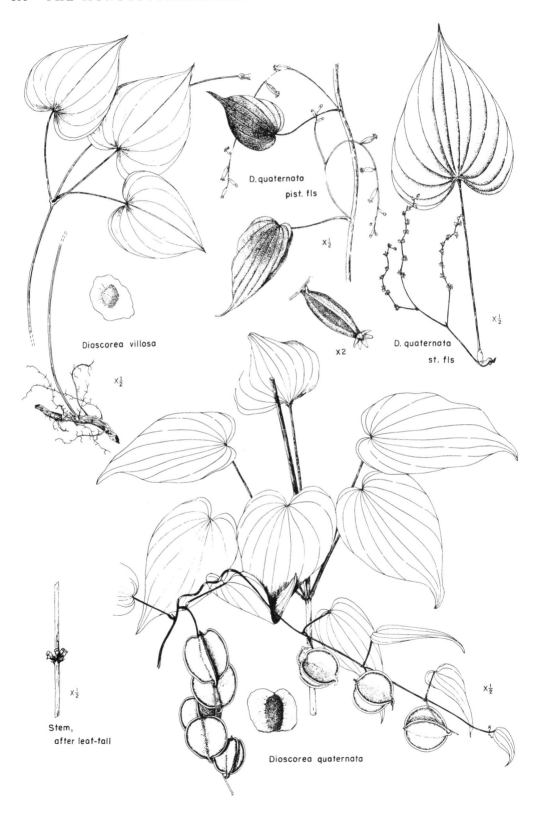

D. quaternata
pist. fls

X½

Dioscorea villosa

X½

X2

D. quaternata
st. fls

X½

Stem,
after leaf-fall

X½

X½

Dioscorea quaternata

8. **Smilax bona-nox** L.

A southern species extending northward into southern Indiana and northern Kentucky; not definitely known from Ohio, but should be looked for on dry, more or less open limestone slopes. Distinguished by whitish scurf near base of stem (best seen in the field) and by leaf-margin—hard and colorless, much like the stronger veins, and (especially on vigorous shoots) spinulose-denticulate. Leaves thick, ovate to triangular, often constricted at middle, truncate at base.

DIOSCOREACEAE. Yam Family

A family of some 650 species of twining herbaceous or woody plants, most of which occur in tropical America; only a few species of one genus indigenous to the United States. Plants with thickened rhizomes or tubers (often large); leaves usually cordate, strongly palmately ribbed, often whorled; petioles often with joint near base; flowers in axillary racemes or panicles, unisexual; perianth of 6 similar parts; stamens 6 in staminate flowers, absent in pistillate; ovary inferior; fruit, usually, a 3-angled or 3-winged capsule.

DIOSCOREA L. YAM

By far the largest genus of the family, containing over nine-tenths of its species; two occur in Ohio, a third has long been in cultivation and occasionally escapes. Dioecious, pistillate inflorescence a raceme, staminate a panicle; flowers minute; fruit a 3-winged, 3-valved capsule, the valves splitting along the apical edge of wing; seeds flat, winged, usually 2 in each locule. Ohio species may be distinguished from other herbaceous climbers, even in the vegetative state, by the strong primary veins extending from base to apex, and by the slightly enlarged and bent base of petiole.

a. Leaves cordate-ovate, the sides convex; plants without axillary tubers; rhizomes straight or branched, 5–15 mm in diam.; native.
  b. Leaves all alternate, or at lowest node in whorl of 3; petioles glabrous; stems usually much twining; capsules about 2 cm long ....1. *D. villosa*
  bb. Leaves at 1 or more lower nodes in whorls of 4–7, alternate above; petioles pubescent at apex; stems erect or leaning, twining above; capsules about 2.5 cm long or longer ....2. *D. quaternata*
aa. Leaves cordate, broad at base, their sides constricted above; plants usually with small potato-like axillary tubers, and much enlarged subterranean tubers; Asiatic ....
                                         3. *D. batatas*

1. **Dioscorea villosa** L. WILD YAM.

A common and widespread species, found almost throughout Ohio. Stems more noticeably climbing than the next, a character apparent in many herbarium specimens; leaves either glabrous beneath, or more or less pubescent, either all alternate or more often the lowest in whorl of 3; petioles glabrous; valves of capsule varying from semi-circular to slightly broader toward apex.

2.   **Dioscorea quaternata** (Walt.) J. F. Gmel.

More southern in range than the last, and in Ohio apparently almost confined to the area south of the limits of Wisconsin glaciation. Similar to *D. villosa*, but generally larger—larger leaves, longer petioles, larger capsules and seeds; distinguished by the 1 or 2 whorls of 4–7 leaves, those above mostly alternate; by the marked but fine pubescence on petiole just below blade and sometimes also at base; and by its larger capsules and larger seeds. The pubescence on petiole is slightly below the lower limits of pubescence on pubescent forms of *D. villosa*. The joint near base of petiole is more prominent than in the last, and the short basal segments of petiole remain attached to stem when the leaves break off in the fall, resulting in a whorl of flat-topped stubs. Leaves usually glabrous beneath, rarely finely pubescent, sometimes glaucous beneath, a character used to segregate *D. glauca* Muhl. which is now included in *D. quaternata.*

3.   Dioscorea batatas Dene.   Chinese Yam.   Cinnamon Vine.

An occasional escape, reported from 4 Ohio counties; formerly sometimes planted, perhaps because of the odd potato-like tubers borne in leaf-axils; extensively cultivated in southeast Asia. The large subterranean tubers, 6–9 dm long, are said to have taste of potato when cooked.

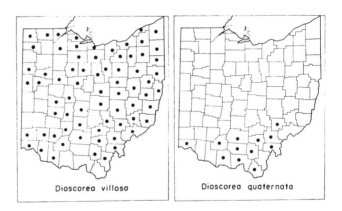

Dioscorea villosa

Dioscorea quaternata

AMARYLLIDACEAE.   Amaryllis Family

A large family of over 80 genera and 1,200 species, poorly represented in the United States where only 6 genera have representatives in the eastern half of the country, 2 of which occur in Ohio. Similar to the Liliaceae in floral characters, from which it is distinguished chiefly by the inferior ovary. Plants mostly with bulbs or corms (except in *Agave*), scapose, the scapes naked or with bracts; leaves basal, mostly linear; inflorescence umbellate, racemose, or paniculate, or flowers solitary; flowers regular or nearly so, sepals and petals similar, separate or united at base, coherent with ovary; stamens 6, inserted on perianth;

ovary 3-locular, style 1, stigmas 3 or 3-lobed or capitate; fruit a loculicidal capsule. In a few genera (as *Narcissus* and *Hymenocallis*) a crown or corona— a showy appendage of the perianth, or corolla-like connection between filaments— is present.

The family contains many showy flowers often cultivated for ornament; most familiar are species of *Narcissus—N. poeticus* L., Poet's Narcissus with short crown, *N. pseudo-narcissus* L., Daffodil with crown often as long as perianth-segments, *N. jonquilla* L., Jonquil with crown of medium length and narrow rush-like leaves, *N. tazetta* L., the Polyanthus Narcissus (including the familiar Paper White of indoor culture) with several flowers in terminal cluster, and innumerable horticultured forms; *Amaryllis* or *Hippeastrum* with strap-shaped leaves and large, usually red or reddish lily-like flowers; *Lycoris*, Magic-Lily from eastern Asia, with strap-shaped leaves in early spring, and pink flowers in umbel on solid scape in late summer; *Hymenocallis*, Spider-lily, mostly tropical or subtropical American plants with large flowers bearing corona, one species, *H. occidentalis* (LeConte) Kunth, native northward in western Kentucky, southwestern Indiana, and southern Illinois; *Galanthus*, Snowdrop, and *Leucojum*, Snowflake, hardy spring bulbous plants with white flowers marked with green. Of the above-mentioned, Daffodil and Poet's Narcissus are sometimes seen on roadsides or about old house-sites, and Snowdrop (*Galanthus nivalis* L.) and Snowflake (*Leucojum aestivum* L.) rarely escape (one report of each, both in extreme northeastern Ohio).

Species of *Agave*, Century-plant, are grown for the fibers contained in the leaves, used in the manufacture of cordage, particularly sisal; some, also, are used (in Latin American countries) in making fermented or distilled beverages, as pulque and mescal. Plants of several genera (among them *Narcissus* and *Amaryllis*) contain toxic alkaloids.

Plants with basal rosette of thick fleshy leaves 1.5–3 dm long, 2–3 cm wide; flowers greenish yellow, tubular-funnelform, almost sessile in a loose elongate spike on tall bracted scape; capsule more or less globose, slightly 3-lobed, 1–1.5 cm in diam. ......
1. *Agave*
Plants with narrow grass-like leaves 1.5–2.5 dm long at flowering time, 2–6 mm wide; flowers in loose umbel on filiform scape, yellow; perianth when open flat, star-like; fruit inconspicuous ....2. *Hypoxis*

1. AGAVE L. American Aloe

A large genus of about 300 species, most numerous in arid and semi-arid regions of the Mexican tableland and the West Indies; a number of species in southwestern United States; one, in a distinct subgenus, *Manfreda*, is south-eastern in range and extends into our area. Leaves in large basal rosette, thick, fleshy or hard, often spiny-margined and spine-tipped; inflorescence a spike or panicle, often very large. In the large western species, during the period of rapid growth of the inflorescence, the large supply of food stored for many years in the leaves is utilized, and after fruiting, the plant dies, or sometimes produces small off-sets. (This is the origin of the name Century-plant.) Flowers tubular

x½

x½

x½

Hypoxis hirsuta

Agave virginica

to funnelform, stamens long-exserted; capsules with hard often thick walls, many-seeded.

1. **Agave virginica** L.   FALSE ALOE
Manfreda virginica (L.) Salisb.

So unlike all other Ohio vascular plants that it can be recognized at once by its large basal rosette of succulent leaves and its tall (1–2 m) stem with widely spaced small leafy bracts and long loose terminal spike. Flowers greenish yellow, very fragrant, especially toward evening when new flowers open; anthers linear, versatile, long-exserted on slender filaments; style, in newly opened flowers much shorter than stamens, much elongating after anthers have shed pollen—an excellent example of protandry, the maturation of anthers before stigmas. A southern species, Florida to Texas, north in the interior to West Virginia, Ohio, Indiana, Illinois, and Missouri, occurring locally in southern Ohio in dry soil in full sun, most often in dry prairie-openings where soil is thin and calcareous rock exposed or near surface.

2. HYPOXIS L.   STARGRASS

One of the largest genera of Amaryllidaceae, with about 100 species, widespread in the southern hemisphere, several in the southeastern states, only one ranging far northward—to Maine and Manitoba. Small plants with narrow, usually hairy, grass-like leaves, slender scape with terminal few-flowered cluster of yellow to whitish flowers, and tuft of fibrous roots from small more or less globular corm, or corm-like rhizome. Where a number of species occur, the seeds furnish good specific characters; some are roughened with short sharp points, some have pebbled or reticulate surface.

1. **Hypoxis hirsuta** (L.) Coville.   YELLOW STARGRASS.

Flowering stems filiform, usually 1–2 dm tall, much surpassed by the narrow leaves; flowers yellow, 1–2 cm across, the perianth-tube completely coherent with ovary, sepals and petals above divergent, similar, but sepals greenish on back, assuming an erect position after flowering; capsule indehiscent and

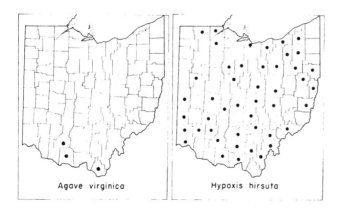

Agave virginica   Hypoxis hirsuta

crowned by the connivent perianth-segments; seeds black, about 1 mm in diam., densely covered with short sharp points. Found throughout Ohio in a variety of habitats; prairie-openings, open oak woods, calcareous ledges, etc.

IRIDACEAE.   Iris Family

Sometimes named IXIACEAE; four characters will generally suffice to distinguish members of this family: leaves 2 ranked, equitant (bases astride one another, or astride stem), flowers (individuals or clusters) subtended by 2 or more spathe-like bracts, stamens 3 placed opposite sepals, and withered perianth deciduous from summit of capsule. Flowers perfect, regular or zygomorphic (bilaterally symmetrical as in *Gladiolus*), usually showy; perianth-parts 6, sepals and petals similar or unlike; stamens 3, opposite sepals, capsule 3-locular, loculicidal or indehiscent, seeds (in most) on axile placentae.

A large family of nearly 1,500 species classified into about 60 genera, over half of which are restricted to the Old World, well over a third to Africa, about one-fourth to the New World, of which *Sisyrinchium* with about 80 species is the largest; only 5 genera contain species indigenous to the United States, 2 of them in our area. The family contains many showy, horticulturally valuable species, the best known of which are in the 2 largest genera, *Iris* (widespread in the Northern Hemisphere) and *Gladiolus* (Old World, and best represented in South Africa); other more or less familiar genera are *Crocus*, a Eurasian genus (with solitary flowers and long slender perianth-tube suggesting a scape), *Tigridia*, native from Mexico to Chile, and *Freesia*, from South Africa.

a. Sepals and petals similar, outwardly spreading.
    b. Plants tall (to 1 m), Iris-like; stems usually branching and loosely flowered; flowers orange with dark spots; capsules 2.5–3 cm across; capsule-lobes, on ripening, widely spreading, later deciduous, exposing shiny black seeds in blackberry-like cluster ....1. *Belamcanda*
    bb. Plants lower, grass-like (but leaves equitant); stems 2-edged or 2-winged; flowers white, bluish, or blue; capsules small (about 5 mm) ....2. *Sisyrinchium*
aa. Sepals and petals unlike, sepals spreading to recurved, petals erect to arching; style-branches petal-like, curving outward over sepals ....3. *Iris*

1. BELAMCANDA Adans. BLACKBERRY-LILY

Similar in growth-form to *Iris*; plants with rhizomes, erect stems, and sword-like leaves; perianth-segments similar, outwardly spreading; stamens 3, attached at base of perianth opposite sepals; capsule completely dehiscent, exposing blackberry-like cluster of shiny black seeds attached to central column. A small eastern Asian genus, one species naturalized in much of eastern United States.

1. BELAMCANDA CHINENSIS (L.) DC. BLACKBERRY-LILY
   Gemmingia chinensis (L.) Kuntze
   Except when in bloom, often mistaken for an *Iris*; stems tall (to 1 m), usually branched above; leaves sword-shape, 3 dm or more long, 2–3 cm wide, the upper

smaller; flowers orange spotted with dark red-purple, perianth spreading, sepals and petals slightly cupped. Unmistakable in fruit, when the blackberry-like cluster of shiny black seeds is exposed. Naturalized from eastern Asia; in dry usually calcareous soil of thickets, roadsides, and open places, often remote from dwellings.

2. SISYRINCHIUM L. Blue-eyed-Grass

Low grass-like plants with linear, mostly basal, equitant leaves, flattened 2-edged or 2-winged stems often with leaf-like bracts; flowers delicate and soon withering, white to blue (in eastern American species), on filiform pedicels in umbel-like clusters subtended by a spathe of 2 bracts; sepals and petals spreading, usually abruptly contracted to a sharp tip; capsules more or less globular, small.

a. Spathes terminating unbranched stems.
  b. Flowers white or pale blue; leaves and stems pale green, glaucous; spathes usually 2 together with 1 outer foliar bract; bracts with wide membranous margins; margins of outer bract free to base ....1. S. albidum
  bb. Flowers bright blue; leaves and stems greener, slightly glaucous; spathes solitary; bracts with narrow membranous margins; margins of outer bract united near base.
    c. Spathes usually red-purple; outer bract with margins united a little above base, 2–7 cm long, inner 1–2 cm long; stems slender, 0.5–1.5 mm wide ....
        2. S. mucronatum
    cc. Spathes green, often purple-tinged; outer bract with margins united 2–6 mm above base, 2–6.5 cm long, inner 1–3 cm long; stems 1.5–3 mm wide ....
        3. S. montanum var. crebrum
aa. Spathes on peduncles from axil of leaf-like bract; flowers blue.
  b. Leaves and stems green (darkening in drying), not glaucous, 2–6 mm wide; stems broadly winged (except sometimes near base); peduncles broadly winged ....
        4. S. angustifolium
  bb. Leaves and stems pale green, glaucous, 1–3 mm wide; stems wiry, slender, narrowly winged; peduncles very slender ....5. S. atlanticum

1. **Sisyrinchium albidum** Raf.
Plants often densely tufted, pale, the narrowly winged stems mostly about twice the length of leaves; outer bract of spathe about twice length of inner; a second spathe (the tip of at least one of its bracts visible within the longer outer pair) included in terminal inflorescence. The presence of 2 spathes (in what may at first appear to be only 1) accounts for the larger number of flowers (and hence showier clusters) in this species. Plants of prairies and dry sandy or rocky, usually calcareous soil, mostly west of the Appalachians.

2. **Sisyrinchium mucronatum** Michx.
Stems slender, very narrowly winged, 0.5–1.5 mm wide, from longer than to twice length of leaves; leaves very little wider than stems. Plants with the aspect of S. albidum but distinguished by solitary spathe, the slightly united (instead of separated) margins of basal part of outer bract, red-purple spathe-bracts, and darker foliage. A northern species.

3. **Sisyrinchium montanum** Greene, var. **crebrum** Fern.
  S. angustifolium of Gray, ed. 7, and Gleason, 1952.
  Similar in general appearance to S. angustifolium; distinguished from it chiefly by the simple, unbranched stems with terminal spathe, unequal spathe-

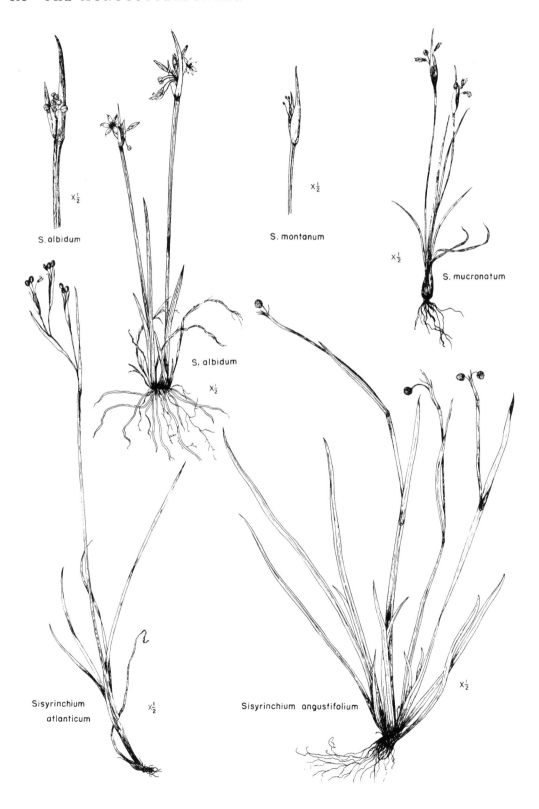

S. albidum

S. montanum

S. mucronatum

S. albidum

Sisyrinchium
atlanticum

Sisyrinchium angustifolium

bracts (outer about 2× inner), and more or less ascending pedicels rarely exceeding inner bract. Both species have margins of outer bracts united above base; both have winged stems 3–4 mm wide. A very local and not well understood taxon.

4. **Sisyrinchium angustifolium** Mill.

   S. gramineum Curtis; *S. graminoides* Bickn.

   Our most abundant and widespread species; stems broadly winged, 2–6 mm wide; leaves 2–6 mm wide; spathes on winged peduncles from axils of leaf-like bracts (sometimes 2–3 from 1 axil); spathe-bracts subequal. Ranging from Newfoundland to Ontario and Minnesota, south to Florida and Texas; roadsides, meadows, and open woods.

5. **Sisyrinchium atlanticum** Bickn.

   Slender pale green or glaucous plants forming dense tufts; stems much longer than leaves, narrowly winged, 1–3 mm wide; leaves pale, 1–3 mm wide, rather stiff; peduncles slender, several from axil of leafy bract high on stem, a feature which gives a distinctive aspect to this species. Most abundant on the Atlantic slope (Florida northward to Maine and Nova Scotia) and also occurring inland in the Mississippi Valley to Missouri, southwestern and northwestern Indiana, and Michigan, a distribution pattern apparently related to glacial outlets of the Great Lakes; local elsewhere.

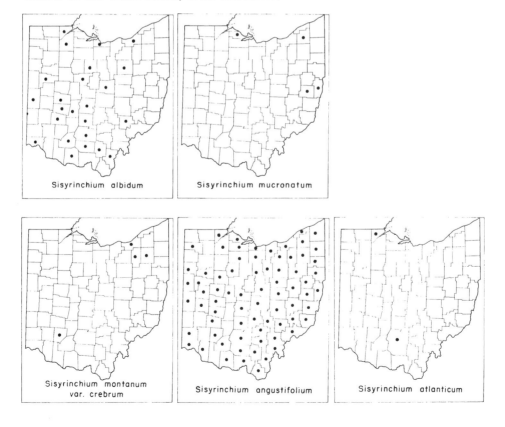

Sisyrinchium albidum

Sisyrinchium mucronatum

Sisyrinchium montanum var. crebrum

Sisyrinchium angustifolium

Sisyrinchium atlanticum

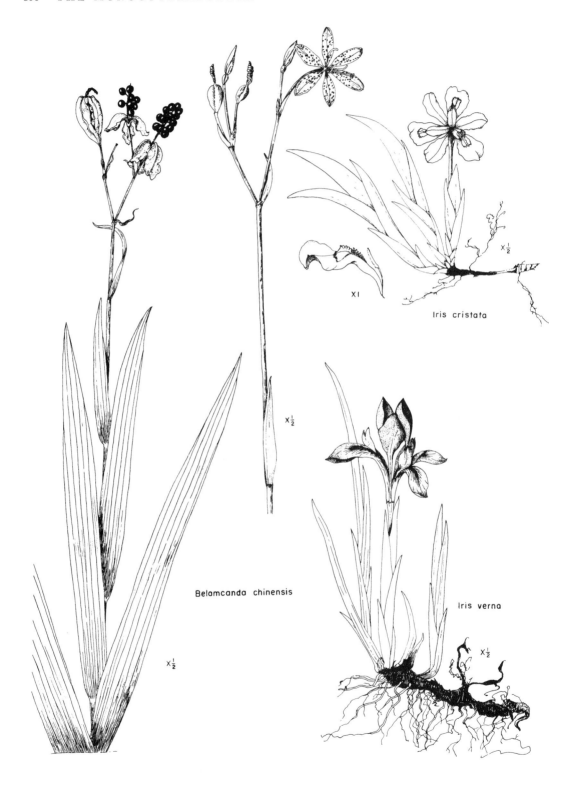

Iris cristata

$X I$

$X \frac{1}{2}$

$X \frac{1}{2}$

Belamcanda chinensis

$X \frac{1}{2}$

Iris verna

$X \frac{1}{2}$

3. IRIS L. IRIS

Perennials with (in eastern American species) horizontal rhizomes and equitant linear to sword-shaped leaves. Flowers showy, both outer and inner perianth-segments colored but unlike; perianth-tube more or less prolonged beyond ovary, raising the 6 clawed and expanded segments of the perianth above subtending spathe; sepals spreading or recurved, petals ascending to erect; style-branches petaloid, 2-lobed at apex, spreading outward above or parallel to sepals, thus arching over the stamens; each style-branch bearing a stigma—a thin plate or lip—under the apex at base of lobes; capsule 3-locular, dehiscent or indehiscent.

A large genus, wide-ranging in the northern hemisphere, with many species indigenous to the United States, where they are particularly abundant in the Mississippi delta; Small, in his Southeastern Flora (1933) distinguished over 90 species, a number greatly reduced by most students of *Iris*. Innumerable horticultural forms and hybrids are in cultivation, representing several of the distinct sections of the genus. Most familiar in American gardens are the tall bearded Irises, often referred to *Iris germanica* L. According to Anderson (1952), "German flags, to use the name for *Iris germanica* which your grandmother used, are a group of earlier-flowering, old-fashioned varieties, most of which have been around for at least some hundreds of years, and Linnaeus gave them this scientific name in the eighteenth century. Our modern bearded irises are not directly related to them and they resemble *Iris germanica* only superficially. . . . " None of these is naturalized in Ohio—probably nowhere in this country—although various color-forms are sometimes seen along roadsides.

The swamp species of Iris should be carefully distinguished from the Sweet Flag (*Acorus*) if rootstocks of the latter are being gathered for making candied flag-root, or the inner portion of its young shoots pulled for salad; the rootstocks of the Irises are very poisonous.

a. Plants low; leaves of flowering stem (spathe-bracts) several, overlapping; flowers usually blue or blue-violet, 1–2 (–3) from one spathe; flowering stem and perianth-tube together less than 2 dm tall.
   b. Leafy shoots erect, leaves (at flowering time) about 1 cm wide, 1–2 dm long, stiffish; sepals blue-violet with pubescent orange band extending from claw to middle of blade and bordered by outwardly pointing white streaks ....1. *I. verna*
   bb. Leafy shoots arching; sepals without pubescent band, crested; crest in a more or less rectangular white, centrally orange patch, 3-ridged, toothed.
      c. Leaves of leafy (sterile) shoots mostly 1–2 cm wide; perianth-tube very slender, 4–6 cm long, much exceeding spathe .... 2. *I. cristata*
      cc. Leaves of leafy (sterile) shoots mostly 1 cm or less wide; perianth-tube less than 2 cm long ....3. *I. lacustris*
aa. Plants tall, with elongate basal leaves; flowering stems simple or branched, mostly 2–10 dm tall, stem-leaves distinctly separated, exposing stem.
   b. Flowers yellow, copper-colored, or reddish.
      c. Ovary and capsule 6-angled; flowers coppery to orange or brick-red; sepals and petals spreading, similar ....4. *I. fulva*
      cc. Ovary and capsule obtusely 3-angled; flowers bright yellow; petals erect, much shorter than sepals ....5. *I. pseudacorus*
   bb. Flowers prevailingly blue or violet.
      c. Ovary and capsule 6-angled or 6-keeled; petals ascending to spreading, smaller than sepals; stems weak, leaning, much shorter than basal leaves ....
                          6. *I. brevicaulis*

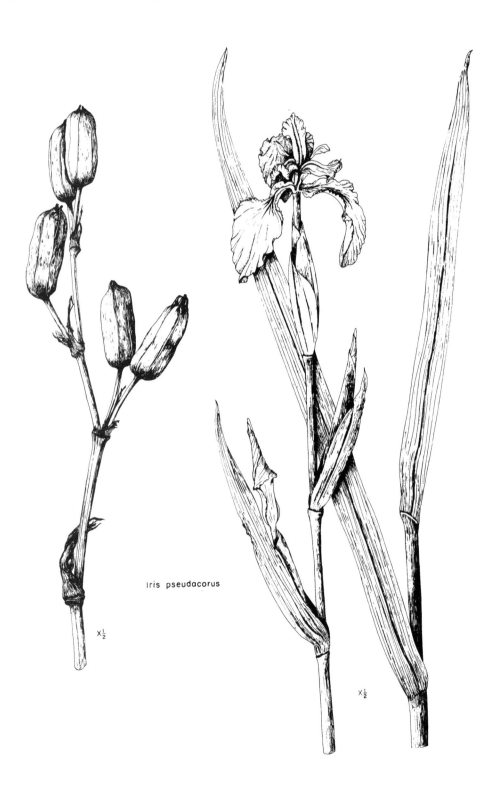

Iris pseudacorus

X½

X½

cc. Ovary and capsule 3-angled; stems ascending, equalling or taller than leaves.
  d. Sepals with pubescent bright yellow blotch at base of blade; petals some-
    what shorter than sepals; ovary at flowering time 2–4 cm long; capsules
    6–11 cm long ....7. *I. shrevei*
  dd. Sepals without blotch, or with green or greenish-yellow blotch at base of
    blade, pubescence inconspicuous; petals ½–⅔ length of sepals; ovary at
    flowering time 1–2 cm long; capsules 3.5–5.5 cm long ....8. *I. versicolor*

1. **Iris verna** L. Violet Iris
   Neubeckia verna (L.) Alef.
   This dwarf Iris has 2 distinct varieties, the typical, a plant of the Atlantic
Coastal Plain and adjacent areas of the Piedmont, and var. *smalliana* Fern. of
the Appalachian Upland, to which Ohio plants belong. The latter is distinguished
by its stout, dark rhizomes (instead of cord-like and whitish), its tendency to
form more compact clumps, its more numerous (5–8) and greener spathe-bracts,
its wider leaves with more numerous veins, and its larger capsules. Perianth-tube
long and slender; sepals and petals about equal in size, the former spreading
and slightly recurved, with an orange, pubescent, median band extending from
claw to middle of blade and bordered laterally by outwardly and apically
pointing prominent white streaks; capsule about 2.5 cm long, short-stalked,
tapering above to slender beak (the dried remnants of perianth-tube); leaves,
at fruiting time, much elongated. In Ohio, confined to acid soils of narrow
ridge-tops of southern Scioto County and adjacent Adams County (not shown
on map).

2. **Iris cristata** Ait. Crested Dwarf Iris
   Neubeckia cristata (Ait.) Alef.
   Often forming large patches, the leaves of the sterile shoots (fans) grace-
fully arching; rhizomes shallow, much branched, with thick knotty joints, branching before terminal flowering shoot; perianth-tube very slender, 4–7 cm long,
exceeding spathe; sepals larger than petals, curved downward, bearing on basal
half of blade a 3-ridged crest in a golden-yellow band surrounded by rectangular
white area outwardly bordered by deep blue; capsule about 1 cm long, hidden
by spathe-bracts. Variable in color, usually blue, varying to pale or deep violet,
occasionally violet-pink or white. A plant of rich woods, in circumneutral soils,
on ravine flats, steep banks or ledges; Appalachian to Ozarkian in range, but
absent from extensive limestone areas of the interior.

3. Iris lacustris Nutt. Dwarf Lake Iris
   Iris cristata var. *lacustris* (Nutt.) Dykes
   Circumscribed in range, confined to "gravelly shores and cliffs in calcareous
soil, around Lakes Superior, Michigan, and Huron"; is not a member of the
Ohio flora, although Waller (1931) maps it as occurring in Cuyahoga County
on basis of specimen so labeled in the Herbarium of the Berlin Botanic Garden,
and Foster (1937), following Waller, includes Ohio in the range of the species.
   Similar to *I. cristata*, but differing in its much shorter perianth-tube, 1–2 cm
long, its more cuneate sepals, its more scarious and shorter spathe-bracts, and its
different chromosome number ($2n = 42$, instead of $2n = 32$), as well as in soil
preference.

4. Iris fulva Ker   Copper Iris

A plant of swamps, swampy woods, sloughs, and muddy shores from the Gulf coast northward in the Mississippi Valley to southeastern Missouri, southern Illinois, and western Kentucky. Waller (1931) reported plants in a pasture on a farm near Mechanicsburg, Champaign County, where they had been for at least 15 years; these were said to have been brought from Clark County; it is probable that they were originally brought to Ohio from the Mississippi Valley. Distinguished by flower color (but not to be confused with similarly colored bearded irises), widely spreading to drooping sepals and petals, and hexagonal capsule. Capsules similar to those of *I. brevicaulis*, from which it can be distinguished in fruit by stems surpassing leaves.

5. Iris pseudacorus L.   Yellow Water Flag

A European species abundantly established in shallow water or moist soil on lake and stream borders, creek flats, and in roadside ditches. A tall Iris, to more than 1 m, with basal leaves of about same height as stem, the leaves somewhat brighter green than those of the native blue flags; flowers bright yellow, the sepals lined with brown, the petals erect—thus distinct from all native species; capsules 5–8 cm long, about 1.5 cm in diameter, the valves at maturity widely spreading to recurved at apex.

6. Iris brevicaulis Raf.   Leafy Blue Flag

I. foliosa Mackenz. & Bush; *I. hexagona* of Gray, ed. 7.

Distinguished from other Blue Flags by its flexuous often leaning stems much shorter than the leaves, spreading perianth-segments, and 6-angled short plump, ribbed capsule. A plant of swampy stream terraces, floodplains, and shores, interior in range.

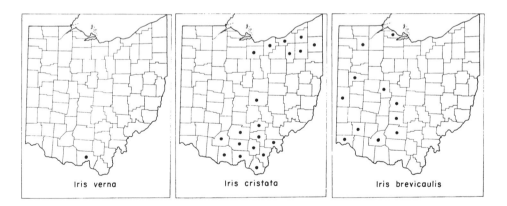

Iris verna   Iris cristata   Iris brevicaulis

7. Iris shrevei Small   Southern Blue Flag

I. virginica L. var. *shrevei* (Small) E. Anders.

This Iris was not distinguished from *I. versicolor* in the seventh edition of Gray's Manual (1908), and probably almost all reports of that species except from northeastern Ohio should be referred to *I. shrevei*. *I. virginica*, of which this is often considered a variety, is confined to the Coastal Plain, from south-

$\times\frac{1}{2}$

$\times\frac{1}{2}$

Iris brevicaulis

eastern Virginia to Texas; *I. shrevei* is largely confined to the drainage basin of the Mississippi River, and lower Great Lakes. See maps by Anderson (1936) of these and of *I. versicolor*. Stems tall, to 1 m, usually with one lateral branch nearly equaling main axis. Distinguished from the next by pubescent orange blotch toward base of sepal, larger petals only somewhat shorter than sepals, long prismatic-cylindric capsule (6–11 cm long), and lower number of chromosomes ($2n = 72$). A plant of swamps, wet flats near streams, and open swampy woods, sometimes forming large colonies; fairly widespread in Ohio, but now extinct in many localities because of drainage of swamps and urban growth.

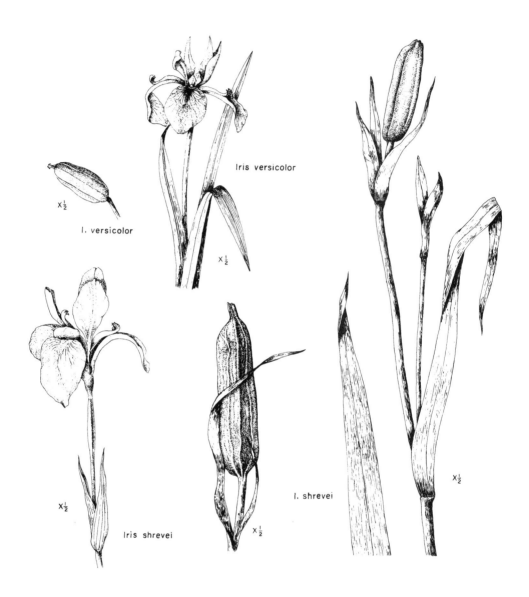

Iris versicolor

I. versicolor

x½

Iris versicolor

x½

Iris shrevei

x½

I. shrevei

x½

x½

x½

8. Iris versicolor L.   Northern Blue Flag

A species of northern range, throughout much of the Great Lakes area, westward to Manitoba, northeastward throughout much of the St. Lawrence Valley to Labrador, and east and southeast to the coastal area from Maine to Chesapeake Bay; in swamps, meadows, and on wet shores. Stems 2–6 dm tall, with 1–2 lateral branches not equaling inflorescence. Distinguished from the last by more or less indistinct greenish or greenish yellow minutely papillose basal area of sepals, smaller petals, shorter capsule (to about 5 cm long), and larger number of chromosomes (2n = 108).

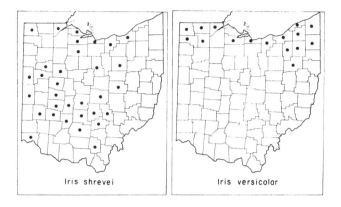

Iris shrevei   Iris versicolor

ORCHIDACEAE. Orchid Family

One of the largest, or perhaps the largest family of flowering plants (the Compositae is sometimes said to be larger), with 450–500 genera and 15,000–20,000 or 25,000 species; world-wide in range, from tropical to arctic latitudes, and from sea-level to alpine summits. Orchids are most abundant in the rain-forests of tropical and sub-tropical regions, where the majority are epiphytes; in arid and semi-arid districts, orchids are poorly represented. In temperate regions, very few epiphytes occur, most of the species are terrestrial, and in the arctic regions all are terrestrial. Only about 45 genera with less than 200 species occur in North America north of Mexico, over half of which may be found in the Southeast; 2 species of orchids, 1 Asiatic, 1 European, are introduced and have become well established: *Zeuxine strateumatica* (L.) Schltr. in Florida, first seen in 1936 and now thoroughly established throughout the state; and *Epipactis helleborine* (L.) Crantz, first found in 1879 near Syracuse, N. Y., and now ranging from Quebec and New England westward to Michigan, Indiana, and Missouri, very abundant in the Northeast, local westward where more recently established.

Orchids inhabit a wide variety of habitats. Only in warm climates are epiphytes found; these often have some leaf-modification which favors retention of water, or have specialized air-roots with a highly absorptive specialized epidermis, the velamen. Many of the orchids of temperate latitudes are woodland species, growing in dry or mesophytic woods, tamarac or arbor vitae bogs, meadows, swamps, and bogs, occasionally in dry open sites.

Greenhouse-grown tropical orchids and orchid hybrids are both aesthetically and commercially valuable. Most of the vanilla of commerce is derived from the long (to 25 cm) narrowly cylindric capsules of *Vanilla planifolia* Andrews, a climbing orchid of Central and South America, Mexico, and extreme southern Florida. This is the only orchid of economic importance except those of the florists' trade.

The literature on orchids is extensive, comprising both technical and popular works. The reader is referred to *Our Wild Orchids* (Morris and Eames, 1929), for photographs of orchids in their native haunts, and for interesting information on orchid habitats and orchid hunting; to *Native Orchids of North America* (Correll, 1950), for orchid illustrations and habitat and life-history information as well as botanical descriptions of all North American (north of Mexico) orchids.

The ORCHIDACEAE is the most advanced of the monocotyledonous plant families in evolutionary development. The flowers are highly specialized, with modification of some of the floral parts.

Plants perennial, epiphytic or terrestrial, these either chlorophyll-bearing (with green leaves) or non-green (without chlorophyll) and saprophytic and living on dead organic matter; roots with mycorrhizal fungi on which orchids are at least in part dependent for nutrition; leaves mostly alternate, rarely opposite or whorled, or reduced to bracts or scales, entire, coriaceous, membranaceous, or succulent; inflorescence a raceme, spike, or panicle, or flowers solitary or few; flowers zygomorphic (bilaterally symmetrical); sepals 3 (or 2 by fusion), the lateral often different from upper; petals 3, the lateral similar (in species-text referred to as petals), the third highly modified and called the lip (labellum) often larger and different in color from others and extended at base into a spur; ovary inferior, of 3 carpels but unilocular, it or the pedicel usually twisted 180°, causing the dorsal sepal to become (in position) the upper sepal and the lip to be downward, the flower thus resupinate (turned upside-down); style, stigmas, and stamens (1 or 2) are united and form a central "column"; if 1 stamen, anther terminal on column, if 2, then lateral on column; stigma low on anterior face of column; pollen coherent in granular or waxy masses, pollinia, in each locule of anther; fruit a unilocular capsule containing thousands of seeds, the 3 valves remaining attached to one another at apex.

A capsule of *Cypripedium acaule* is estimated to contain over 54,000 seeds— the number based on weight of seeds, and number of seeds per milligram. Seedling development, in many species, is exceedingly slow, several years elapsing between germination and the growth of the first scale-like leaves on the subterranean protocorm, a structure produced by enlargement of embryo (see Stoutamire, 1964).

The species of two of the Ohio genera (*Hexalectris* and *Corallorhiza*) are saprophytes and devoid of chlorophyll, i.e., plants living on dead organic matter; those of three genera (*Arethusa, Aplectrum, Tipularia*) are without leaves at flowering time and might be mistaken for saprophytes; some species of *Spiranthes* lose their leaves by the time the flowers mature, but their stems are green and bear green bracts. All other Ohio orchids are terrestrial green plants. A few Ohio orchids have winter-green leaves, those of *Aplectrum* and *Tipularia* developing in fall and persisting until April or May, those of *Goodyera* persisting throughout the year.

Almost all species of orchids are becoming rare; none should be disturbed; few wild orchids succeed in cultivation, except possibly for a few years. Plants offered for sale have almost always been dug from the wild, a practice which has exterminated some of the showier species from many localities. Culture of orchids from seed is a slow and tedious undertaking.

The sequence of genera departs (in a few cases) from that followed by Fernald, in order to bring genera within the same tribe or series (as outlined by Lawrence, 1960) into juxtaposition.

a. Flowers large, few (1–3); lip a large inflated sac or pouch 1.5 cm long or more; leaves and stem glandular pubescent .... 1. *Cypripedium*
aa. Flowers smaller, few or many; lip, if pouch-like then small or in part expanded and flattened; leaves glabrous, stem glabrous or puberulent.
   b. Flowers solitary or few and axillary.
      c. Leaves 5 or 6 in a whorl at top of stem; flower solitary, terminal .... 5. *Isotria*
      cc. Leaves not in a whorl, solitary or alternate.
         d. Flowers 1–6 (usually 3), axillary, white to pink; leaves alternate, broad-ovate, small .... 6. *Triphora*
         dd. Flower solitary (rarely 2), terminal; leaf solitary.
            e. Floral bract large, foliaceous; leaf about half-way up stem, more or less elliptic, well developed at flowering time; flower pink (rarely white), lip fringed and bearded .... 4. *Pogonia*
            ee. Floral bract minute; leaf from uppermost sheath, linear-lanceolate, grass-like, immature or absent at flowering time; flower rose-purple and white, lip with ragged margin and fringed crests .... 8. *Arethusa*
   bb. Flowers in a spike or raceme, sometimes few.
      c. Plants without green leaves at flowering time; over-wintering leaves in nos. 15 and 16.
         d. Stem, reduced leaves, and bracts green; flowers white, small, in a more or less spiral spike .... 10. *Spiranthes*
         dd. Stems and sheaths or bracts not green.
            e. Flowers with long slender spur; over-wintering leaf solitary, elliptic to broad-ovate, purple beneath .... 15. *Tipularia*
            ee. Flowers without free spur.
               f. Lip white, spotted with purple or unspotted, 3–8 mm long; rhizome coral-like .... 18. *Corallorhiza*
               ff. Lip white, streaked with purplish, 10–12 mm long, with 3 low longitudinal ridges on lower part; stem from one of chain of subglobose corms; overwintering leaf plicate .... 16. *Aplectrum*
               fff. Lip pale yellowish striped with red-purple, 14–17 mm long, with 5–7 crests or longitudinal ridges from base almost to apex; rhizome coral-like, annulate .... 17. *Hexalectris*
      cc. Plants with green leaves at flowering time.
         d. Leaves with white or whitish veins or reticulations, basal or almost basal, persistent throughout the year .... 11. *Goodyera*
         dd. Leaves without white veins or reticulations, absent in winter.
            e. Leaves all basal or nearly basal, cauline, if present, reduced to scales.
               f. Basal leaves 2, elliptic to orbicular, large.
                  g. Lip prolonged into a spur; floral bracts more or less conspicuous.
                     h. Flowers yellowish green or whitish; floral bracts shorter than flowers, linear to lanceolate .... 3. *Habenaria*
                     hh. Flowers pink or lavender and white; floral bracts foliaceous, equal to or longer than flowers, lanceolate to elliptic .... 2. *Orchis*
                  gg. Lip not prolonged into a spur; floral bracts minute ....
                                       14. *Liparis*
               ff. Basal leaves several, usually shriveled by flowering time, elliptic, small; or, basal and lower stem leaves linear to linear-lanceolate; flowers white, small, in a more or less spiral spike ....
                                       10. *Spiranthes*

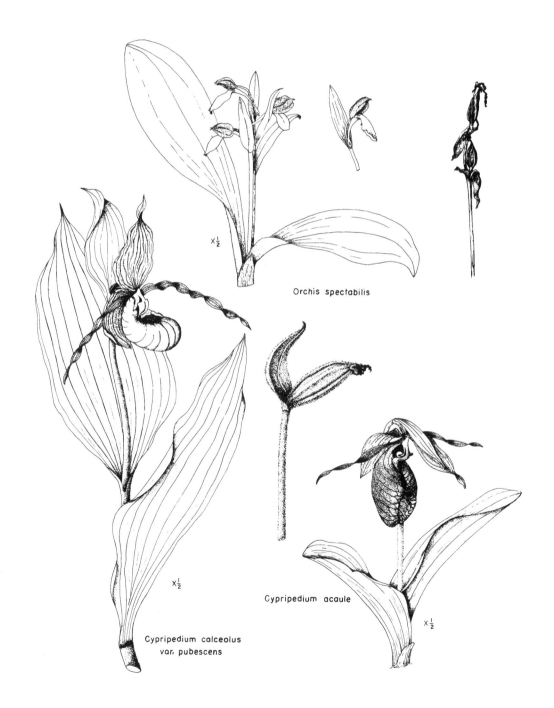

X½

Orchis spectabilis

X½

Cypripedium acaule

X½

Cypripedium calceolus
var, pubescens

ee. Leaves cauline, or cauline and basal.
   f. Leaves 1 or 2 at about middle of stem, broad; flowers small.
      g. Leaves 2, opposite, round-ovate; flowers purplish or greenish ....12. *Listera*
      gg. Leaf solitary; flowers minute, whitish ....13. *Malaxis*
   ff. Leaves several, alternate, or if only 1 or 2, then on lower half of stem.
      g. Lip directed upward, widening toward apex, bearded; flowers large, pink ....7. *Calopogon*
      gg. Lip directed downward.
         h. Spur at base of lip ....3. *Habenaria*
         hh. Spur absent.
            i. Flowers white, small, in crowded more or less spiral spike; leaves low on stem, linear or linear lanceolate, mostly reduced to sheathing bracts above; floral bracts small ....10. *Spiranthes*
            ii. Flowers greenish, veined with purple, in loose bracted raceme; stem leafy almost to inflorescence; floral bracts large, foliaceous ....9. *Epipactis*

1. CYPRIPEDIUM L.   Lady's-slipper.   Moccasin-flower

Widespread, in tropical, temperate, and boreal regions of the northern hemisphere; the only genus of its Tribe in temperate latitudes, although others with the distinctive sac-like lip and with 2 fertile anthers occur in the tropics and may be seen in greenhouse collections. Perennials with coarse fibrous roots; leaves basal or cauline, strongly ribbed and usually plicate; stems erect, 1–3 flowered (in eastern species); flowers large, showy, pink, white, or yellow; perianth of 3 sepals—the 2 lower usually united and thus apparently opposite the erect sepal, 3 petals—the 2 lateral very unlike the inflated sac-like (or slipper-like) lip; stamens 2, lateral on the column just below its enlarged petaloid tip (a staminodium, or greatly modified sterile stamen); column declined forward, partially shielding the orifice of sac-like lip.

The leaves and stems of Ohio species are more or less densely covered with glandular hairs; these are poisonous to the touch; if handled, they may produce a severe case of dermatitis similar to that caused by poison ivy.

a. Leaves 2, basal; scape bearing 1 pink (rarely white) flower ....1. *C. acaule*
aa. Leaves alternate on stem; flowers 1–3.
   b. Leaves erect to strongly ascending, crowded near middle of stem, stem below middle with several bladeless sheaths; sepals and petals greenish yellow, lip waxy white ....2. *C. candidum*
   bb. Leaves more evenly distributed, sheaths below only toward base of stem.
      c. Sepals and lateral petals white, petals oblong-elliptic, flat, blunt; lip suffused with pink and darker streaks; pubescence of pale hairs ....3. *C. reginae*
      cc. Sepals and lateral petals brown, brownish, or green; petals narrow, acuminate, usually spirally twisted; lip yellow with darker veins; pubescence of brown hairs.
         d. Petals 5–9 cm long, lip 3–5 cm long; sepals greenish yellow and darker streaked ....4. *C. calceolus* var. *pubescens*
         dd. Petals 3.5–5 cm long, lip 2–3 cm long; sepals madder purple ....
            5. *C. calceolus* var. *parviflorum*

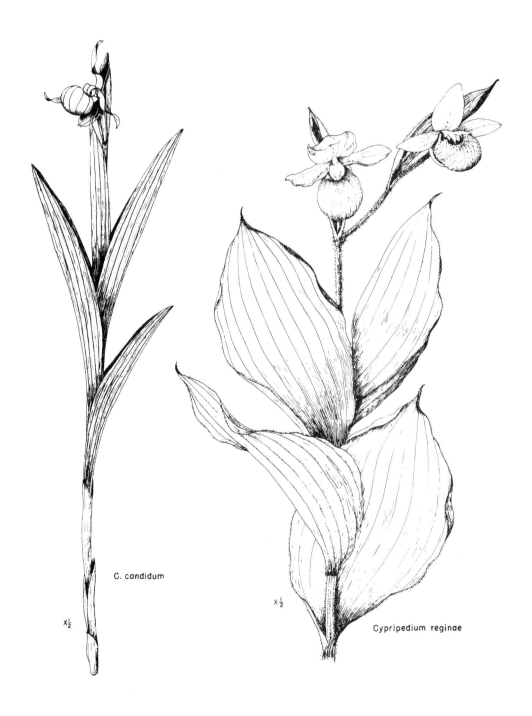

C. candidum

X½

X½

Cypripedium reginae

1. **Cypripedium acaule** Ait. PINK MOCCASIN-FLOWER
   Fissipes acaulis (Ait.) Small
   Distinguished from other species by growth-habit—2 (rarely 3) large (1–2.5 dm), glandular pubescent, oval to narrowly elliptic basal leaves and naked stem (scape) 2–4 (–5) dm tall, with 1 foliaceous bract subtending solitary pink flower. Sepals and lateral petals yellowish to greenish brown with darker streaks; lip pink with red veins, inrolled along narrow frontal fissure. Albino forms occur, with white lip, and all floral parts pale. Plants of acid, often sandy soil in dry oak or oak-pine woods, or sometimes in bogs. Ranging from Newfoundland westward to Alberta, southward to New England, New Jersey, and northern Indiana, and in the Appalachian area, to Georgia, Alabama, and Tennessee.

2. **Cypripedium candidum** Muhl. WHITE LADY'S-SLIPPER
   Very rare in Ohio, and doubtless extinct in some counties from which there are records; the Montgomery County specimen bears the date 1878. Distinguished from all other Ohio species by its waxy white lip, or, in the absence of flowers, by its rigidly erect habit and erect or stiffly ascending leaves crowded at about middle of stem so that the bases overlap, the lower half of stem with several prominent sheaths. Ranging from Ontario, west-central New York, and northern New Jersey, westward across the Lake States into eastern North and South Dakota, northeastern Nebraska, and Missouri, with an early record from "the barrens of Ky."; in marly or other lime-rich soil of wet meadows and swamps, or rich lower ravine slopes.

3. **Cypripedium reginae** Walt. QUEEN LADY'S-SLIPPER; BIG PINK-AND-WHITE
   C. hirsutum of Gray, ed. 7; *C. spectabile* Salisb.
   The largest and showiest of the Lady's-slippers, the whole flower sometimes more than 1 dm across; "lip pouch-shaped, with shallow vertical and evenly distributed white furrows, white, crimson-magenta or rose-pink in front, often with purple or rose veins, 2.5–5 cm long" (Correll, 1950); sepals and petals white. Stems 3–8 dm tall, densely glandular hirsute with pale hairs; leaves with veins and margins prominently pale long-pubescent. Wide-ranging, but very local, from Newfoundland to Manitoba and North Dakota through the Lake States and southward to Missouri, and in the mountains to Georgia and Tennessee. Local in Ohio, and long extinct in its southernmost station (Hamilton County specimens dated 1836 and 1837).
   The most poisonous species of *Cypripedium*.

4. **Cypripedium calceolus** L. var. **pubescens** (Willd.) Correll LARGE YELLOW
   LADY'S-SLIPPER
   C. pubescens Willd.; *C. parviflorum* Salisb. var. *pubescens* (Willd.) Knight.
   The typical variety is Eurasian, and American varieties were long considered to be specifically distinct. This and the next may be distinguished from other Lady's-slippers by the yellow flowers or, after flowering, by the brown rather than pale glandular pubescence of stems, leaves, and developing capsules. Flowers 1–2, each with foliaceous bract; lip 3–5 cm long, lateral petals 5–9 cm long, usually spirally twisted, greenish yellow with dark streaks. Plants of rich

to dry woods in circumneutral to slightly alkaline soil with incorporated humus. Somewhat more southern than the next.

5.   **Cypripedium calceolus** L. var. **parviflorum** (Salisb.) Fern.   SMALL YELLOW LADY'S-SLIPPER

C. parviflorum Salisb.

Not always distinguished from the last, from which it differs in its smaller size, smaller flowers, lip 2–3 cm long, lateral petals 3.5–5 cm long, darker color of sepals and petals, and in its habitat—very wet and boggy places. However, the varieties are not always distinct, and there are occasional exceptions in habitat. Wider ranging than the last, somewhat more northern, and extending farther west.

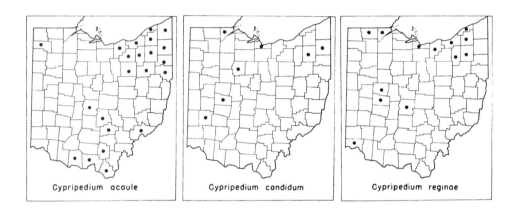

Cypripedium acaule   Cypripedium candidum   Cypripedium reginae

2. ORCHIS L.   ORCHIS

A genus of about 100 species of the northern hemisphere, only 3 of which occur in North America; the name is the ancient Greek name, adopted by Linnaeus in 1753. More or less succulent plants with thickened fibrous or tuberous roots, stems less than 4 dm tall, leafy (in an Alaskan species) or naked, with 1–2 (–3) basal leaves; inflorescence a loose raceme with foliaceous bracts; sepals and petals similar, but petals smaller and connivent with sepals to form a hood over column; lip large, with prominent spur at base; pollen collected into 2 large masses, one in each locule of anther.

1.   **Orchis spectabilis** L.   SHOWY ORCHIS

Galeorchis spectabilis (L.) Rydb.

Plants glabrous, with 2 large (6–20 cm long) lustrous and succulent basal leaves; raceme with foliaceous bracts; sepals and petals lavender to roseate (rarely white); lip white, broad ovate to rhombic, entire; spur white, about as long as lip. An orchid of rich mesophytic woods, ranging almost throughout the Deciduous Forest, flowering (in mid-latitudes) in May.

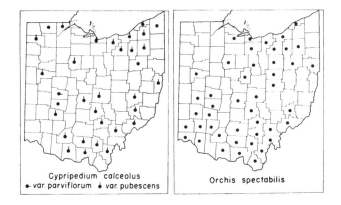

Cypripedium calceolus
• var. parviflorum ♦ var. pubescens

Orchis spectabilis

3. HABENARIA Willd.

A large genus of some 400–500 species, most abundant in warm regions; frequently divided on basis of technical floral features, especially of column, into a number of genera. All, except the few species with basal leaves and scapose stems, are sufficiently similar in appearance so that it is usually possible to recognize a *Habenaria* at any stage of growth. Plants glabrous, stems of most species leafy or leafy-bracted; flowers in racemes or spikes, often showy, sometimes inconspicuous; sepals and petals more or less alike in form and color; lip various, undivided, entire or toothed, or fringed, or 3-parted; spur longer or shorter than lip; column short.

a. Plants with 2 large, subopposite, often orbicular leaves spreading flat on the ground; stems scapose.
    b. Stem without bracts (except in inflorescence); flowers yellowish green, sessile, erect or nearly so; spur tapering to sharp point ....1. *H. hookeri*
    bb. Stem with 1-several lanceolate bracts; flowers greenish white, pedicellate, ascending to spreading; spur cylindric and slightly enlarged toward tip (clavate) ....
                        2. *H. orbiculata*
aa. Plants with alternate cauline leaves, these gradually or abruptly reduced in size upward; stems not scapose.
    b. Lip 3-parted.
        c. Divisions of lip not fringed, finely erose on margins, middle lobe notched; flowers rose-purple; spur 2–3 cm long; lip 1–2 cm long ....3. *H. peramoena*
        cc. Divisions of lip fringed.
            d. Flowers lilac- to rose-purple; spur strongly recurved, equaling to 2 × length of ovary.
                e. Spur 15–25 mm long; lip 6–16 mm broad ....4. *H. psycodes*
                ee. Spur 20–35 mm long; lip 18–30 mm broad ....
                        5. *H. psycodes* var. *grandiflora*
            dd. Flowers white to whitish, or yellow-green.
                e. Spur 2–5 cm long; lip 1.5–3 cm long; flowers white ....
                        6. *H. leucophaea*
                ee. Spur 1.5–2 cm long; lip 1–1.5 cm long; flowers yellow-green ....
                        7. *H. lacera*
    bb. Lip not 3-parted.
        c. Lip fringed; flowers about 2 cm long, exclusive of spur; spur more than 15 mm long.
            d. Flowers white ....8. *H. blephariglottis*

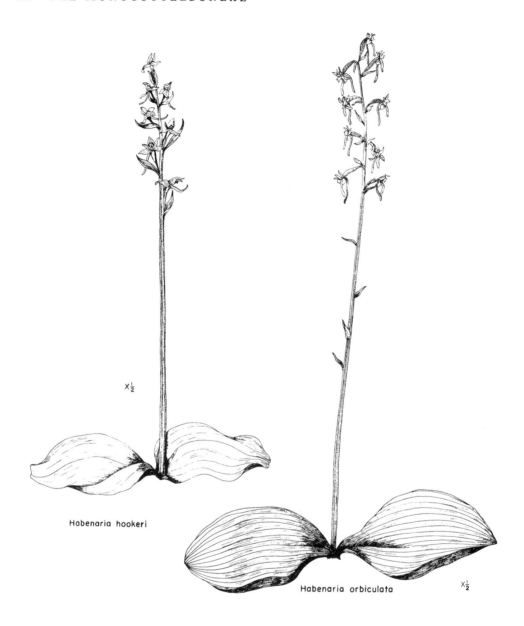

X½

Habenaria hookeri

Habenaria orbiculata   X½

dd. Flowers orange ....9. *H. ciliaris*

cc. Lip entire or toothed; flowers smaller, about 1 cm long, exclusive of spur; spur usually less than 10 mm long.

d.   Floral bracts shorter than ovary; flowers twisted to one side, lip and spur thus appearing lateral; lip cuneate-oblong, with 3 short apical teeth; plants usually with only 1 well developed cauline leaf ..... 10.   *H. clavellata*

dd. Floral bracts longer than ovary, at least the lower exceeding flowers; plants usually with 2 or more well developed cauline leaves.

e.   Floral bracts conspicuous, divergent, at least the lower leaf-like; inflorescence loose, flowers somewhat divergent from stem.

f.   Lip 2–3 toothed at apex; spur saccate, 2–3 mm long; floral bracts
2–4 × length of flower; flowers green ....
11.   *H. viridis* var. *bracteata*

ff. Lip without apical teeth, often with one lateral tooth on each
side near base, and with a median tubercle below middle; lower
floral bracts much exceeding flowers; flowers greenish yellow ......
12.   *H. flava* var. *herbiola*

ee. Floral bracts inconspicuous, ascending to incurved, not leaf-like;
inflorescence compact, flowers erect or appressed, green or greenish
yellow; lip lanceolate, not much broadened at base ....
13.   *H. hyperborea*

1. **Habenaria hookeri** Torr.   HOOKER'S ORCHID
Lysias hookeriana (Gray) Rydb.

A northern woodland species ranging from Nova Scotia and Quebec to
northern Minnesota and southward to New England, Pennsylvania, Ohio, and
northern Indiana. Flowers about 1.5–2 cm across, yellowish green, sessile or
nearly so in narrow raceme; upper sepal triangular-ovate, and with the narrow
attenuate petals forming a hood over column; lateral sepals reflexed; lip narrow,
upcurved, tapering to apex, and with lateral margins rolled under; spur 1.5–2.5
cm long, tapering to pointed apex. Distinguished from the next by its sessile
flowers, tapering spur and usually bractless stem, and from other 2-leaved
orchids by floral characters and position of leaves—flat on ground. Ohio plants
belong to the typical variety.

2. **Habenaria orbiculata** (Pursh) Torr.   LARGE ROUND-LEAVED ORCHID
Lysias orbiculata (Pursh) Rydb.

Northern in range, extending almost across the continent in Canada, southward to New England, Ohio, northern Indiana, Minnesota, and in the mountains
to West Virginia, Virginia, and Georgia. Flowers about 2 cm across, greenish
white, pedicellate in loose raceme; upper sepal roundish, erect; lateral sepals
narrow-ovate, obtuse, somewhat reflexed to drooping; petals ascending; lip
strap-shaped, pendent; spur 1.5–3 cm long, slender-clavate. Distinguished from
the last by leaf-color (shining above, silvery beneath), pedicellate flowers,
clavate spur, and the 1-several lanceolate bracts on stem. Ohio plants belong to
the typical variety.

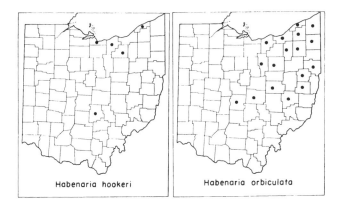

Habenaria hookeri   Habenaria orbiculata

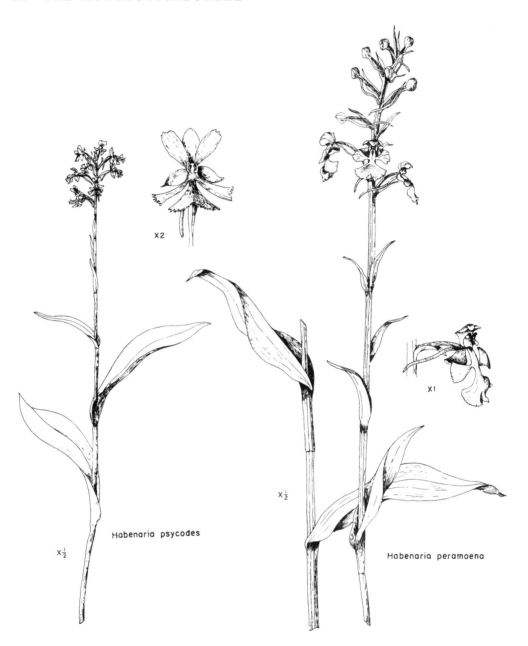

X2

X1

Habenaria psycodes

X½

X½

X½

Habenaria peramoena

3.   **Habenaria peramoena** Gray   PURPLE FRINGELESS ORCHID
Blephariglottis peramoena (Gray) Rydb.
One of the most beautiful summer wildflowers of southern Ohio, usually growing in wet grassy meadows or swamps; formerly rather frequent on the Illinoian Till Plain of southwestern Ohio. Stems 3–10 dm tall; leaves on lower

half of stem 1–2 dm long, upper much reduced; raceme dense, many flowered, 10–20 cm long and 5–6 cm in diam.; flowers phlox-purple (color and inflorescence-shape from a distance resembling that of *Phlox maculata*). Distinguished from all other species by the deeply 3-parted but not fringed lip, the middle division broad wedge-shaped and notched at apex, the lateral truncate, apical margins finely erose. A southern species, ranging from the Appalachian Upland of Alabama and Tennessee northward to Delaware, Maryland, southern Ohio, Indiana, and Illinois and westward to Arkansas and southeastern Missouri, chiefly south of the limits of Wisconsin glaciation.

4. **Habenaria psycodes** (L.) Spreng. SMALL PURPLE FRINGED ORCHID
   Blephariglottis psycodes (L.) Rydb.
   Wider ranging and more northern than the last, extending to Newfoundland, Quebec, and northwest Ontario. Distinguished from the last by its fringed lip, and from the next by flower size (as stated in key); fringe of lip cut to about ⅓ way to base. Usually in wetter situations than the last.

5. **Habenaria psycodes** var. **grandiflora** (Bigel.) Gray. LARGE PURPLE FRINGED
   ORCHIS
   H. fimbriata (Ait.) R. Br.; *Blephariglottis grandiflora* (Bigel.) Rydb.
   This differs from the last only in its larger size—longer and wider raceme, longer and wider lip, and somewhat deeper fringe (to more than ⅓ way to base), characters which do not seem to warrant designation as a species, although extremes are well marked. In Ohio, known only from Ashtabula and Portage counties.

6. **Habenaria leucophaea** (Nutt.) Gray. PRAIRIE WHITE FRINGED ORCHIS
   Blephariglottis leucophaea (Nutt.) Farw.
   The white flowers, 3-parted fringed lip (about 2 cm long) and very long spur (about 4 cm long) distinguish this species. A rare species of sphagnum bogs of the Northeast and Lake States, and in wet depressions in the Prairie Region westward to North and South Dakota, Nebraska, Kansas, and Louisiana. The Franklin County record is based on a specimen in the Darlington Herbarium at West Chester State College, collected by Wm. S. Sullivant prior to 1842; a photograph furnished by Robert B. Gordon shows the very long spur.

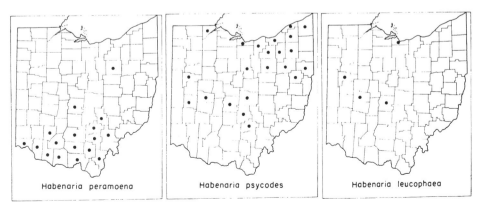

Habenaria peramoena   Habenaria psycodes   Habenaria leucophaea

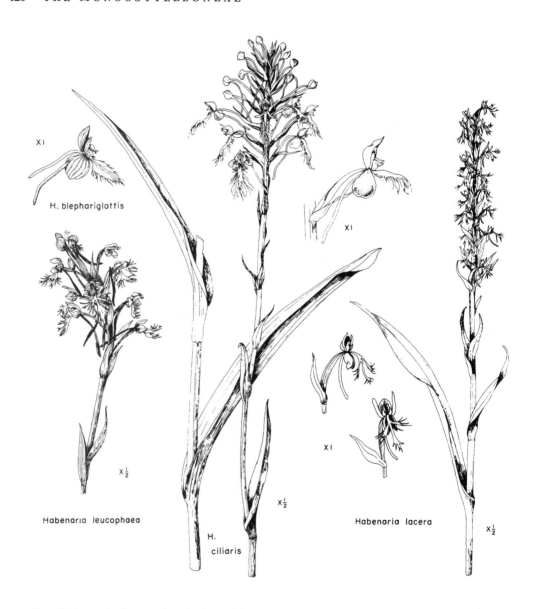

H. blephariglottis

X1

X1

X1

X½

Habenaria leucophaea

H. ciliaris

X½

Habenaria lacera

X½

7.   **Habenaria lacera** (Michx.) Lodd.   RAGGED FRINGED ORCHID
Blephariglottis lacera (Michx.) Farw.

More frequent in Ohio than any other species of *Habenaria* and less exacting
as to habitat—in dry or wet meadows or clearings, open woods or woods-borders,
or occasionally in bogs or boggy lake-borders. Inconspicuous because of its
yellow-green flowers with ragged lip; lip 3-parted, the lateral lobes often cut
almost to base into filamentous sections, the middle lobe narrow but expanding
apically and there irregularly fringed. Leaves paler and duller green than those
of other species. Local throughout most of the Deciduous Forest region. Hybri-

dizes with *H. psycodes* producing ×*H. andrewsii* White, reported from Lorain County by Correll (1950); this has purple to white flowers tinged with purple, and its characters are usually intermediate between those of the parents.

8.   **Habenaria blephariglottis** (Willd.) Hook.   WHITE FRINGED ORCHID
      Blephariglottis blephariglottis (Willd.) Rydb.

A wide-ranging white bog orchid in which 2 or 3 varieties may be segregated: (1) var. *blephariglottis*, northern in range, to which Ohio plants belong; (2) var. *conspicua* (Nash) Ames, of the south Atlantic and Gulf Coastal Plain, with larger flowers and looser racemes; (3) var. *integrilabia* Correll, most abundant on the Cumberland Plateau of Kentucky and Tennessee, with entire or nearly entire lip and petals. The typical variety is distinguished by its narrowly ovate-oblong copiously fringed lip (about 10–12 mm long). Found in sphagnum and sedge bogs, and tamarack swamps.

9.   **Habenaria ciliaris** (L.) R. Br.   ORANGE OR YELLOW FRINGED ORCHID
      Blephariglottis ciliaris (L.) Rydb.

A very showy species occurring, in suitable locations, almost throughout the Deciduous Forest, from southern New England to Ontario and Wisconsin, south to Florida and east Texas; growing in sun or partial shade in bogs, swamps, wet banks, dry grassy slopes—in a wide variety of sites but always in acid, often sandy soil. Easily recognized when in bloom by its rich golden yellow or orange flowers with copiously fringed lip. Similar in foliage and floral structure to the last, and best separated from dried and discolored specimens of it by the longer, finer fringe of lip.

10.   **Habenaria clavellata** (Michx.) Spreng.   CLUB-SPUR ORCHID.   GREEN WOOD-
       LAND ORCHID
       Gymnadeniopsis clavellata (Michx.) Rydb.

A slender species with 1, rarely 2 lower leaves large, to 15 cm long and 3.5 cm wide, the upper much reduced and sometimes all bract-like. Flowers few to many in loose or dense spike, outwardly arching, the peculiar twist of the ovary making them appear askew; lip with 3 short apical teeth; spur clavate. Widespread, occurring locally almost throughout the Deciduous Forest, usually in

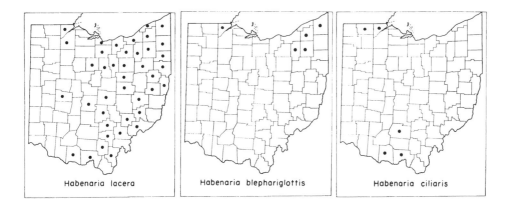

Habenaria lacera   Habenaria blephariglottis   Habenaria ciliaris

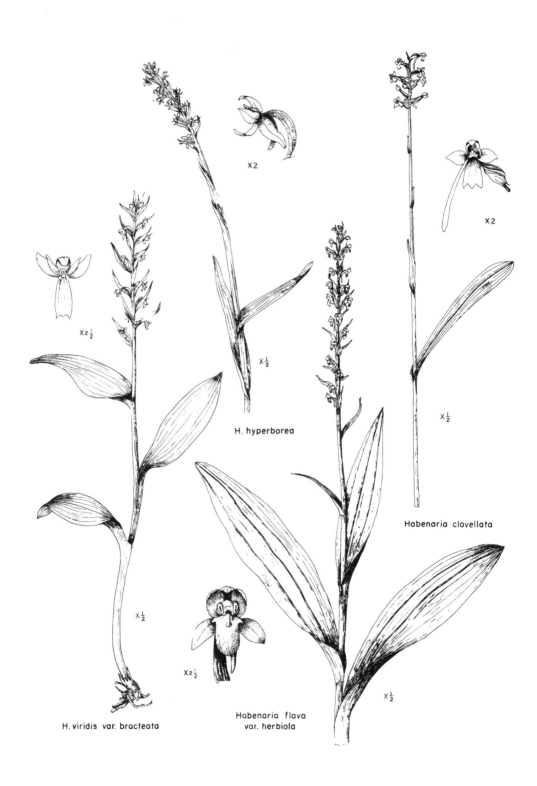

X2

X2½

X½

H. hyperborea

X2

X½

Habenaria clavellata

X½

X2½

H. viridis var. bracteata

Habenaria flava
var. herbiola

X½

moist to wet acid soil of stream or lake borders and mossy slopes in sun or shade.

11. **Habenaria viridis** (L.) R. Br. var. **bracteata** (Muhl.) Gray.   LONG-BRACTED ORCHID

   H. bracteata (Willd.) R. Br.; *Coeloglossum bracteatum* (Willd.) Parl.

   This small-flowered green *Habenaria* is distinguished from other species by its large, divergent floral bracts, its short (2–3 mm) pouch-shaped spur, and its narrowly oblong lip 5–10 mm long and 3-toothed at apex, the middle tooth short or lacking. The species is circumboreal in distribution; the typical variety is Eurasian, var. *bracteata* North American, transcontinental in the North (Newfoundland to British Columbia and Alaska) and ranging south to mid-temperate latitudes.

12. **Habenaria flava** (L.) R. Br. var. **herbiola** (R. Br.) Ames & Correll.   TUBER-CLED REIN ORCHID

   Two varieties of *H. flava* (*Perularia flava* (L.) Farw.) are distinguished; the typical variety is southern, and its larger leaves are all below middle of stem and its floral bracts only as long as flower; the more northern var. *herbiola*, to which Ohio plants belong, is more leafy, and has much longer floral bracts, characters which lend a very different aspect to the growing plants. Both varieties have a median tubercle on lip, but lip is broader in the typical variety. Flowers green or greenish yellow, sessile, small (about 5 mm across), very fragrant.

13. **Habenaria hyperborea** (L.) R. Br.   TALL NORTHERN GREEN ORCHID

   Limnorchis hyperborea (L.) Rydb.

   A tall green-flowered species of bogs and wet woods; 2 intergrading varieties of similar range are sometimes segregated. One of the few orchids occurring within the Arctic Circle; transcontinental in the far North. Very rare in Ohio, confined to a few northern bogs. Distinguished from other green-flowered Ohio species by the narrow inflorescence with erect or appressed flowers, the long-pointed lanceolate ascending or suberect bracts, and the lanceolate lip. Varies greatly over its extensive range, and, following Gleason (1952) and Correll (1950), varieties are not here distinguished.

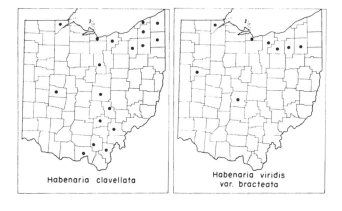

Habenaria clavellata

Habenaria viridis
var. bracteata

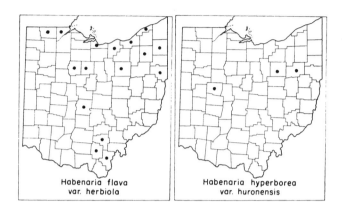

Habenaria flava
var. herbiola

Habenaria hyperborea
var. huronensis

4. POGONIA Juss.

Slender-stemmed plants with widespreading fleshy fibrous roots; stem with 1 leaf near middle and a leafy bract subtending the 1–3 terminal flowers; long-petioled basal leaves arise from spreading stolons; flowers pink or white; perianth parts distinct, sepals and petals similar (the latter shorter), spreading; lip narrow below, widening above middle, fringed and crested; column projecting forward, its terminal anther deflexed. A small genus of eastern Asia and eastern North America; in earlier manuals, more inclusive than as now understood.

1. **Pogonia ophioglossoides** (L.) Ker.   Rose Pogonia.   Snake-mouth

A delicate orchid of deep sphagnum bogs, wet mossy meadows, and, on the Coastal Plain, pine-barren swamps and flatwoods; wide-ranging, but local, in some northern bogs said to occur by the thousand. Distinguished from other Ohio orchids by its slender stem bearing a single leaf at about middle, leafy-bracted terminal pink flower with spreading sepals and petals, tongue-shaped or spatulate lip "crested on the face with a three-ranked brush of fleshy white hairs tipped with yellow-brown," veined with rose, and lacerate-toothed at apex; column about 1 cm long, projecting forward.

Pogonia ophioglossoides

5. ISOTRIA Raf. WHORLED POGONIA

A genus of 2 eastern American woodland species; similar in floral structure to *Pogonia*, in which genus its species were included in earlier manuals; sepals and petals distinct but unlike, the sepals linear and much longer than petals; stems slender but fleshy, glabrous, hollow, with a whorl of 5–6 leaves subtending terminal flower or flowers, and a few short scales near base.

1. **Isotria verticillata** (Willd.) Raf. WHORLED POGONIA
   Pogonia verticillata (Willd.) Nutt.

Distinguished from all other Ohio orchids by the whorl of usually 5 leaves subtending the single erect, long-peduncled flower with 3 linear brown-purple sepals more than twice length of petals. Sterile plants may be mistaken for *Medeola*, but can be distinguished from it by the hollow, fleshy, glabrous stem instead of wiry, loosely pubescent stem of *Medeola*. The whorled Pogonia is a spring-flowering species usually found in acid soil of dry oak woods, often with Mountain Laurel. It first appears above ground as a slender brown cone (the sepals pressed closely together), then as a small ovoid dark green body (the tightly enfolded still small leaves); at flowering time (late April—early May in southern Ohio) the leaves are small, sometimes less than 2 cm long, but grow rapidly and before the flower fades may be 4–5 cm long, later attaining a length of 8–10 cm; the capsule, about 3 cm long, is erect, elevated on long peduncle, and with the withered leaves, persistent into the next year. The densely hairy roots extend outward near the surface of the ground and are said to reach a length of several feet (Correll, 1950). Eastern United States west to Michigan, Missouri, and Arkansas.

6. TRIPHORA Nutt.

A small genus of tropical and east-temperate North America, formerly included in *Pogonia*, and differentiated on technical characters of column, anthers, and pollen-grains. Inconspicuous stoloniferous plants bearing fleshy tubers; stems usually slender, succulent, with a few small alternate leaves; flowers few to several in axils of uppermost leaves.

1. **Triphora trianthophora** (Sw.) Rydb. THREE BIRDS ORCHID. NODDING POGONIA
   Pogonia trianthophora (Sw.) BSP.

A delicate inconspicuous orchid; stems fragile, succulent, green-brown; leaves alternate, ovate to almost orbicular, small (6–18 mm long); flowers 1–6, usually 3, borne singly from axils of upper leaves, pale pink, white, or rarely deep pink, 1–1.5 cm long; sepals and petals lanceolate, lip with 3 longitudinal green ridges, 3-lobed, sinuate on margin. Stolons and tubers entangled in leaf-litter; flowering stem grows rapidly, the first (lowest) flower opening about 2 weeks after stem appears above ground, its position changing from a nodding bud to erect flower, later (if capsule matures) again nodding; stem above first flower bent down, straightening as second flower opens, and so on to stem tip. Flowers retain their freshness and wide-spreading sepals only a few hours; by the second

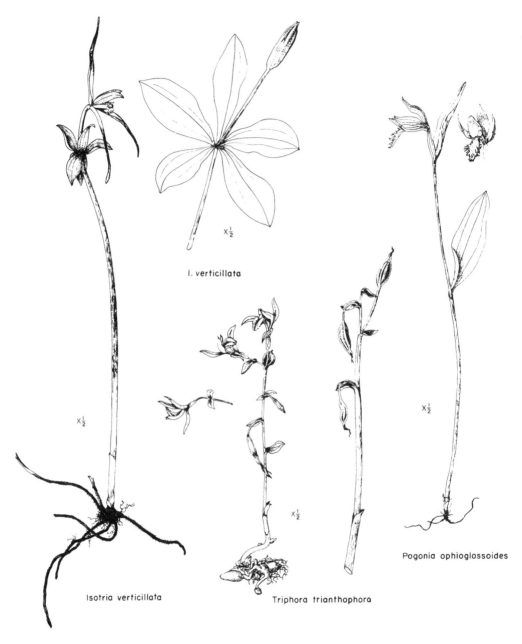

I. verticillata

×½

×½

×½

×½

Isotria verticillata

Triphora trianthophora

Pogonia ophioglossoides

morning, sepals are drooping and the delicate beauty passed. Local, in rich woods throughout much of the Deciduous Forest; flowering from early August to early September, but not every year; some clones may produce abundant above-ground stems one year, and few or none the next, the stolons and tubers continuing to develop much as is the habit with the entirely saprophytic orchids.

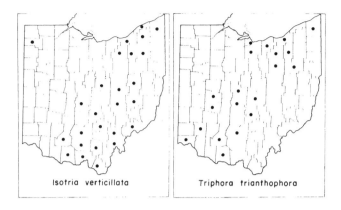

Isotria verticillata   Triphora trianthophora

7. CALOPOGON R. Br. GRASS-PINK

Stem from a small hard corm, slender, 3–8 dm tall, sheathed toward base by 1 (rarely 2) basal, erect, narow leaves; inflorescence a raceme, the flowers apparently inverted, i.e., lip uppermost (not resupinate); perianth-parts wide-spreading, lip bearded. A small genus of eastern United States, southern Canada, and the West Indies.

1. **Calopogon pulchellus** (Salisb.) R. Br. GRASS-PINK
Limodorum tuberosum BSP., not L.
A tall, slender-stemmed orchid of wet peaty or marly soil, bogs, and dry sandy soil of cliff-tops or pine-oak woods; leaves erect, linear, grass-like, strongly veined; flowers in early summer, 3–4 cm across, rose-pink or rose-purple (rarely white), 2–8 in a loose raceme; lip directed upward, fan-shaped above, narrow toward base, bearded with slender white hairs thickened and golden at tip, and nearer apex of lip with shorter, thicker, rust-colored hairs. A variable species; Ohio plants belong to the typical variety.

8. ARETHUSA L. ARETHUSA

A genus of 2 species, one in eastern North America, the other in Japan. Bulbous plants but a few dm tall, the stem bearing a few loose bracts toward base, and 1 leaf arising from axil of upper bract and developing after the flowering season; inflorescence of 1, rarely 2 rose-purple showy flowers; sepals and petals similar, united at base, ascending and arched over column; lip with basal half erect, apical half recurved.

1. **Arethusa bulbosa** L. ARETHUSA; DRAGON'S MOUTH
A very rare orchid growing loosely attached in sphagnum of deep bogs. Stems 6–30 cm tall; flowers in late spring, large and showy, sepals and petals magenta, together forming an arch over column; lip wavy-crisped on margin, strongly veined, crested with glandular capillary-fringed yellow and purple ridges. Northeastern in range, extending south in the mountains.

Calopogon pulchellus

Arethusa bulbosa

Epipactis helleborine

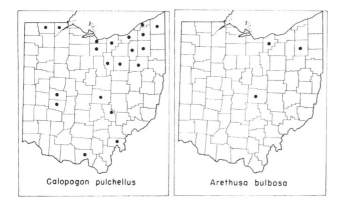

Calopogon pulchellus

Arethusa bulbosa

9. EPIPACTIS Sw. HELLEBORINE

A genus of about 20 species, most of which are in temperate Europe and Asia; 2 occur in North America—one native in the West, another naturalized from Europe in the East. Plants with simple leafy stems and bracted raceme; sepals and smaller petals similar; lip saccate at base, broadly ovate to cordate in apical half; column short and broad.

1. EPIPACTIS HELLEBORINE (L.) Crantz. HELLEBORINE
    E. latifolia (L.) All.
    Stems 2–8 dm tall, glabrous below, pubescent above; leaves alternate, ovate to lanceolate, acute to acuminate, sessile and clasping at base, finely ciliolate; raceme loose or dense; flowers about 1.5 cm across, greenish, tinged with purple, lip darker toward base. Apparently first found in the United States in 1879, and in Canada, near Toronto, in 1890, and now established over a considerable area, occurring as a roadside weed, and in woods where competing with the native flora; the only Ohio specimens, from Summit County, were collected in 1958 and 1959.

10. SPIRANTHES Richard. LADIES' TRESSES

Gyrostachys Pers.; *Ibidium* Salisb.

Slender stemmed plants with a cluster of thick, fleshy or tuberous roots, or rarely with only one tuber; leaves mostly basal, ovate to linear, those of stem usually much reduced and bract-like; inflorescence a spirally twisted spike of 1–2 or more rows of flowers; flowers (in ours) white, more or less tubular or gaping, often with green or yellow blotch or shade on lip; sepals and petals similar, narrow, the upper sepal (or all sepals) and lateral petals connivent, forming a hood which projects over column and lip; lip various in shape, not crested, its basal part inrolled more or less around the short column. Color of lip is an important character, and should be noted from fresh specimens.

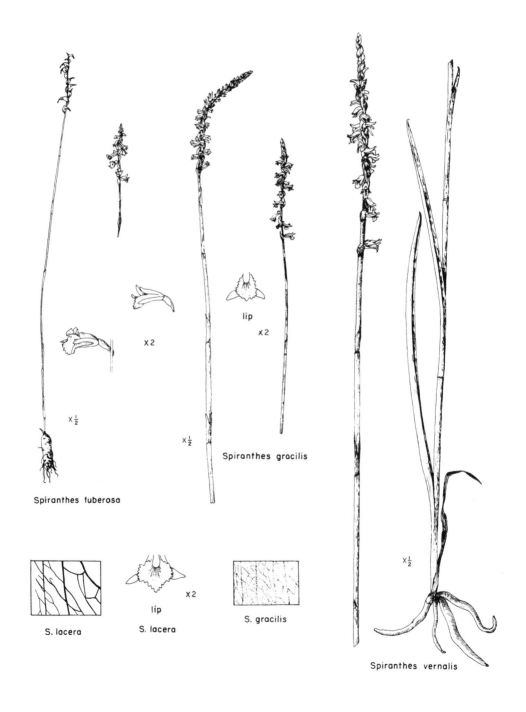

X 2

lip

X 2

X ½

X ½

Spiranthes gracilis

Spiranthes tuberosa

S. lacera

lip

S. lacera

X 2

S. gracilis

X ½

Spiranthes vernalis

a. Spike a single row of flowers, spirally twisted.
    b. Leaves basal, ovate to oval or elliptic, absent or decaying at flowering time; perianth-segments 4 mm long or less; rachis and ovaries glabrous.
        c. Lip white, without green or greenish central blotch; stem very slender; root solitary, tuberous thickened, vertical ....1. *S. tuberosa*
        cc. Lip white with prominent green blotch; stem slender; roots several, fleshy but not tuberous thickened.
            d. Leaves usually present at flowering time, thinnish; veins and veinlets obvious; white margin of lip wide ....2. *S. lacera*
            dd. Leaves usually absent at flowering time, thicker; veins visible only by transmitted light; white margin of lip narrow ....3. *S. gracilis*
    bb. Leaves basal and cauline, linear, usually present at flowering time; perianth-segments 6–10 mm long, white or yellowish; rachis and ovaries pubescent
                                                      4. *S. vernalis*
aa. Spike dense, flowers in 2 or more (usually 3) rows, spiral twisting less pronounced.
    b. Basal leaves oblanceolate, large, to 15 cm long.
        c. Lip yellow; basal leaves 1–2 cm wide; flowering in early summer ....5. *S. lucida*
        cc. Lip white; basal leaves 0.5–1 (–1.5) cm wide; flowering in late summer or fall
                                                      6. *S. ovalis*
    bb. Basal leaves linear to narrowly oblanceolate, 1–3 dm long; flowers fragrant.
        c. Lip fiddle-shaped, narrowed in middle, much widened to a round-ovate entire terminal lobe; lateral sepals connivent with petals and upper sepal, forming a hood; cauline leaves linear-spatulate, gradually reduced upward ....
                                              7. *S. romanzoffiana*
        cc. Lip not fiddle-shaped, oblong to ovate-oblong, its margin toward apex crisped or erose; lateral sepals free; cauline leaves, except 1–3 lowest, reduced to acuminate sheathing bracts ....8. *S. cernua*

1. **Spiranthes tuberosa** Raf.   LITTLE LADIES' TRESSES

S. beckii of auth., not Lindl.; *S. grayi* Ames; *Ibidium beckii* of auth.; *Gyrostachys simplex* (A. Gray) Kuntze

Distinguished from other species by its very slender stem with very slender spirally twisted glabrous spike of a single rank of small flowers (perianth 2–3 mm long). Root solitary, tuberous-thickened, vertical; basal leaves ovate, 1–2 cm long, usually withered at flowering time; bracts small and inconspicuous. Variable in density of spike and closeness of spiralling, characters which have been used to segregate varieties of what is here considered to be a variable species. Dry oak woods, moist shaded slopes and borders of woods, flowering in September and October. Southern in range extending north to Massachusetts, New Jersey, Ohio, Indiana, and Missouri.

2. **Spiranthes lacera** Raf.

S. gracilis (Bigel.) Beck in part.

This species, recognized by Fernald (1946, 1950), included in *S. gracilis* by Correll (1950), and not mentioned by Gleason (1952) is so similar to *S. gracilis* that it is with difficulty separated from it. *S. lacera* is said to be more northern than *S. gracilis*, yet specimens showing its characteristic features have been found in Hamilton, Clermont, and Adams counties. Differences in the lip of the two are illustrated by the enlarged drawings (made from fresh flowers) in which the green area is shaded; differences in venation of basal leaves are shown, adapted from Fernald's half-tones (1946). For further discussion and illustrations of these 2 species, see Fernald (1946).

3.   **Spiranthes gracilis** (Bigel.) Beck.   SLENDER LADIES' TRESSES
Ibidium gracile (Bigel.) House; *Gyrostachys gracilis* (Bigel.) Kuntze.

One of the most frequent species of *Spiranthes*, occurring in open woods, fields, pastures, prairies, in moist or dry, acid or alkaline soils, almost throughout the eastern half of the country (if *S. lacera* is included) and adjacent Canada. Stems 2–6 dm tall, slender; basal leaves not present at flowering time, cauline reduced to bracts; spike from loosely to tightly spirally twisted; lip with prominent green blotch. The spiral spike with a single row of flowers, green blotch on lip, and absence of basal leaves at flowering time distinguish this from other species (except *S. lacera*).

4.   **Spiranthes vernalis** Engelm. & Gray.   NARROW-LEAVED LADIES' TRESSES
Ibidium vernale (Engelm. & Gray) House; *S. praecox* and *I. praecox* of earlier authors, in part.

The common name, Spring Ladies' Tresses (a translation of the specific name), is a misnomer in Ohio, where plants bloom in August or early September. One of Ohio's tallest species, often 6–7 dm tall, with persistent grass-like basal and lower cauline leaves; distinguished from other species by height, long narrow leaves, length of perianth (often about 1 cm), spiral twist of the long (to 2 dm) spike with a single row of flowers, usually yellowish color of lip, and densely pubescent inflorescence. In dry grassy meadows and prairies, in acid or alkaline soil; southern in range, extending from Florida and east Texas northward on the Coastal Plain to New England, and in the interior to Ohio and Missouri.

5.   **Spiranthes lucida** (H. H. Eat.) Ames.   SHINING OR WIDE-LEAF LADIES'
TRESSES
S. plantaginea (Raf.) Torr.; *Ibidium plantagineum* (Raf.) House; *Gyrostachys plantaginea* (Raf.) Britton

The earliest of the Ladies' Tresses, flowering (in southern Ohio) in late May and early June, where it grows in seepage zones at the foot of dolomite cliffs and on rocky stream margins. Distinguished from other species by its large, lustrous, ascending basal leaves often half the height of the stem, its dense spike, yellow lip, and habitat. Ranging from Nova Scotia and Maine to Wisconsin, and southward to Delaware, Ohio, Kentucky, and Missouri.

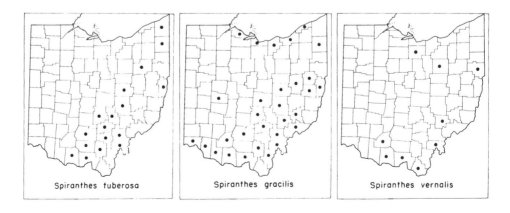

Spiranthes tuberosa   Spiranthes gracilis   Spiranthes vernalis

6. **Spiranthes ovalis** Lindl.   LESSER LADIES' TRESSES
Ibidium ovale (Lindl.) House

An infrequent and inconspicuous species of open woods and woodland borders; southern in range, from the Gulf States northward to West Virginia, Ohio, Indiana, and Missouri. Although slightly resembling the last, it cannot be mistaken for it if color of lip (white), season of bloom (October), and habitat are considered. Leaves narrower than those of the last, and tapering to long petiolar base. From other fall-blooming species, it is distinguished by its small flowers (about 4 mm long) in a 3-ranked compact short spike.

7. **Spiranthes romanzoffiana** Cham.   HOODED LADIES' TRESSES
Gyrostachys romanzoffiana (Cham.) MacM.; *Ibidium strictum* (Rydb.) House

A widespread species, transcontinental in the North (from Newfoundland and Labrador to Alaska), southward to Pennsylvania, northeastern Ohio, Michigan, westward through the northern tier of states, and southward in the western mountains to Colorado, Utah, and California; in swamps and wet meadows, flowering in summer. Distinguished from other species by its fiddle-shaped lip, and "hood" made up of the 3 connivent sepals and 2 petals. Similar

Spiranthes lucida

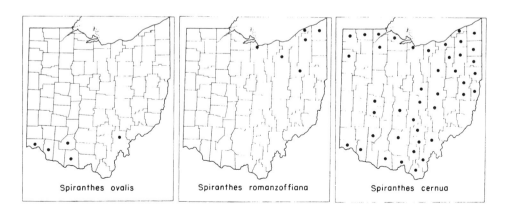

Spiranthes ovalis   Spiranthes romanzoffiana   Spiranthes cernua

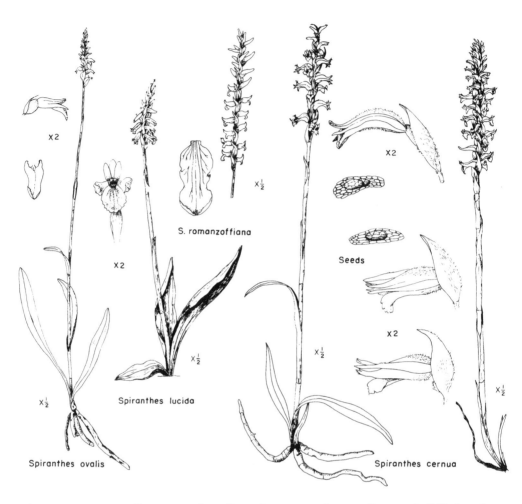

S. romanzoffiana

Seeds

Spiranthes lucida

Spiranthes ovalis

Spiranthes cernua

in appearance to *S. cernua*, but flowering somewhat earlier, and differing in characters noted above.

8. **Spiranthes cernua** (L.) Richard. Nodding Ladies' Tresses
   Ibidium cernum (L.) House; *Gyrostachys cernua* (L.) Kuntze

An extremely variable and wide-ranging species, found throughout the eastern half of the United States and adjacent Canada. Stems slender or stout, sometimes 5 mm or more in diam.; leaves basal and cauline, usually narrowly oblanceolate, those on stem much reduced and mostly bract-like; inflorescence more or less pubescent, bracts from shorter than to equalling or exceeding flowers; perianth downwardly arching, usually about 1 cm in length; lateral sepals attenuate; lip varies from ovate-oblong to oblong, with 2 pubescent basal auricles or callosities, margin crisped; flowers fragrant.

No one or few characters distinguish all individuals of this highly variable species. Varieties have been designated, but are not geographically separated, and are intergrading; var. *odorata* (Nutt.) Correll in the restricted sense is large and almost confined to the Coastal Plain, where it grows in swamps and low woods; its lip is ovate to broadly obcuneate, tapering to a subacute or obtuse apex, its stem near base 3–7 mm thick (a Lawrence County specimen referable to this variety); var. *ochroleuca* (Rydb.) Ames is a rather large variant, very fragrant, and differs from typical *S. cernua* in having only 1 embryo in its seeds instead of 2 or more; such plants have been found in the Lynx Prairie of Adams County.

S. *cernua* is found in a wide variety of habitats; wet meadows, mossy swamps, fields, dry oak woods, in the interstices between prairie bunch-grasses on thin limestone soil; its flowering season extends from early September to late October, early in meadows and swamps, late in prairies where the variant is very fragrant.

11. GOODYERA R. Br.   RATTLESNAKE-PLANTAIN

Distinguished from all other Ohio orchids by the basal rosette of evergreen leaves marked with silvery white reticulations. Flowers small, white or greenish, in spike-like raceme; petals and upper sepal united into a hood extending over lip; lip sac-like or pouch-like, with an apical horizontal or recurved prolongation or beak. A genus of about 30 species of tropical and temperate latitudes of the northern hemisphere.

a. Raceme dense, many flowered, cylindric; lip more or less globose, with short beak;
   leaves clearly and finely reticulate, 4–8 cm long ....1. *G. pubescens*
aa. Raceme not dense, one-sided or spiral; beak of lip elongate; leaves with broad whitish
   reticulations or mottled dark and light green.
   b. Leaves 1–3 cm long; flowers small (perianth 4 mm long), raceme one-sided ....
                                                                2. *G. repens* var. *ophioides*
   bb. Leaves 2–6 cm long; flowers larger (perianth 5–6 mm long), raceme loosely spiral
                                                                3. *G. tesselata*

1. **Goodyera pubescens** (Willd.) R. Br.   DOWNY RATTLESNAKE-PLANTAIN
   Peramium pubescens (Willd.) MacM.; *Epipactis pubescens* (Willd.) A. A. Eat.
   Widely distributed in the Allegheny Plateau section of Ohio and westward near Lake Erie, often in oak woods. Easily recognized by its rosettes of ovate to elliptic evergreen leaves finely reticulate with white, the rosettes often in groups of 2–10, and connected by more or less superficial rhizomes; stems 2–4 dm tall, bracteate, pubescent; inflorescence pubescent; raceme crowded, the perianth more or less spherical; lip subglobose, the tip short. For illustrations of the growth of this orchid from seed, and the early association with fungal hyphae, see Correll (1950).

2. **Goodyera repens** (L.) R. Br. var. **ophioides** Fern.   DWARF RATTLESNAKE-
   PLANTAIN
   Peramium ophioides Rydb.; *Epipactis repens* (L.) Crantz. var. *ophioides* (Fern.) A. A. Eat.
   A northern species extending south in the higher mountains to North Carolina and Tennessee; represented in the Ohio flora by one specimen from

×5

×½

Malaxis unifolia

Listera
cordata

×2½

×½

lip

lip

×2

×½

Goodyera tesselata

Goodyera repens

Goodyera pubescens

Cuyahoga County. The smallest of the 3 species, with stems 1–2 dm tall; raceme one-sided, flowers small; lip deeply saccate with elongate recurved tip.

3.   **Goodyera tesselata** Lodd.   CHECKERED RATTLESNAKE-PLANTAIN
Peramium tesselatum (Lodd.) Heller; *Epipactis tesselata* (Lodd.) A. A. Eat.

A northern species extending southward into the northern tier of states from New England to Minnesota; known in Ohio by a single collection from Ashtabula County. Somewhat larger than the preceding; raceme loosely spiral; lip saccate, its tip slightly recurved. Leaves often mottled dark and light green, hence the common name.

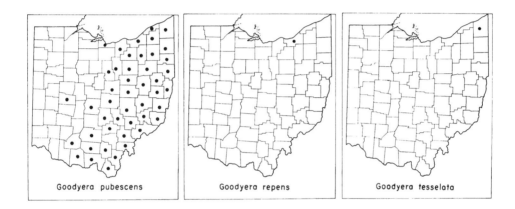

12.   LISTERA R. Br.   TWAYBLADE

Fibrous rooted plants with slender stem bearing a pair of sessile leaves near middle; flowers small, in bracted raceme; sepals and petals similar, lip longer than sepals, drooping to almost horizontal, 2-lobed or 2-cleft. A small genus, most of whose species are northern in range.

1.   **Listera cordata** (L.) R. Br.   HEARTLEAF TWAYBLADE
Ophrys cordata L.

A slender, delicate species; stems (of Ohio specimens) 1–1.5 dm tall, glabrous except near leaves; leaves round-ovate, 1.5 cm long; raceme open, glabrous throughout; sepals and petals about 2 mm long, greenish, lip 4 mm long, watery purple, drooping but upcurved toward tip, 2-cleft, and bearing near base a pair of horn-like teeth. Known in Ohio only from Ashtabula County; transcontinental in range in the North (Greenland to Alaska), usually in damp, mossy coniferous woods.

13.   MALAXIS Sw.   ADDER'S-MOUTH

Small orchids with tiny flowers, stems arising from a hard bulbous base, bearing 1–5 leaves and a terminal bracted raceme of (usually) greenish flowers. A large genus, poorly represented in America.

1. **Malaxis unifolia** Michx.   GREEN ADDER'S-MOUTH

Microstylis unifolia (Michx.) BSP.; *M. ophioglossoides* Eaton; *Acroanthes unifolia* (Michx.) Raf.

Stem bearing a single oval leaf at about the middle and a terminal minutely bracted raceme; raceme at first short and compact, elongating and becoming more open during flowering as the rachis and pedicels elongate; flowers greenish; upper sepal erect, lower similar, spreading, petals narrowly linear, recurved (all about 1–1.5 mm long); lip broad, about 2–3 mm long, 3-toothed at apex (center tooth short), at first drooping, later (because of twist of pedicel) becoming erect; capsule ovoid, 3–5 mm long. Inconspicuous plants, perhaps more frequent than the map indicates; in bud, the plant resembles an Adder's-tongue fern. Said to grow in "damp woods and bogs;" all that I have seen were in dryish oak woods.

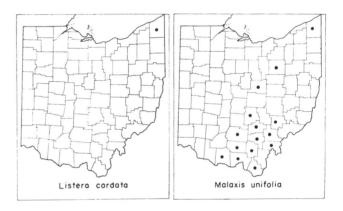

Listera cordata   Malaxis unifolia

14.   LIPARIS Richard.   TWAYBLADE

Plants with two large glossy basal leaves wrapped at their bases with bladeless sheaths, solid almost superficial pseudo-bulbs, and naked scape bearing a bracted raceme; flowers with narrow sepals and petals and broad, entire lip; capsules ribbed or winged. A large genus of chiefly Asiatic and tropical species.

a.   Leaves elliptic to broadly oval, scape strongly angled; flowers on long divergent slender pedicels; lip flat, watery-purple, broadly obovate, subtruncate and emarginate at apex; capsule equalling or shorter than pedicel ....1.   *L. lilifolia*
aa. Leaves lanceolate to elliptic; scape slightly angled above; flowers on ascending pedicels; lip yellowish green, cuneate to obovate, concave; capsule longer than pedicel ....
                                                                    2.   *L. loeselii*

1. **Liparis lilifolia** (L.) Richard.   LARGE TWAYBLADE
Leptorchis lilifolia (L.) Kuntze

Distinguished from our other orchids by its two basal leaves, superficial pseudo-bulb, strongly angled to winged scape, and minute bracts in the inflorescence; from the next by its longer pedicels, and expanded purple, almost translucent lip, 8–12 mm long. A widespread species growing in woods, frequently on steep mossy banks. Southern in range, extending northward locally to Maine and Vermont, and in the interior to the Lake States.

2. **Liparis loeselii** (L.) Richard.   FEN ORCHID
   Leptorchis loeselii (L.) MacM.
   Plants usually smaller than the last, the leaves narrower; flowers yellowish green, the lip concave, with upturned margins, about 5 mm long; capsule longer than pedicel. A northern species extending southward in the Appalachian upland; in moist or wet soil near streams or, more often in bogs.

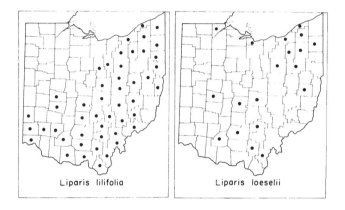

Liparis lilifolia   Liparis loeselii

15. TIPULARIA Nutt.   CRANEFLY ORCHIS

Plants with a single leaf arising from the terminal (youngest) of a horizontal series of hard, spherical corms or solid bulbs; leaf developing in autumn and persisting until late spring; inflorescence a many-flowered raceme maturing in August; flowers with similar sepals and petals, 3-lobed lip produced basally into a long slender spur. A small genus with one eastern American and one eastern Asian species (or 4 Asian species, according to Li, 1952).

1. **Tipularia discolor** (Pursh) Nutt.   CRANEFLY ORCHID
   T. unifolia (Muhl.) BSP.
   Readily recognized in flower or in vegetative condition, for it is very unlike all other Ohio orchids, except *Aplectrum*, from which it differs in all characters except growth cycle. Leaf elliptic to broad-ovate with truncate to subcordate base, dark bronzy green and shining above, rich dark purple beneath, 5–10 cm long, with thick petiole about equaling or shorter than blade. By mid-April, the leaf begins to wither, and for nearly four months there is no above-ground leaf or stem, unless the plant flowered the summer before and the stem of dry capsules persists. In August the slender flower stem arises, elongating rapidly, and bears a loose many-flowered bractless raceme; flowers greenish-sienna, sepals and petals veined with red-purple; perianth asymmetric, one petal partly overlapping dorsal sepal; lip 3-lobed, the terminal lobe long, the lateral short and near base; spur very long and slender, (1.5–2 cm), exceeding the pedicel and

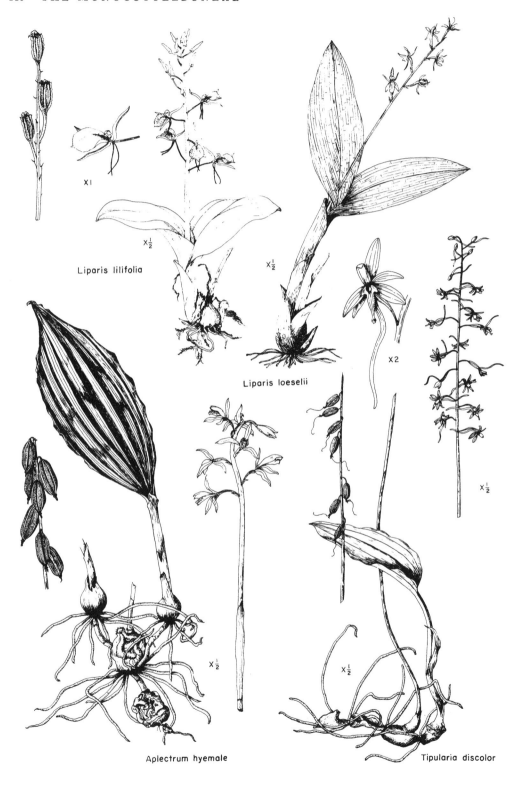

Liparis lilifolia

X1

X½

Liparis loeselii

X½

X½

X2

X½

Aplectrum hyemale

X½

Tipularia discolor

X½

slender ovary; capsules about 1 cm long, ellipsoid, pendant, the shriveled elongate spur frequently dangling from apex. Plants of dryish beech and oak woods (not occurring in rich mixed mesophytic forest); southern in range, extending northward into our area; often fairly abundant in some southern Ohio counties.

16. APLECTRUM (Nutt.) Torr.

A monotypic genus of eastern North America.

1. **Aplectrum hyemale** (Muhl.) Torr.   PUTTY-ROOT.   ADAM-AND-EVE
Plants with a chain of subglobose glutinous corms (putty-like) connected by sections of slender rootstock; leaf solitary, basal, elliptic, 10–20 cm long, dark dull green with whitish veins, longitudinally plicate, gathered at base into an enwrapping sheath. Leaf developing in autumn, usually withering by flowering time in late May or June. Flower-stem 2–6 dm tall, the lower half clothed with tubular sheaths, with an open raceme of greenish or yellowish flowers usually strongly suffused with purple; floral bracts small; sepals and petals spreading, 10–15 mm long; lip 3-lobed, obovate in general outline, pale and marked with purple-red, not spurred; capsules about 2 cm long, plump, pendant. Wide-ranging from Quebec and Vermont to Saskatchewan, southward to the Gulf States and Arkansas; widespread in Ohio in rich woods, but not common.

17. HEXALECTRIS Raf.

A small genus of saprophytic orchids with golden brown or red-brown stems and reduced scale-leaves of the same color; rhizomes coarse, coral-like, annulate; flowers in loose terminal raceme; sepals and petals about equal; lip 3-lobed, crested with longitudinal ridges; pollen-masses 8.

1. **Hexalectris spicata** (Walt.) Barnh.   CRESTED CORAL-ROOT
H. aphylla (Nutt.) Raf.
A handsome orchid, the stem sometimes 8 dm tall, more often 4–5 dm tall, with raceme 1–3 dm long; flowers large, sepals and petals free, sepals slightly

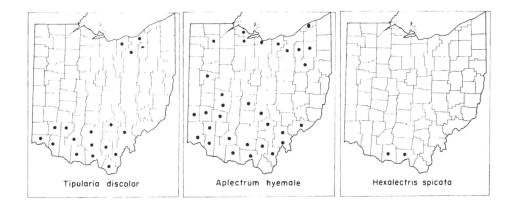

Tipularia discolor   Aplectrum hyemale   Hexalectris spicata

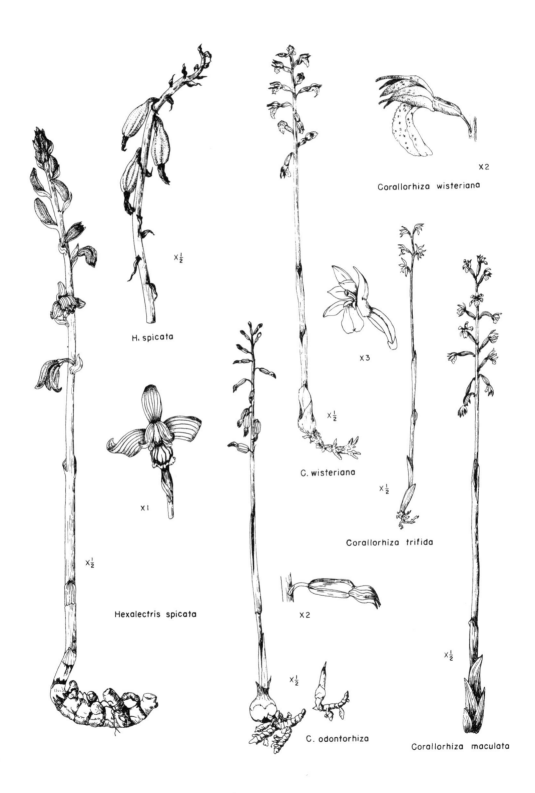

H. spicata

×½

×1

Hexalectris spicata

×½

C. wisteriana

×½

Corallorhiza wisteriana

×2

×3

Corallorhiza trifida

×½

×2

C. odontorhiza

×½

Corallorhiza maculata

×½

longer than petals; lip 3-lobed; veins on all segments except lip are a rich sienna color on a brownish-tawny to yellowish-tawny ground, those on the lip (which has 5–7 nerves modified into longitudinal ridges or crests) are rich red or royal purple on a pale yellow-buff ground. Plants of dry calcareous soil in open woods and thickets, often where rock is close to surface. The coral-like masses of rhizomes develop in the ground for several years before flowering stems arise; a clone may flower one year and not the next. A southern species, ranging northward from Florida, Texas, and northern Mexico to southern Ohio, Indiana and Missouri. In Ohio, confined to Adams and southwestern Scioto counties, where it flowers in August or early September.

18. CORALLORHIZA Chat. CORAL-ROOT

A small genus of the northern hemisphere; plants saprophytic, rhizomes coral-like, flowering stems devoid of chlorophyll, yellowish, brownish, or purplish, with similarly colored scale-like or sheathing reduced leaves, and bearing a terminal raceme of flowers in color much like stem, but spotted or striped with purple; sepals and petals similar; lip 3-lobed or simple; a hump or gibbosity mostly adnate to ovary formed by bases of lateral sepals and lip.

a. Flowering in spring (April or May in Ohio).
    b. Lip not lobed, 5.5–7 mm long; southern in range ....1. *C. wisteriana*
    bb. Lip with 2 short lobes near base, 3–4 (–5) mm long, northern in range ....
                                   2. *C. trifida*
aa. Flowering in mid-summer or fall.
    b. Lip not lobed, 2.5–3 mm long; capsules about 7 mm long; autumnal ....
                                   3. *C. odontorhiza*
    bb. Lip 3-lobed, 5–8 mm long; capsules about 15 mm long; flowering in mid-summer
                                   4. *C. maculata*

1. **Corallorhiza wisteriana** Conrad. SPRING CORAL-ROOT
The only spring-flowering Coral-root in most of Ohio; stems yellowish strongly suffused with purple, somewhat swollen at base; sheaths striate; sepals and petals yellowish within, reddish or purple-backed; lip white with purple dots. A southern species, ranging northward from Florida and Texas into our area.

2. **Corallorhiza trifida** Chat. NORTHERN CORAL-ROOT
C. corallorhiza (L.) Karst.
A circumboreal species flowering in late spring or early summer. Fernald (1950) distinguishes as var. *verna* (Nutt.) Fern. pale yellow-green plants with yellow-green flowers, white often unspotted lip, and greener capsules.

3. **Corallorhiza odontorhiza** (Willd.) Nutt. SMALL-FLOWERED OR FALL CORAL-
    ROOT
The most abundant and latest flowering of the Ohio species of Coral-root; wide-ranging, throughout much of eastern United States, usually in open woods. Flowers very small, perianth only 3–4 mm long, the sepals and petals scarcely spreading; lip white with 2 purple spots and purple margin.

4.   **Corallorhiza maculata** Raf.   SPOTTED CORAL-ROOT
     C. multiflora Nutt.

A summer-blooming woodland species, transcontinental in the North, ranging south to New Jersey, Ohio, Indiana, South Dakota, and in the mountains to North Carolina, Colorado, and California. Stems 2–5 dm tall; entire plant usually strongly suffused with brown-purple; sepals and petals usually suffused with purple, often spotted; lip white, spotted with purple, with 2 lateral lobes toward base.

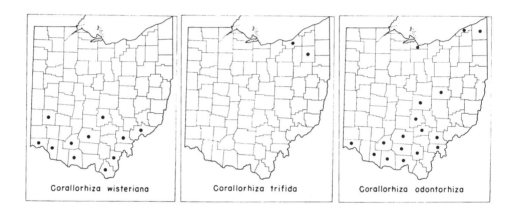

Corallorhiza wisteriana   Corallorhiza trifida   Corallorhiza odontorhiza

Corallorhiza maculata

LITERATURE CITED

ANDERSON, E. 1936. The species problem in Iris. Mo. Bot. Gard. Ann. **23:** 457–509.

ANDERSON, E. 1952. Plants, Man and Life. Little, Brown and Company.

ANDERSON, E. and T. W. WHITAKER. 1934. Speciation in Uvularia. Arn. Arb. Jour. **15:** 28–42.

BALDWIN, J. T., JR. and B. M. SPEESE. 1955. Chromosomes of taxa of the Alismataceae in the range of Gray's Manual. Amer. Jour. Bot. **42:** 406–411.

BEAL, E. O. 1960. Sparganium (Sparganiaceae) in the southeastern United States. Brittonia **12:** 176–181.

BRAUN, E. L. 1961. The Woody Plants of Ohio. Ohio State University Press. Columbus.

BRAUN, E. L. 1964. Erythronium rostratum in southern Ohio. Castanea **29:** 185–186.

BRITTON, N. L. and A. BROWN. 1896. Illustrated Flora of the Northern United States, Canada and the British Possessions. Ed. 1. New York.

CLAUSEN, R. T. 1936. Studies in the genus Najas in the northern United States. Rhodora **38:** 333–345.

CORE, EARL L. 1948. The Flora of the Erie Islands. Franz Theodore Stone Lab., Ohio State Univ., Contr. 9.

CORRELL, D. S. 1950. Native Orchids of North America. Chronica Botanica Co., Waltham, Mass.

CRONQUIST, A. 1963. In Gleason and Cronquist, 1963.

DEAM, C. C. 1940. Flora of Indiana. Ind. Dept. Cons., Div. of Forestry.

DEAN, D. S. 1953. A study of Tradescantia ohiensis in Michigan. The Asa Gray Bull., N.S. II: 379–388.

DEAN, D. S. 1959. Distribution of tetraploid and diploid Tradescantia ohiensis in Michigan and adjoining areas. Amer. Midl. Nat. **61:** 204–209.

DRESS, W. J. 1955. Hemerocallis lilio-asphodelus, the Day-lily. Baileya 3: 107–108.

ERICKSON, R. O. 1941. Mass collections—Camassia scilloides. Mo. Bot. Gard. Ann. **28:** 293–297.

FASSETT, N. C. 1951. Grasses of Wisconsin. Univ. of Wisconsin Press.

FASSETT, N. C. and B. CALHOUN. 1952. Introgression between *Typha latifolia* and *T. angustifolia*. Evolution 6: 367–379.

FERNALD, M. L. 1932. The linear-leaved North American species of Potamogeton, section Axillares. Memoirs Gray Herb. III: 1–183, and Memoirs Amer. Acad. Arts & Sci., 17 (Part 1).

FERNALD, M. L. 1940. What is Arisaema triphyllum? Rhodora **42:** 247–254. (in Some Spermatophytes of eastern North America.)

FERNALD, M. L. 1944a. Specific distinctions between Polygonatum biflorum and P. canaliculatum. Rhodora **46:** 9–12.

FERNALD, M. L. 1944b. Trillium flexipes Raf. Rhodora **46:** 16–17.

FERNALD, M. L. 1944c. The identity of Yucca filamentosa. Rhodora **46:** 5–9.

FERNALD, M. L. 1946. Some species in Rafinesque's "Herbarium Rafinesquianum." Rhodora **48:** 5–9.

FERNALD, M. L. 1950. Gray's Manual of Botany, ed. 8. American Book Company.

FERNALD, M. L. and A. E. BRACKETT. 1929. The representatives of Eleocharis palustris in North America. Rhodora **31:** 57–77.

FOSTER, R. C. 1937. A cyto-taxonomic survey of the North American species of Iris. Contr. Gray Herb. 119: 1–83.

GALE, S. 1944. Rhynchospora, section Eurhynchospora, in Canada, the United States, and the West Indies. Rhodora **46:** 89–134, 159–197, 207–249, 255–278.

GLEASON, H. A. 1952. Illustrated Flora of the Northeastern United States and Adjacent Canada. New York Botanical Garden.

GLEASON, H. A. and A. CRONQUIST. 1963. Manual of Vascular Plants of Northeastern United States and Adjacent Canada. D. Van Nostrand Company.

HERMANN, F. J. 1940. The genus Carex in Indiana, in Deam's Flora of Indiana, pp. 212–276.

HICKS, L. E. 1932. Flower production in the Lemnaceae. Ohio Jour. Sci. 32: 115–131.

HICKS, L. E. 1937. The Lemnaceae of Indiana. Amer. Midl. Nat. 18: 774–789.

HITCHCOCK, A. S. 1951. Manual of the Grasses of the United States. Second edition, revised by A. Chase. U.S.D.A. Miscellaneous Publication No. 200.

HOTCHKISS, N. and H. L. DOZIER. 1949. Taxonomy and distribution of North American cattails. Amer. Midl. Nat. 41: 237–254.

JONES, G. N. 1951. On the nomenclature of Luzula saltuensis. Rhodora 53: 242–244.

JONES, QUENTIN. 1951. A cytotaxonomic study of the genus Disporum in North America. Contr. Gray Herb., no. CLXXIII.

LAWRENCE, G. H. M. 1960. Taxonomy of Vascular Plants. Macmillan Co.

LI, HUI-LIN. 1952. Floristic relationships between eastern Asia and eastern North America. Am. Phil. Soc. Trans., n.s. 42: 371–429.

LÖVE, A. and D. LÖVE. 1957. Drug content and polyploidy in Acorus. Proc. Genetics Soc. Can. 2: 14–17.

LÖVE, A. and D. LÖVE. 1958. Biosystematics of Triglochin maritimum agg. Naturaliste Canadien 85: 156–165.

MACKENZIE, K. K. 1931. Cyperaceae: Carex. North American Flora 18. New York Bot. Garden.

MACKENZIE, K. K. 1940. North American Cariceae. 2 vol. New York Botanical Garden.

MILLER, P. 1731. The Gardeners Dictionary, ed. 1.

MORRIS, F. and E. A. EAMES. 1929. Our Wild Orchids. Charles Scribner's Sons.

OGDEN, E. C. 1943. The broad-leaved species of Potamogeton of North America north of Mexico. Rhodora 45: 57–105, 119–163, 171–214.

OGDEN, E. C. 1947. Potamogeton tennesseensis new to the Manual range. Rhodora 49: 255–256.

OGDEN, E. C. 1953. Key to the North American species of Potamogeton. New York State Museum, Circular 31.

OWNBEY, M. 1953. The chromosomes of Disporum maculatum. Rhodora 55: 61–62.

OWNBEY, R. P. 1944. The liliaceous genus Polygonatum in North America. Ann. Mo. Bot. Gard. 31: 373–413.

PARKS, C. R. and J. W. HARDIN. 1963. Yellow Erythroniums of the eastern United States. Brittonia 15: 245–259.

POHL, R. W. 1947. A taxonomic study on the grasses of Pennsylvania. Amer. Midl. Nat. 38: 513–604.

SHINNERS, L. H. 1944. Notes on Wisconsin grasses—IV. Leptoloma and Panicum. Amer. Midl. Nat. 32: 164–180.

SMALL, J. K. 1933. Manual of the Southeastern Flora. Publ. by author, New York.

SMITH, J. G. 1895. A revision of the North American species of Sagittaria and Lophotocarpus. Mo. Bot. Gard. Ann. Rept. 6: 27–64, Plates 1–29.

STOUTAMIRE, W. P. 1964. Seeds and seedlings of native orchids. Mich. Bot. 3: 107–119.

SVENSON, H. K. 1929–45. Monographic studies in the genus Eleocharis. Rhodora 31: 121–135, 152–163, 167–191, 199–219, 224–242; 34: 193–203, 215–227; 36: 377–389; 39: 210–231, 236–273; 41: 1–19, 43–77, 90–110; 47: 273–302, 363–388.

WALLER, A. E. 1931. The native Iris of Ohio and bordering territory. Ohio Jour. Sci. 31: 29–43.

WHERRY, E. T. 1942. The relationship of Lilium michiganense. Rhodora 44: 453–456.

GENERAL INDEX

Technical terms which are defined in the text, either by statement or parenthetical phrase, are listed alphabetically under *Definitions*. In a number of places, mention is made of chromosomes, introgression, and cytogenetic investigations (or desirability of such studies); these are entered under *Cytotaxonomic Studies*, and are listed alphabetically by name of genus.

INDEX TO PLANT NAMES

The page on which a species is illustrated is in **boldface**. Forms are not indexed. Synonyms are in *italics*. Names occurring in Introduction are not indexed.